Annual Reports in Organic Synthesis—1981

Annual Reports in Organic Synthesis

ANNUAL REPORTS IN ORGANIC SYNTHESIS—1970
John McMurry and R. Bryan Miller, Eds.

ANNUAL REPORTS IN ORGANIC SYNTHESIS—1971
John McMurry and R. Bryan Miller, Eds.

ANNUAL REPORTS IN ORGANIC SYNTHESIS—1972
John McMurry and R. Bryan Miller, Eds.

ANNUAL REPORTS IN ORGANIC SYNTHESIS—1973
R. Bryan Miller and Louis S. Hegedus, Eds.
John McMurry, Series Editor

ANNUAL REPORTS IN ORGANIC SYNTHESIS—1974
Louis S. Hegedus and Stephen R. Wilson, Eds.
R. Bryan Miller, Series Editor

ANNUAL REPORTS IN ORGANIC SYNTHESIS—1975
R. Bryan Miller and L. G. Wade, Jr., Eds.

ANNUAL REPORTS IN ORGANIC SYNTHESIS—1976
R. Bryan Miller and L. G. Wade, Jr., Eds.

ANNUAL REPORTS IN ORGANIC SYNTHESIS—1977
R. Bryan Miller and L. G. Wade, Jr., Eds.

ANNUAL REPORTS IN ORGANIC SYNTHESIS—1978
L. G. Wade, Jr., and Martin J. O'Donnell, Eds.

ANNUAL REPORTS IN ORGANIC SYNTHESIS—1979
L. G. Wade, Jr., and Martin J. O'Donnell, Eds.

ANNUAL REPORTS IN ORGANIC SYNTHESIS—1980
L. G. Wade, Jr., and Martin J. O'Donnell, Eds.

ANNUAL REPORTS IN ORGANIC SYNTHESIS—1981
L. G. Wade, Jr., and Martin J. O'Donnell, Eds.

Annual Reports in Organic Synthesis—1981

edited by

L. G. Wade, Jr.
Department of Chemistry, Colorado State University, Ft. Collins, Colorado

Martin J. O'Donnell
Department of Chemistry, Indiana University and Purdue University, Indianapolis, Indiana

ACADEMIC PRESS   1982
A Subsidiary of Harcourt Brace Jovanovich, Publishers
NEW YORK   LONDON
PARIS   SAN DIEGO   SAN FRANCISCO   SÃO PAULO   SYDNEY   TOKYO   TORONTO

Academic Press Rapid Manuscript Reproduction

ACADEMIC PRESS, INC.
111 Fifth Avenue, New York, New York 10003

United Kingdom Edition published by
ACADEMIC PRESS, INC. (LONDON) LTD.
24/28 Oval Road, London NW1 7DX

LIBRARY OF CONGRESS CATALOG CARD NUMBER: 17-167779
ISBN: 0-12-040812-0

PRINTED IN THE UNITED STATES OF AMERICA

82 83 84 85   9 8 7 6 5 4 3 2 1

CONTENTS

PREFACE

One of the most difficult problems facing chemists today is that of "keeping up with the literature." For several reasons, the problem is particularly severe for the synthetic organic chemist. Bits of information of potential use are scattered throughout common chemistry journals and can be found in any paper, not just those dealing strictly with synthesis. Thus, a synthetic chemist must read a large number of journals and must organize and index what he reads to make the information available for future reference. All synthetic chemists do this; but the task is becoming more difficult each year as the flow of information increases.

The problem, however, is shared to some extent by all. Most organic chemists are at some time faced with the problem of synthesizing a desired material, and for many the problems are formidable. Nonspecialists faced with a synthetic problem are not likely to have kept pace with the developments in synthetic chemistry that may well solve their problems and will not have the necessary information in their files.

Thus, we felt that an organized annual review of synthetically useful information would prove beneficial to nearly all organic chemists, both specialist and nonspecialist in synthesis. It should help relieve some of the information-storage burden of the specialist and should enable the nonspecialist who is seeking help with a specific problem to become rapidly aware of recent synthetic advances. Ideally also, it should appear as promptly as possible after the close of the abstracting period. This year we have placed particular emphasis on keeping the abstracts as concise as possible, while indicating the generality of the reactions involved. We have tried to combine similar publications into inclusive abstracts, particularly in Chapter I. This practice has allowed us to include a larger number of references without a substantial increase in the book's length.

In producing *Annual Reports in Organic Synthesis—1981*, we have abstracted 48 primary chemistry journals, selecting useful synthetic advances. We have tried to present the information in an organized manner, emphasizing rapid visual retrieval. Only the common journals received by our libraries have been abstracted. Any journal received after March 1, 1982 will be covered in the next volume. We have also exercised selectivity in choosing which papers to abstract. Our general guidelines have been to include all reactions and methods that are new, synthetically useful, and reasonably general. Each entry is comprised primarily of structures, accompanied by very few comments. The purpose of this emphasis is to aid the reader in scanning the book. The mind is capable of absorbing a whole picture in an instant, but is considerably slowed by having to read sentences. If the pictures presented catch the reader's interest, he or she should then seek details from the original paper.

For the sixth year we have included a principal author index to aid the user. No subject index is included because to do so would greatly increase both the cost of the book and the lead time for publication. Instead, we have chosen to use an extensive table of contents. Chapters I–III are organized by reaction type and constitute a major part of the book. The organization of these sections is self-explanatory; thus, there should be no difficulty in locating a new method of oxidation or a new cyclo-propanation procedure. Chapter IV deals with methods of synthesizing heterocyclic systems and Chapter V covers the use of new protecting groups. Chapter VI is divided into three main parts and covers those synthetically useful transformations that do not fit easily into the first three chapters. The first part deals only with functional group synthesis; the second covers ring expansion and contraction; and the third involves useful multistep sequences, the individual steps of which may be well known. Future volumes of this series will maintain the present table of contents as much as possible. If no entry is found for a particular section, the last volume in which one appears will be cited in the table of contents.

Any undertaking of this type involves a series of compromises. We have chosen to emphasize reasonable cost, rapid publication, and rapid visual retrieval of information at the admitted expense of detail and beauty.

The arduous task of drawing the multitude of structures appearing in this review was carried out by Ms. Audrey Maakestad and Ms. Katy Krupa. We thank them very much for their efforts. We also thank William Bruder, Dr. Arlene Courtney, Sharon Fricke, Lydia Milne, Dr. Forrest Sheffy, and Ron Wilde for aid in proofreading the manuscript.

L. G. WADE, JR.

MARTIN J. O'DONNELL

JOURNALS ABSTRACTED

Accounts of Chemical Research
Acta Chemica Scandinavica
Aldrichimica Acta
Angewandte Chemie International Edition in English
Australian Journal of Chemistry
Bulletin of the Chemical Society of Japan
Bulletin de Societes Chimiques Belges
Bulletin de la Société Chimique de France
Canadian Journal of Chemistry
Chemical Communications
Chemical and Pharmaceutical Bulletin
Chemical Reviews
Chemical Society Reviews
Chemische Berichte
Chemistry and Industry
Chemistry Letters
Collection of Czechoslovakian Chemical Communications
Comptes Rendus Hebdomadaires de Seances de l'Academie des Sciences (C)
Doklady Chemistry
Gazzetta Chimica Italiana
Helvetica Chimica Acta
Indian Journal of Chemistry
Israel Journal of Chemistry
Journal of the American Chemical Society
Journal of Chemical Research
Journal of the Chemical Society (Perkin I)
Journal of the Chemical Society (Perkin II)
Journal of General Chemistry (USSR)
Journal of Heterocyclic Chemistry
Journal of Medicinal Chemistry
Journal of Organic Chemistry
Journal of Organic Chemistry (USSR)
Journal of Organometallic Chemistry
Journal für Praktische Chemie
Liebig's Annalen der Chemie
Monatshefte für Chemie
Organic Preparations and Procedures International
Organic Syntheses
Pure and Applied Chemistry
Recueil des Travaux Chimiques des Pays-bas
Russian Chemical Reviews
Steroids
Synthesis
Synthetic Communications
Tetrahedron
Tetrahedron Letters
Topics in Current Chemistry
Zeitschrift für Chemie

GLOSSARY OF ABBREVIATIONS

Ac   acetyl
AIBN   Azobisisobutyronitrile
Ar   aryl
9-BBN   9-borabicylo[3.3.1] nonane
BOC (t-Boc)   t-butyloxycarbonyl
Bu   Butyl
Bz   benzyl
Cbz   Benzyloxycarbonyl
COD   1,5-cyclooctadiene
Cp   cyclopentadienyl
DBN   1,5-Diazabicyclo[4.3.0] non-3-ene
DBU   1,5-diazabicyclo[5.4.0] undecene-5
DCC   dicyclohexylcarbodiimide
DDQ   2,3-cichloro-5,6-dicyanobenzoquinone
DEAD   diethyl azodicarboxylate
DIBAH (DIBAL)   diisobutylaluminum hydride
DMAD   dimethyl acetylenedicarboxylate
DMAP   4-N,N-dimethylaminopyridine
DME   1,2-dimethoxyethane
DMF   dimethylformamide
DMSO   dimethyl sulfoxide
E+   general electrophile
Et   ethyl
Hex   hexyl
HMPA, HMPT   hexamethylphosphoramide (hexamethylphosphoric triamide)
hv   irradiation with light
KAPA   potassium 3-aminopropylamide
L   triphenylphosphine ligand
LAH   lithium aluminum hydride
LDA   lithium diisopropylamide
LICA   lithium isopropylcyclohexylamide

MCPBA   *meta*-chloroperbenzoic acid
Me   methyl
MEM β-methoxyethoxymethyl
Ms   methanesulfonyl
NBS   N-bromosuccinimide
NCS   N-chlorosuccinimide
Ni (R)   Raney Nickel
Ⓟ   polymeric backbone
PCC   pyridinium chlorochromate
Ph   phenyl
(Phen)   1,10-phenanthroline
Phth   phthaloyl
PPA   polyphosphoric acid
PPE   polyphosphate ester
Pr   propyl
Py, pyr   pyridine
Q^+ quaternary ammonium
RT   room temperature
Tf   trifluoromethane sulfonate
TFA   trifluoroacetic acid
TFAA   trifluoroacetic anhydride
THF   tetrahydrofuran
THP   tetrahydropyranyl
TMEDA   tetramethylethylenediamine
TMP   2.2,6,6-tetramethylpiperidine
TMS   trimethylsilyl
Ts, Tos   *p*-toluenesulfonyl
Z   benzyloxycarbonyl; also used for electron-withdrawing groups such as -CN, -COOR, etc.
Δ heat
ϕ phenyl
18-C-6   18-crown-6

I

CARBON—CARBON BOND FORMING REACTIONS

I.A.   Carbon-Carbon Single Bonds
        (see also: I.E, I.F, I.G).

I.A.1. Alkylation of Aldehydes, Ketones and Their Derivatives

I.A.1-1 D. Fishman, J. T. Klug and A. Shani, <u>Synthesis</u>, 137
(1981); J. A. Miller and G. Zweifel, <u>J. Amer. Chem. Soc.</u>, 103
6217 (1981).

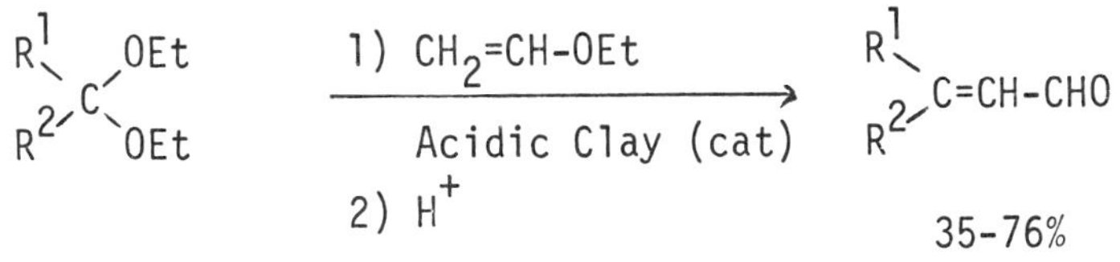

$$\begin{array}{c} R^1 \diagdown \\ \qquad C \diagup OEt \\ R^2 \diagup \diagdown OEt \end{array} \qquad \xrightarrow[\substack{\text{Acidic Clay (cat)} \\ 2) \; H^+}]{1) \; CH_2=CH-OEt} \qquad \begin{array}{c} R^1 \diagdown \\ \qquad C=CH-CHO \\ R^2 \diagup \end{array}$$

35-76%

1

I.A.1-2 P. F. Hudrlik and A. K. Kulkarni, $\underline{J.\ Amer.\ Chem.\ Soc.}$, $\underline{103}$, 6251 (1981); H. L. Goering and C. C. Tseng, $\underline{J.\ Org.}$ $\underline{Chem.}$, $\underline{46}$, 5250 (1981); D. Schinzer and C. H. Heathcock, $\underline{Tetrahedron\ Lett.}$, $\underline{22}$, 1881 (1981).

$$CH_3CH=N-\bigcirc \quad \xrightarrow[\substack{2)\ ^tBuMe_2SiCl,\ -78°C. \\ 3)\ LDA/THF,\ -78°C. \\ 4)\ nC_6H_{11}Br \\ 5)\ HOAc/H_2O/CH_2Cl_2}]{\substack{1)\ LDA \\ THF,\ 0°C.}} \quad {}^tBuMe_2Si\diagdown\diagup-CHO$$

nC_6H_{11}

63%

I.A.1-3 A. Umani-Ronchi et al, $\underline{J.\ Organometal.\ Chem.}$, $\underline{204}$, 281 (1981); T. A. Shustrova et al, $\underline{J.\ Org.\ Chem.\ (USSR)}$, $\underline{17}$, 277 (1981); R. T. Logan et al, $\underline{J.\ Chem.\ Soc.,\ Perkin\ I}$, 2306 (1981).

$$\bigcirc\!\!=\!\!O \quad \xrightarrow[\substack{2)\ nBuBr}]{\substack{1)\ K/Al_2O_3 \\ Hexane,\ 25°C}} \quad \text{(2-nBu cyclohexanone)}$$

73%

Also, alkylation of nitriles, ketimines and aldehyde N,N-dimethylhydrazones.

I.A.1-4 W. C. Still and I. Galynker, $\underline{Tetrahedron}$, $\underline{37}$, 3981 (1981).

$$\text{(2-Me cyclooctanone)} \quad \xrightarrow[\substack{2)\ MeI}]{\substack{1)\ LDA \\ THF,\ -60°C.}} \quad \text{(2,2-Me product)} \quad +\ 2,2\text{-}Me_2$$

> 95% trans   15%

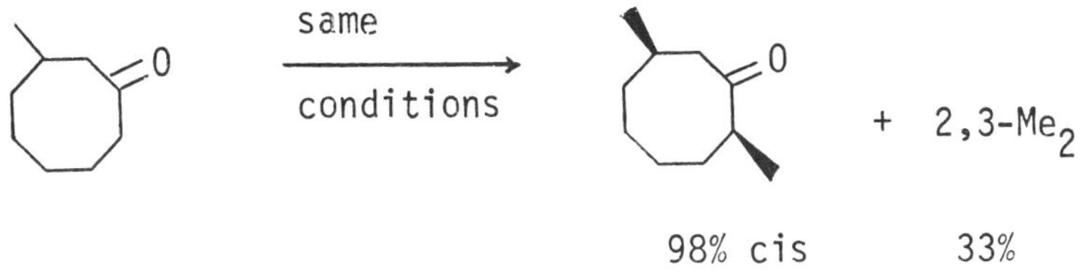

98% cis   33%

I.A.1-5 R. Gompper and H. H. Vogt, Chem. Ber., 114, 2866, 2884 (1981); A. S. Narula, Tetrahedron Lett., 22, 4119 (1981).

"Substituent Effects in the Methylation of Enolate Anions."

I.A.1-6 J. Tsuji et al, Tetrahedron Lett., 22, 1359, 1363, 2651 (1981).

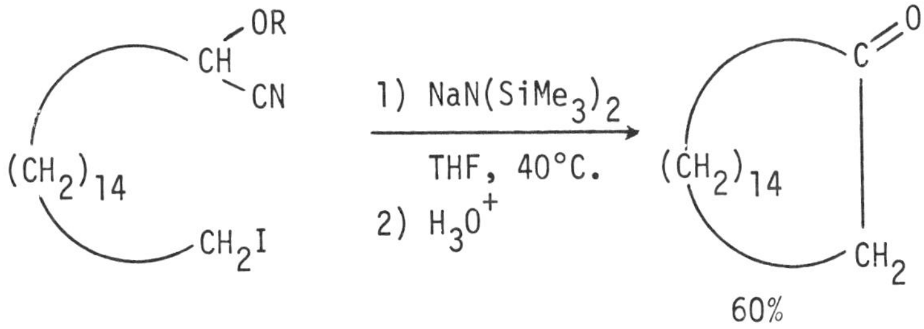

60%

I.A.1-7   J. C. Fiaud and J. L. Malleron, Chem. Commun., 1159 (1981).

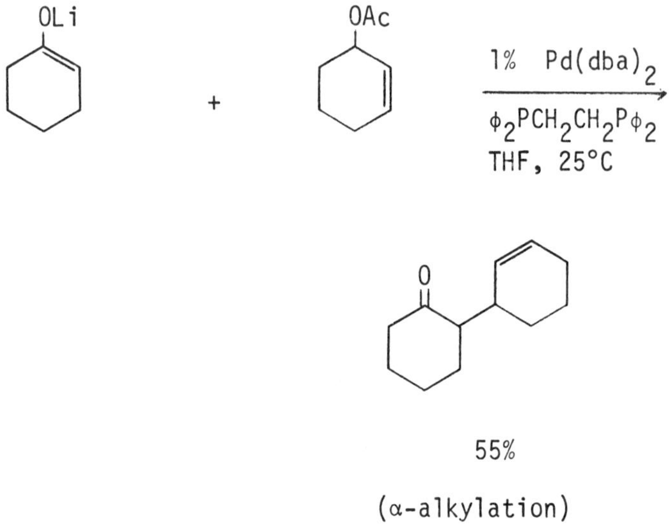

55%

(α-alkylation)

dba = dibenzylideneacetonato

I.A.1-8   G. A. Russell, B. Mudryk and M. Jawdosiuk, J. Amer. Chem. Soc., 103, 4610 (1981).

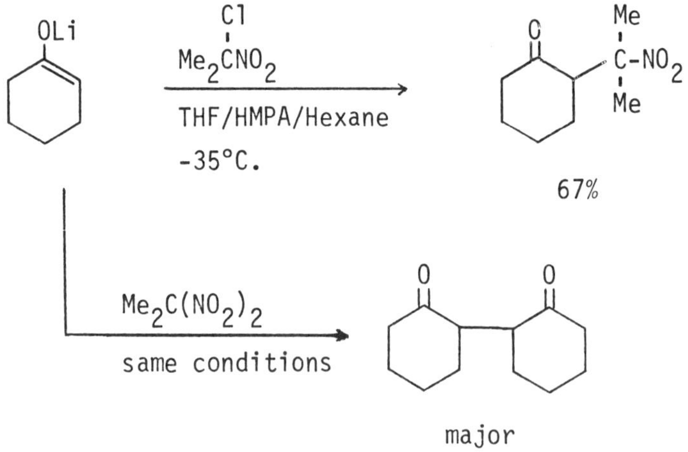

67%

major

I.A.1-9 P. S. Mariano et al, *J. Org. Chem.*, **46**, 4643 (1981);
M. Koreeda and Y. P. L. Chen, *Tetrahedron Lett.*, **22**, 15
(1981); D. Seebach et al, *Justus Liebigs Ann. Chem.*, 2272
(1981).

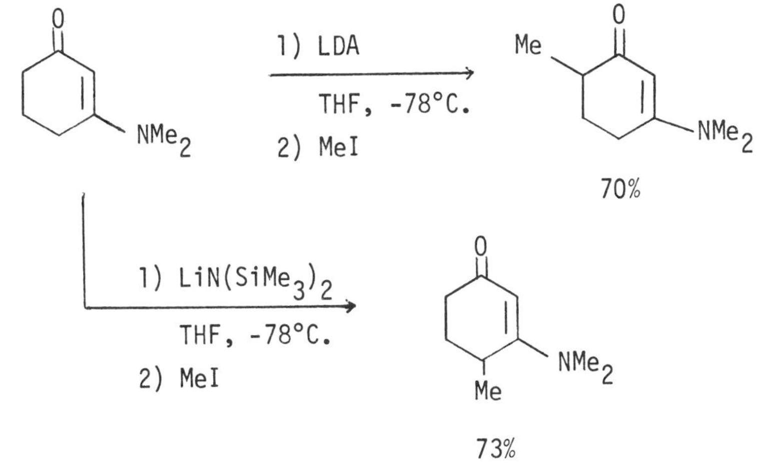

I.A.1-10 T. C. T. Chang and M. Rosenblum, *J. Org. Chem.*, **46**,
4103 (1981).

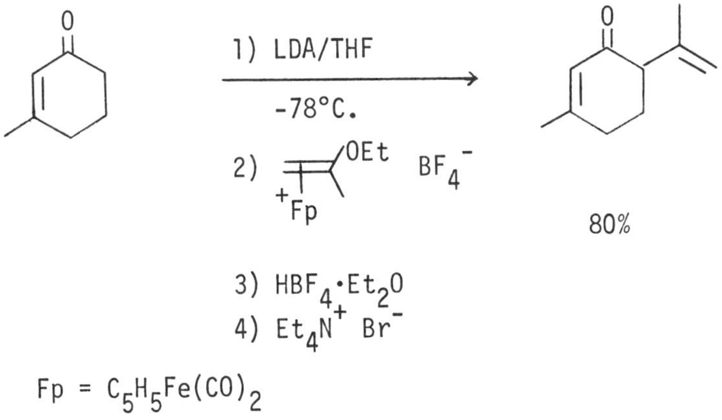

$Fp = C_5H_5Fe(CO)_2$

I.A.1-11 P. Beak and L. G. Carter, J. Org. Chem., 46, 2363
(1981).

$$\text{Ar-C-CH}_2\text{CH}_3 \quad \xrightarrow[\begin{array}{c}\text{THF, -78°C.}\\ \text{2) E}^+\end{array}]{\text{1) s-BuLi/TMEDA}} \quad \text{Ar-C-CHCH}_3$$

31-87%

E^+ = RX, RCHO, R$_2$CO, MeOD, Bu$_3$SnCl, Me$_3$SiCl.

I.A.1-12 F. W. Sum and L. Weiler, Tetrahedron, 37 (Suppl. 1),
303 (1981); J. S. Hubbard and T. M. Harris, J. Org. Chem.,
46, 2566 (1981).

$$\underset{\text{CO}_2\text{Me}}{\overset{\text{O}}{\|}} + \underset{\text{OTHP}}{\overset{\text{Br}}{\|}} \quad \xrightarrow[\begin{array}{c}\text{2) NaH/ClPO(OEt)}_2\\ \text{3) 2 eq. LiMe}_2\text{Cu}\end{array}]{\text{1) THF}}$$

$$\text{THPO} \diagdown\diagup\diagdown\diagup \diagdown \text{CO}_2\text{Me}$$

62%

I.A.1-13 T. Fujisawa and K. Sakai, Chem. Lett., 55 (1981).

$$\xrightarrow[\begin{array}{c}\text{CH}_2\text{Cl}_2\text{, -15°C.}\\ \text{2) aq. NaOH, 55°C.}\end{array}]{\text{1) Br}_2}$$

62%

I.A.1-14 K. Suzuki and M. Sekiya, Synthesis, 297 (1981); R.
Shabana, J. B. Rasmussen and S. O. Lawesson, Bull. Soc. Chim.
Belg., 90, 75 (1981); I. Rico, D. Cantacuzene and C.
Wakselman, Tetrahedron Lett., 22, 3405 (1981).

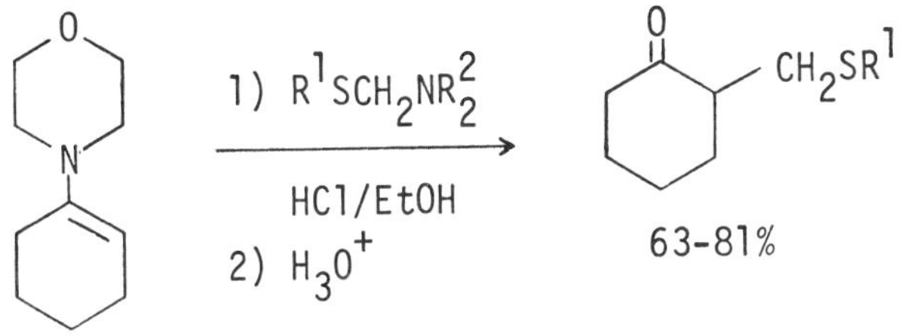

I.A.1-15 J. K. Smith, D. E. Bergbreiter and M. Newcomb, J.
Org. Chem. 46, 3157 (1981); K. Takebe et al, Chem. Ind.,
540 (1981).

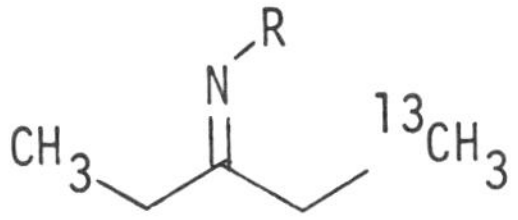

Study of regioselectivity in
deprotonation of alkyl ketimines.

I.A.1-16 H. Sakurai et al, J. Org. Chem., 46, 4631 (1981).

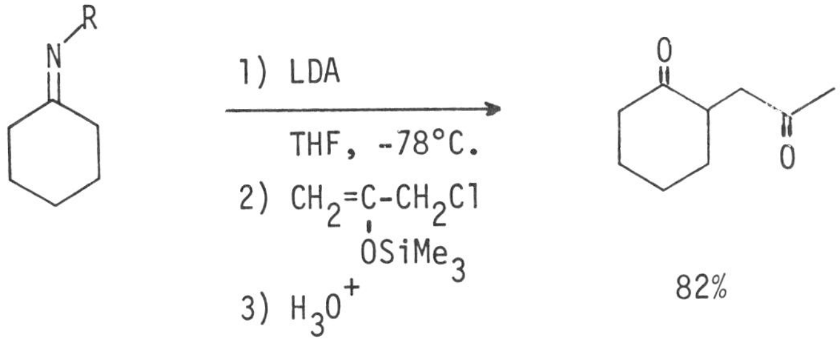

I.A.1-17 R. Lidor and S. Shatzmiller, J. Amer. Chem. Soc.,
103, 5916 (1981).

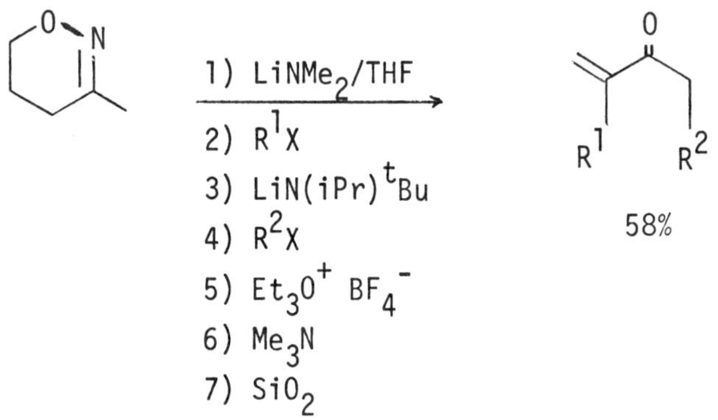

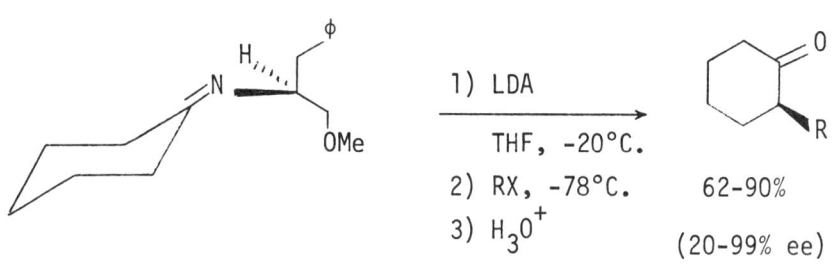

1) LiNMe$_2$/THF
2) R^{1}X
3) LiN(iPr)^{t}Bu
4) R^{2}X
5) Et$_3$O$^+$ BF$_4^-$
6) Me$_3$N
7) SiO$_2$

58%

I.A.1-18 D. Enders, CHEMTECH, 504 (1981).

Review: "Forming Asymmetric C-C Bonds."

I.A.1-19 A. I. Meyers et al, J. Amer. Chem. Soc., 103, 3081,
3088 (1981).

1) LDA

THF, -20°C.

2) RX, -78°C.   62-90%

3) H$_3$O$^+$

(20-99% ee)

I.A.1-20 M. T. Reetz, M. Sauerwald and P. Walz, <u>Tetrahedron</u>
<u>Lett.</u>, <u>22</u>, 1101 (1981); M. T. Reetz, S. Huttenhaïn and F.
Hubner, <u>Synth. Commun.</u>, <u>11</u>, 217 (1981).

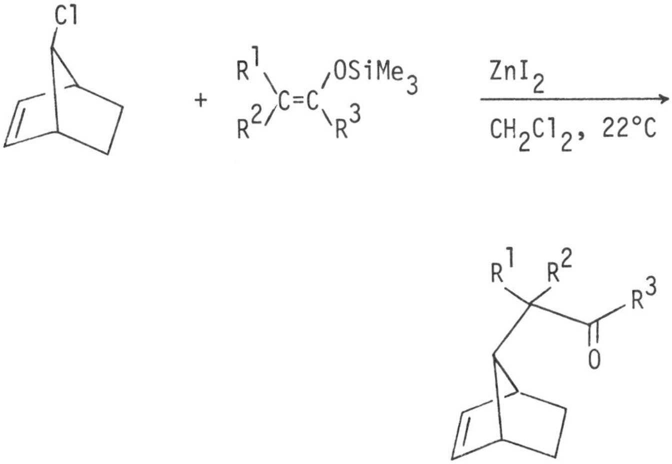

95% (GC)

$R^3 = {}^tBu$, Et or OMe.

I.A.1-21 I. Fleming and T. V. Lee, <u>Tetrahedron Lett.</u>, <u>22</u>,
705 (1981); N. Ono, H. Miyake and A. Kaji, <u>Synthesis</u>, 1003
(1981).

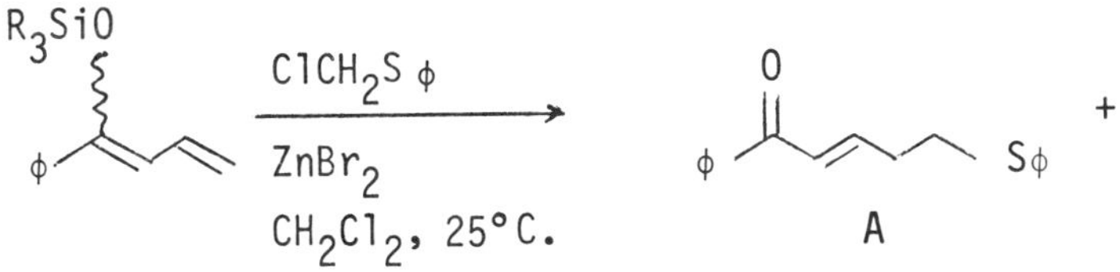

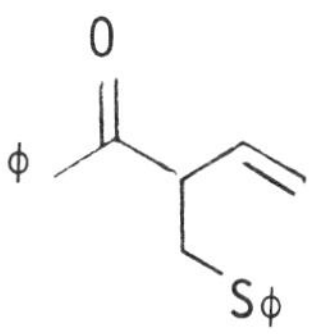

R_3	Yield	%A	%B
ϕ_3	59%	85	15
Me_3	71%	45	55
$^tBuMe_2$	32%	23	77

I.A.1-22 L. Blanco, P. Amice and J. M. Conia, Synthesis, 289, 291 (1981).

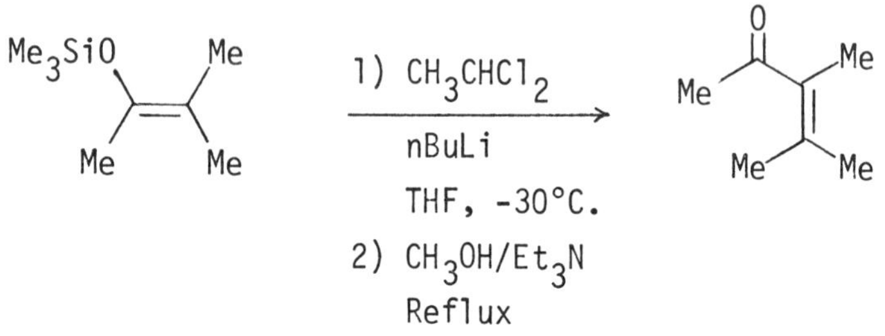

Me3SiO
Me
Me
Me
1) CH3CHCl2
nBuLi
THF, -30°C.
2) CH3OH/Et3N
Reflux
O
Me
Me
Me
Me

Cyclic enol ethers give ring expanded α-methylcycloalkenones.

I.A.1-23 S. M. Makin et al, J. Org. Chem. (USSR), 17, 630 (1981); M. T. Reetz and A. Giannis, Synth. Commun., 11, 315 (1981).

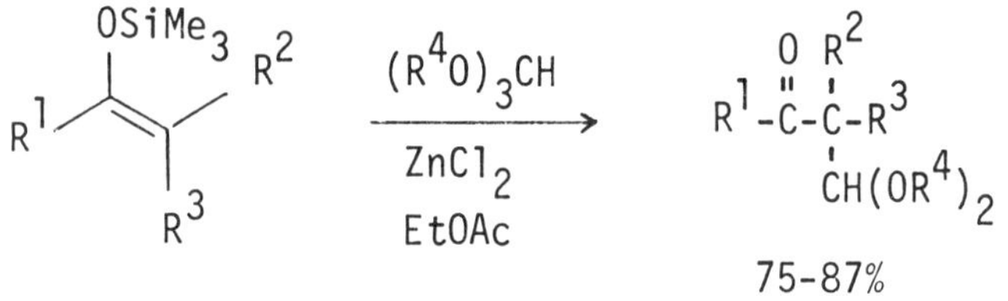

OSiMe3
R2
R1
R3
(R4O)3CH
ZnCl2
EtOAc
O R2
R1-C-C-R3
CH(OR4)2
75-87%

I.A.1-24 K. Ikeda, Y. Terao and M. Sekiya, Chem. Pharm. Bull., 29, 1156, 1747 (1981); T. Shono, Y. Matsumura and K. Tsubata, J. Amer. Chem. Soc., 103, 1172 (1981).

OSiMe3
R-N
N-R
N-R
(1/3 eq.)
CH3COCl/TiCl4
CH2Cl2, 0°C.
O
O
CH2-N-C-CH3
R
32-48%

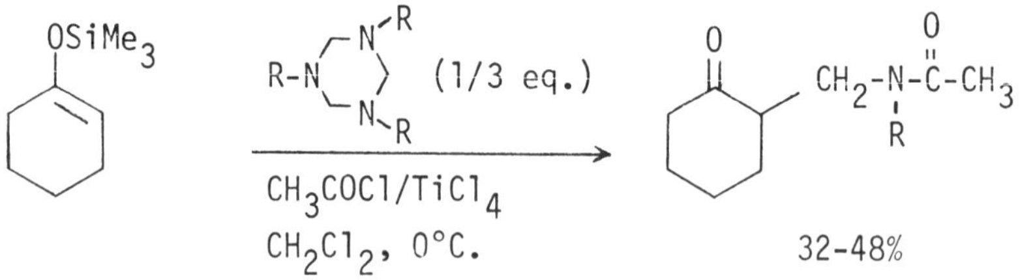

I.A.1-25   A. J. Birch et al, <u>Tetrahedron Lett.</u>, <u>22</u>, 1433
(1981); <u>J. Chem. Soc., Perkin I</u>, 1006 (1981); <u>Tetrahedron</u>,
<u>37</u> (Suppl. 1), 289 (1981); <u>J. Organometal Chem.</u>, <u>208</u>, C31
(1981).

I.A.1-26   W. L. Mock and H. R. Tsou, <u>J. Org. Chem.</u>, <u>46</u>, 2557
(1981); <u>ibid</u>, 2795.

I.A.1-27   E. Seoane et al, <u>Chem. Ind.</u>, 157 (1981); <u>Tetrahe-
dron Lett.</u>, <u>22</u>, 1733 (1981); R. T. Logan, R. G. Roy and G. F.
Woods, <u>J. Chem. Soc., Perkin I</u>, 2631 (1981); L. A. Paquette,
E. Farkas and R. Galemmo, <u>J. Org. Chem.</u>, <u>46</u>, 5434 (1981); N.
DeKimpe et al, <u>Tetrahedron Lett.</u>, <u>22</u>, 1837 (1981).

I.A.2. Alkylations of Nitriles, Acids and Acid Derivatives

I.A.2-1   R. Grigg et al, Tetrahedron Lett., 22, 4107 (1981).

$$ArCH_2CN \quad + \quad ROH \quad \xrightarrow[\substack{\phi_3P \\ Na_2CO_3 \\ ROH, \ Reflux}]{RhCl_3} \quad \overset{R}{\underset{|}{ArCHCN}}$$

$$30-78\%$$

I.A.2-2   J. H. Poupaert et al, J. Chem. Res. (S), 192 (1981).

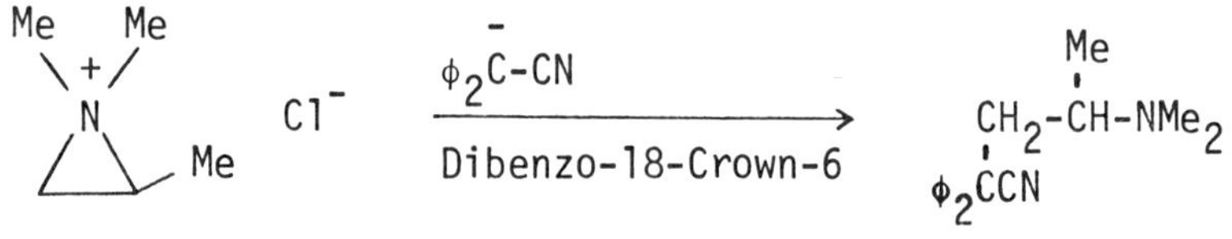

I.A.2-3   K. Takabe, S. Ohkawa and T. Katagiri, Chem. Lett., 489 (1981).

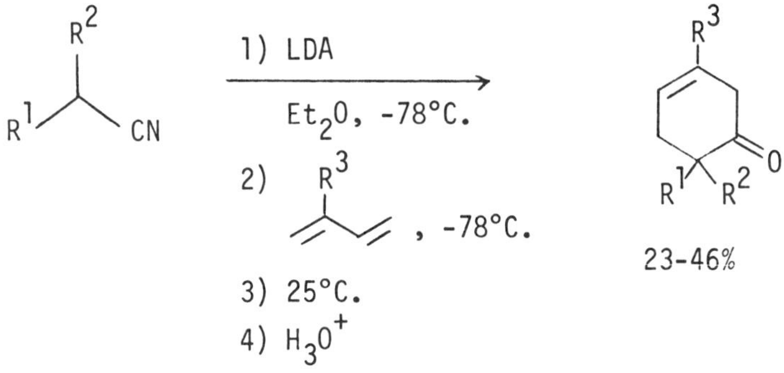

I.A.2-4 A. I. Meyers and J. P. Lawson, <u>Tetrahedron Lett.</u>, <u>22</u>, 3163 (1981); M. W. Anderson, R. C. F. Jones and J. Saunders, <u>ibid</u>, 261; D. J. Brunelle, <u>ibid</u>, 3699; S. P. McManus et al, <u>J. Org. Chem.</u>, <u>46</u>, 3097 (1981); B. H. Lipshutz and R. W. Hungate, <u>ibid</u>, 1410.

$$CH_3 \overset{OMe}{=} N \overset{OK}{\underset{CO_2Me}{}} \quad \begin{array}{l} 1)\ ^tBuLi \\ 2)\ E^+ \\ 3)\ BF_3 \cdot Et_2O,\ 25°C \end{array} \longrightarrow$$

$$E-CH_2 - \overset{O}{\underset{N}{\diagdown}} \underset{CO_2Me}{}$$

38-56%

E^+ = ArCHO, RCHO, ϕCH_2Br.

I.A.2-5 Y. H. Chang and W. T. Ford, <u>J. Org. Chem.</u>, <u>46</u>, 3756, 5364 (1981).

$$\phi CH_2CH_2CO_2CH_2\!\!-\!\!\textcircled{P} \quad \begin{array}{l} 1)\ \phi_3CLi/THF \\ 2)\ E^+ \\ 3)\ aq.\ KOH \\ \quad Q^+\ OH^-/\triangle \\ 4)\ H_3O^+ \end{array} \longrightarrow \quad \phi CH_2\underset{E}{CHCO_2H}$$

40-87%

E^+ = RX or RCOCl.

I.A.2-6 L. N. Mander and R. J. Hamilton, Tetrahedron Lett., 22, 4115 (1981); A. Casares, J. M. L. Cardoso and L. A. Maldonado, Synth. Commun., 11, 223 (1981).

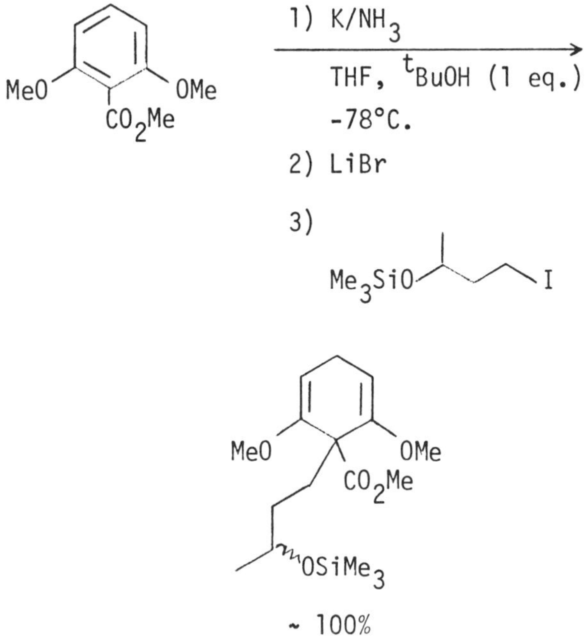
MeO
OMe
CO2Me
1) K/NH3
THF, tBuOH (1 eq.)
-78°C.
2) LiBr
3)
Me3SiO
I
MeO
OMe
CO2Me
OSiMe3
~ 100%

I.A.2-7 S. Djuric, J. Venit and P. Magnus, Tetrahedron Lett., 22, 1787 (1981).

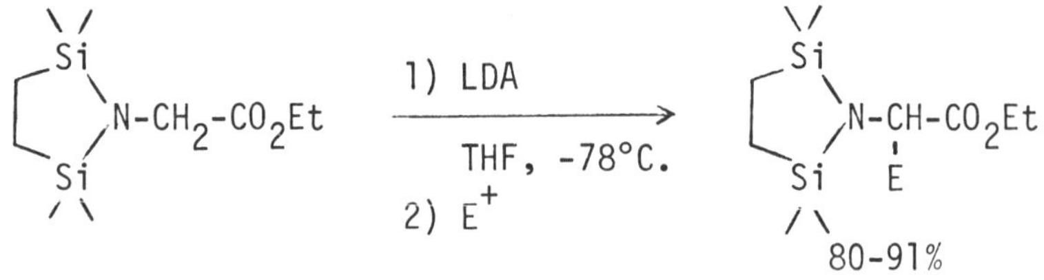
Si
N-CH2-CO2Et
Si
1) LDA
THF, -78°C.
2) E+
Si
N-CH-CO2Et
Si
E
80-91%

E^+ = RX, ϕCHO, Me_3Si-≡-CH_2Br

I.A.2-8 L. C. Yu and P.Helquist, J. Org. Chem., 46, 4536
(1981); Synth. Commun., 11, 591 (1981).

Me$_2$N~~~~CO$_2$Me   1) LDA   →   CH$_2$=C(CO$_2$Me)(R)

                        THF/-78°C.

                        2) RX/HMPA   51-74%

                        3) MeI/MeOH

                        4) DBN

I.A.2-9 C. Lion and J. E. Dubois, Tetrahedron, 37, 319 (1981).

^{t}BuCH=C(OSiMe$_3$)(OMe) ^{t}BuCl   → ^{t}Bu$_2$CH-CO$_2$Me

                              ZnCl$_2$

                                                      58%

I.A.2-10 I. Bohm, E. Hirsch and H. U. Reissig, Angew. Chem.,
Int. Ed. Engl., 20, 574 (1981).

Me$_3$SiO—△(H, CO$_2$Me)(R^1)(R^2)—R^3 1) LDA   →

                                                 THF, -78°C.

                                                 2) R^{4}X

                                                 3) Et$_3$NH$^+$F$^-$

                                                 THF, 25°C.

R^1-C(=O)-C(R^2)(R^3)-CH(R^4)-CO$_2$Me

                        63-93%

I.A.2-11 N. R. Long and M. W. Rathke, <u>Synth. Commun.</u>, <u>11</u>, 687 (1981); K. G. Bilyard and P. J. Garratt, <u>Tetrahedron Lett.</u>, <u>22</u>, 1755 (1981).

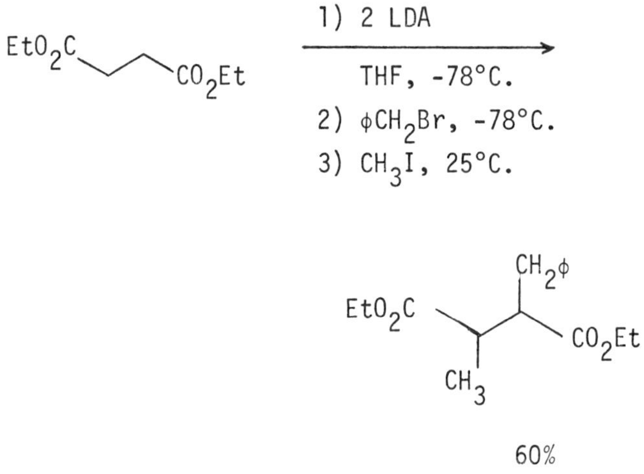

EtO_2C $\diagdown\diagup\diagdown$ CO_2Et

1) 2 LDA
   ⟶
   THF, -78°C.

2) φCH₂Br, -78°C.

3) CH₃I, 25°C.

EtO_2C — 〈 CH₂φ / CO₂Et / CH₃ 〉

60%

I.A.2-12 G. Frater, <u>Tetrahedron Lett.</u>, <u>22</u>, 425 (1981); Z. Yoshida et al, <u>ibid</u>, 3413.

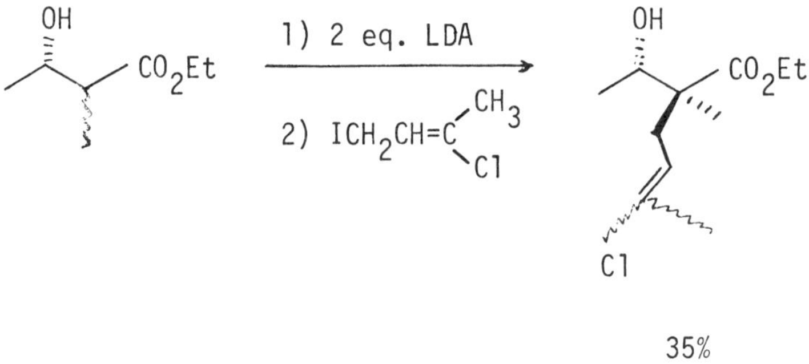

OH 〈 CO₂Et 〉

1) 2 eq. LDA
   ⟶
2) ICH₂CH=C〈CH₃ / Cl〉

OH 〈 CO₂Et / Cl 〉

35%

Product converted in several steps to 4,4- and 6,6-Disubstituted Cyclohex-2-en-1-ones (86% ee).

I.A.2-13 P. M. Savu and J. A. Katzenellenbogen, J. Org. Chem., 46, 239 (1981); V. Snieckus et al, ibid, 2029; K. Itoh et al, Tetrahedron Lett., 22, 1691 (1981).

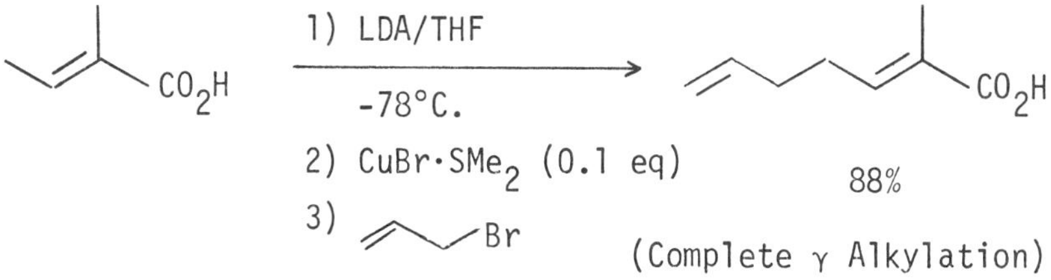
CO2H
1) LDA/THF
-78°C.
2) CuBr·SMe2 (0.1 eq)
3) Br
CO2H
88%
(Complete γ Alkylation)

I.A.2-14 I. Patterson and L. G. Price, Tetrahedron Lett., 22, 2829, 2833 (1981).

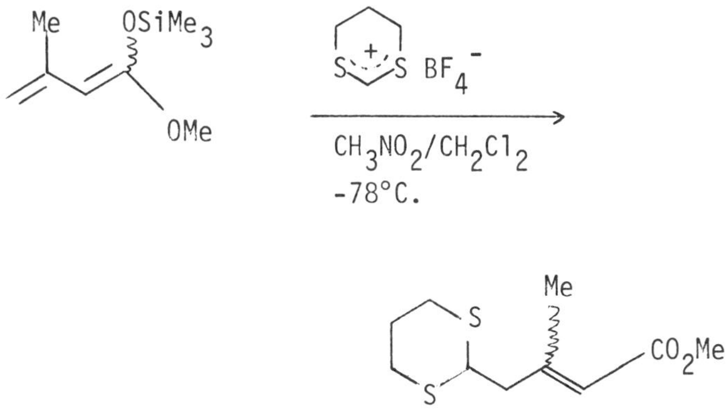
Me
OSiMe3
OMe
S + S BF4-
CH3NO2/CH2Cl2
-78°C.
Me
S
S
CO2Me

92% (94% γ)

I.A.2-15 F. Sauriol-Lord, T. B. Grindley, J. Org. Chem., 46, 2831 (1981).

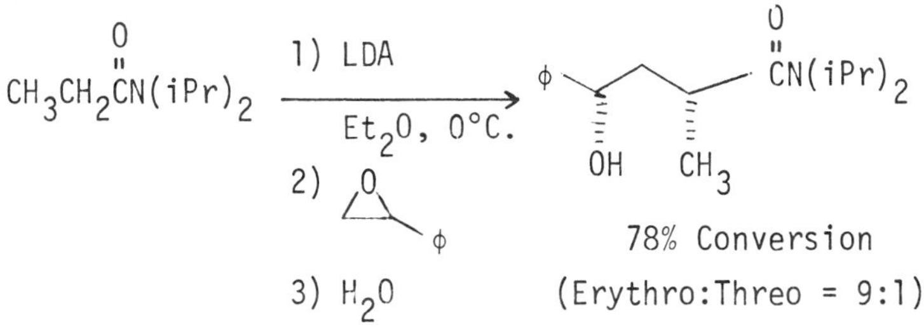
CH3CH2CN(iPr)2
1) LDA
Et2O, 0°C.
2) O φ
3) H2O
φ
OH
CH3
CN(iPr)2
78% Conversion
(Erythro:Threo = 9:1)

I.A.2-16   J. Mulzer and T. Kerkmann, <u>J. Amer. Chem. Soc.</u>, <u>103</u>, 3620 (1980).

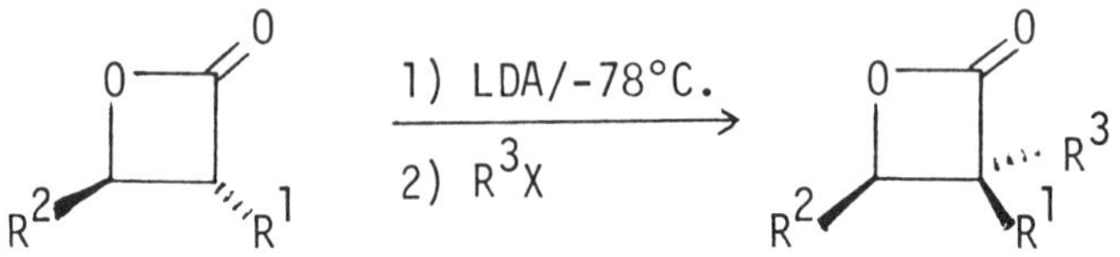

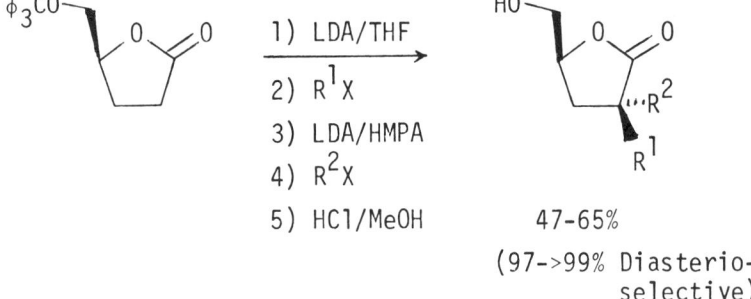

75-99%

"Forbidden" elimination on warming anion to room temperature forms acrylic acid anion.

I.A.2-17   K. Koga et al, <u>Chem. Lett.</u>, 1621 (1981); P. A. Bartlett and C. F. Pizzo, <u>J. Org. Chem.</u>, <u>46</u>, 3896 (1981); H. M. Shieh and G. D. Prestwich, <u>ibid</u>, 4319; G. Helmchen et al, <u>Angew. Chem., Int. Ed. Engl.</u>, <u>20</u>, 207 (1981).

1) LDA/THF
2) R^1X
3) LDA/HMPA
4) R^2X
5) HCl/MeOH

47-65%
(97->99% Diasterio-
selective)

I.A.2-18 R. Naef and D. Seebach, Angew. Chem., Int. Ed. Engl.,
20, 1030 (1981); D. Seebach and D. Wasmuth, ibid, 971; D.
Seebach and R. Naef, Helv. Chim. Acta, 64, 2704 (1981); G.
Frater, U. Muller and W. Gunther, Tetrahedron Lett., 22, 4221
(1981).

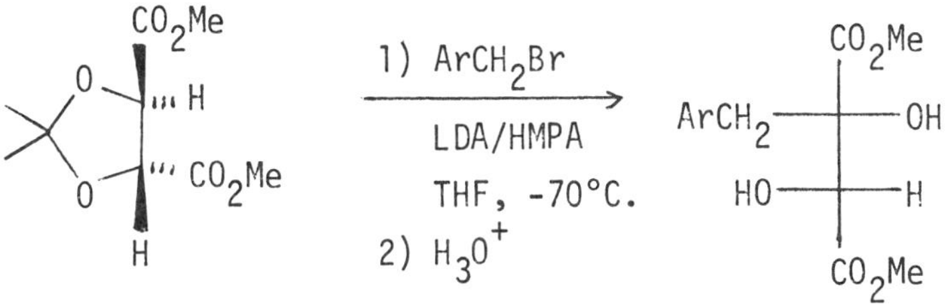

54-77%

(Yield before Hydrolysis)

I.A.3. Alkylation of β-Dicarbonyl, β-Cyanocarbonyl Systems
and Other Active Methylene Compounds

I.A.3-1 F. Freeman, Synthesis, 925 (1981).

Review: "Reactions of Malononitrile Derivatives."

I.A.3-2 H. Stamm and V. Gailius, Chem. Ber., 114, 3599 (1981).

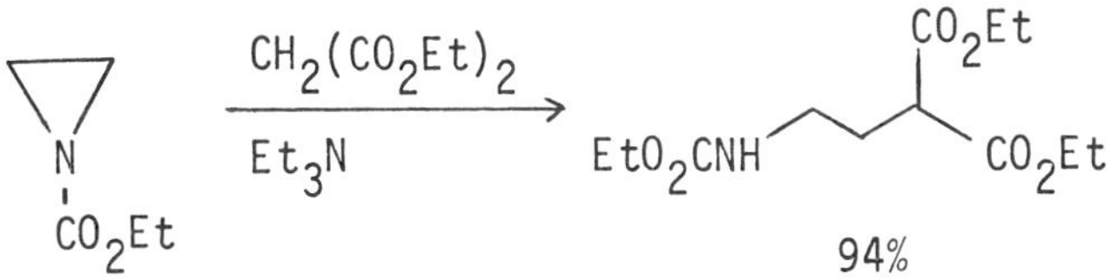

94%

Products readily converted to pyrrolines and pyrroles.

I.A.3-3 O. Mitsunoba et al, Bull. Chem. Soc. Jpn., 54, 2107
(1981); J. Marquet and M. Moreno-Manas, Chem. Lett., 173
(1981); T. L. Ho, Synth. Commun., 11, 237 (1981).

$$CH_2 \overset{CN}{\underset{CO_2Et}{<}} \quad \xrightarrow[EtO_2C-N=N-CO_2Et]{ROH/\phi_3P} \quad R-CH \overset{CN}{\underset{CO_2Et}{<}}$$

43-83%

I.A.3-4 J. P. Marino and M. P. Ferro, J. Org. Chem., 46, 1828
(1981).

$$R^1 \overset{O}{\underset{R^2 \overset{-}{\cdots} CO_2Me}{\Vert}} \quad Na^+ \quad \xrightarrow[THF, Reflux]{iPrS \overset{+}{\diagup} P\phi_3 \quad BF_4^-} \quad R^1 \diagup \overset{SiPr}{\underset{R^2 \quad CO_2Me}{\bigcirc}}$$

80-90%

Hydrolysis of vinyl sulfide via iodolactonization.

I.A.3-5 B. M. Trost and D. P. Curran, J. Amer. Chem. Soc.,
103, 7380 (1981).

1) I $\diagup$ SiMe$_3$

KH/DME, 25°C.

2) MCPBA

3) Q$^+$ F$^-$

40%

I.A.3-6 A. K. Saksena, A. T. McPhail and K. D. Onan, Tetrahedron Lett., 22, 2067 (1981).

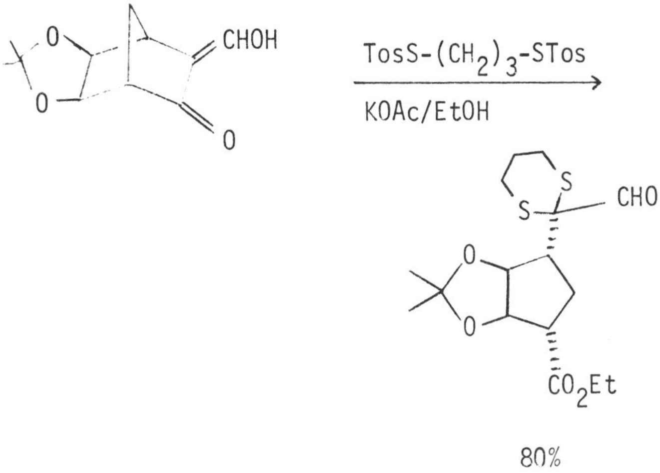
CHOH
O
O
O
$TosS-(CH_2)_3-STos$
KOAc/EtOH
S
S
CHO
O
O
CO_2Et
80%

I.A.3-7 D. J. Goldsmith and J. K. Thottathil, Tetrahedron Lett., 22, 2447 (1981); T. L. Ho, Synth. Commun., 11, 7 (1981).

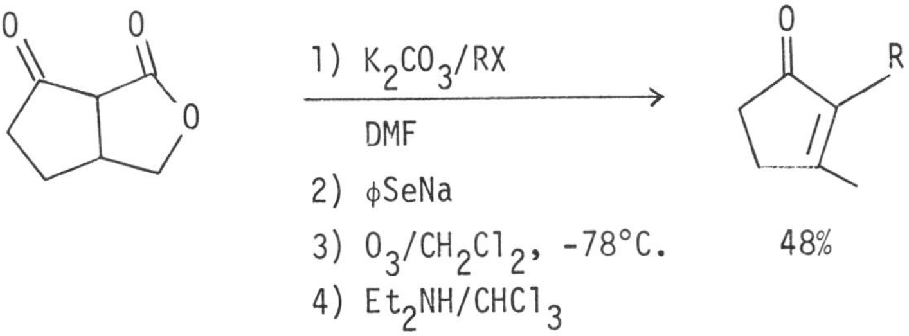
O
O
O
1) K_2CO_3/RX
DMF
2) ϕSeNa
3) O_3/CH_2Cl_2, -78°C.
4) $Et_2NH/CHCl_3$
O
R
48%

I.A.3-8 E. Stamm and R. Keese, Synthesis, 231 (1981).

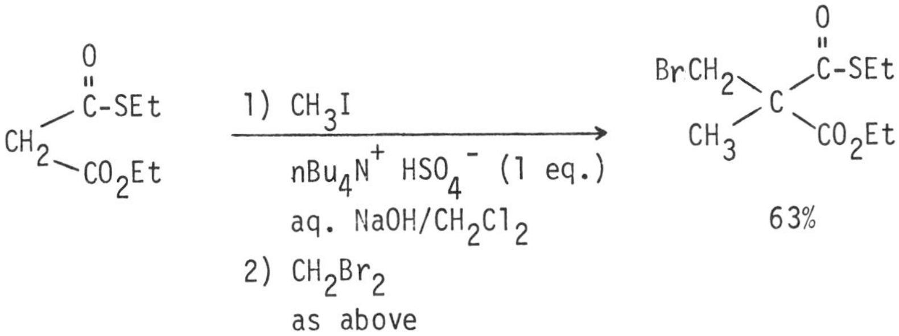
O
C-SEt
CH_2
CO_2Et
1) CH_3I
nBu_4N^+ HSO_4^- (1 eq.)
aq. $NaOH/CH_2Cl_2$
2) CH_2Br_2
as above
O
$BrCH_2$
C-SEt
C
CH_3
CO_2Et
63%

I.A.3-9 M. A. Casadei, C. Galli and L. Mandolini, <u>J. Org.</u>
<u>Chem.</u>, <u>46</u>, 3127 (1981); P. M. Warner, B. L. Chen and E.
Wada, <u>ibid</u>, 4795; R. G. Eilerman and B. J. Willis, <u>Chem.</u>
<u>Commun.</u>, 30 (1981).

$$Br(CH_2)_{n-1}CH \overset{CO_2Et}{\underset{CO_2Et}{\diagup}} \quad \xrightarrow[\substack{18\text{-Crown-6} \\ DMSO/80°C. \\ High\ dilution}]{KOEt} \quad (CH_2)_{n-1} C \overset{CO_2Et}{\underset{CO_2Et}{\diagup}}$$

n	yield	n	yield
6	72%	11	1%
7	70%	12	34%
8	22%	13	52%
9	3%	17	62%
10	0.1%	21	55%

I.A.3-10 G. A. Russell, B. Mudryk and M. Jawdosiuk, <u>Synthesis</u>,
62 (1981).

$$CH_3\overset{O}{\overset{\|}{C}}CH_2CO_2Et \quad \xrightarrow[\substack{DMF,\ 25°C \\ 2)\ (CH_3)_2C \diagdown_{NO_2}^{Cl} \\ hv,\ 35°C. \\ 3)\ H_2O}]{1)\ NaH} \quad CH_3\overset{O}{\overset{\|}{C}}-\overset{}{\underset{\underset{CH_3}{\overset{\|}{C}}\diagup^{CH_3}}{C}}-CO_2Et$$

77%

I.A.3-11 O. S. Wolfbeis, Monat. Chem., 112, 369 (1981); Chem. Ber., 114, 3471 (1981); R. Block and P. Orvane, Synth. Commun. 11, 913 (1981).

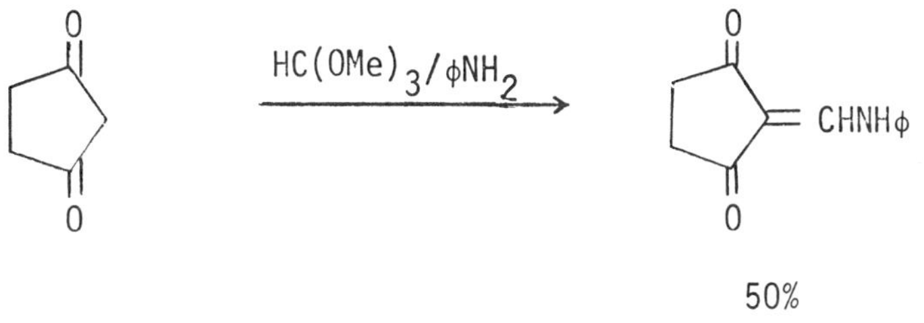

HC(OMe)$_3$/ϕNH$_2$

CHNHϕ

50%

NR with 6-membered rings.

I.A.3-12 A. J. Pearson et al, J. Amer. Chem. Soc., 103, 6686 (1981); Tetrahedron Lett., 22, 1645 (1981); ibid, 2929; J. Chem. Soc., Perkin I, 1614 (1981); A. J. Birch and G. R. Stephenson, Tetrahedron Lett., 22, 779 (1981); F. Effenberger and M. Keil, ibid, 2151; B.F.G. Johnson et al, J. Organometal. Chem., 204, 221 (1981).

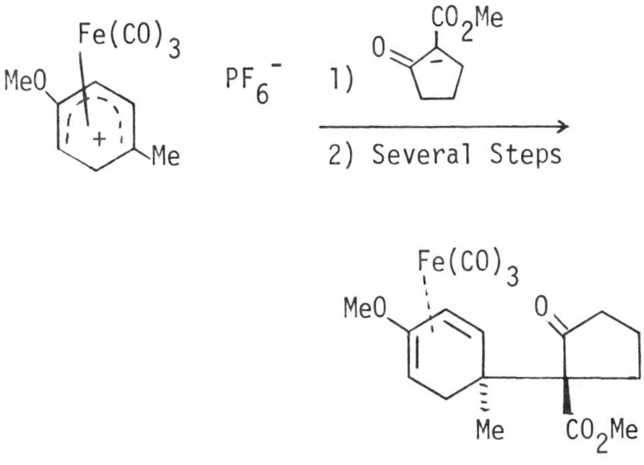

Fe(CO)$_3$

MeO

PF$_6^-$

Me

1)

CO$_2$Me

O

2) Several Steps

Fe(CO)$_3$

MeO

O

Me   CO$_2$Me

I.A.3-13 B. Akermark, A. Ljungqvist and M. Panunzio, Tetrahedron Lett., 22, 1055 (1981); B. Akermark and A. Jutand, J.
Organometal. Chem., 217, C41 (1981).

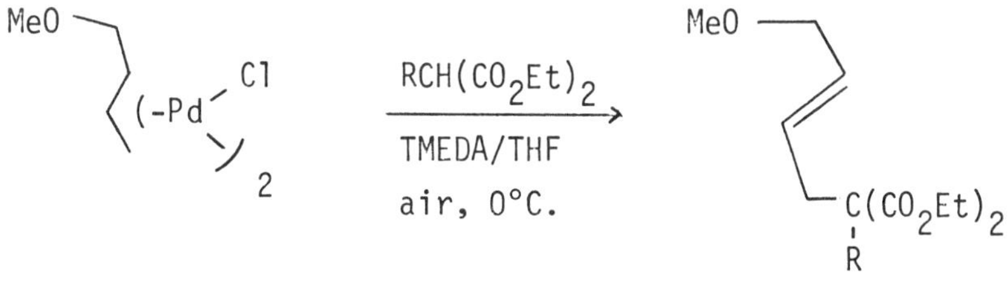

R = H, Me, φ, NHCHO.

15-90%

I.A.3-14 U. Schollkopf et al, Synthesis, 646 (1981); Justus
Liebigs Ann. Chem., 439, 709 (1981).

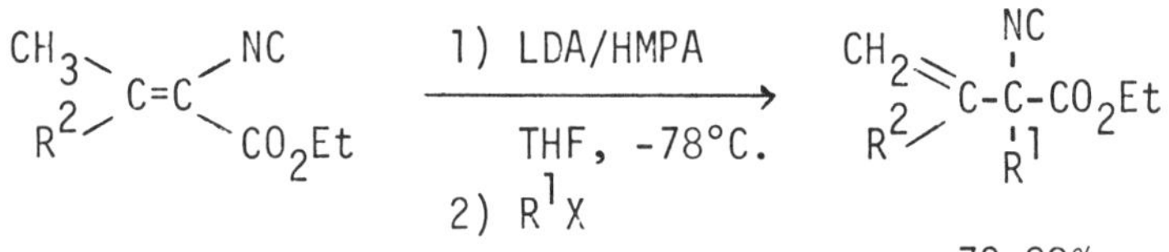

72-88%

I.A.3-15 U. Schollkopf et al, Synthesis, 966, 969 (1981);
Justus Liebigs Ann. Chem., 696, 1378 (1981); Angew. Chem. Int.
Ed. Engl., 20, 798 (1981); R. Gompper and U. Heinemann, ibid,
296.

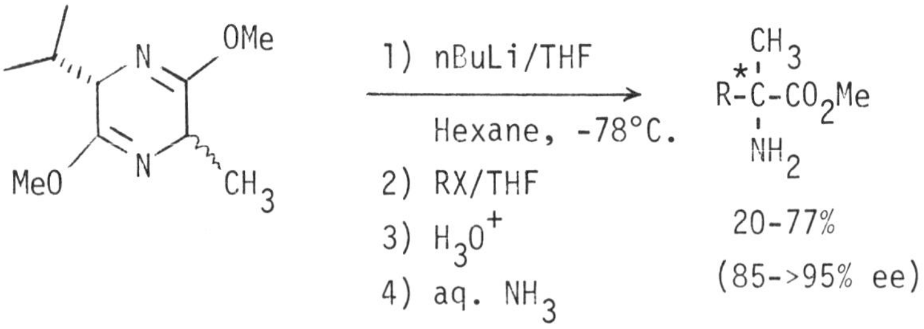

20-77%

(85->95% ee)

.A.3-16 V. Sunjic et al, Helv. Chim. Acta, 64, 1145 (1981).

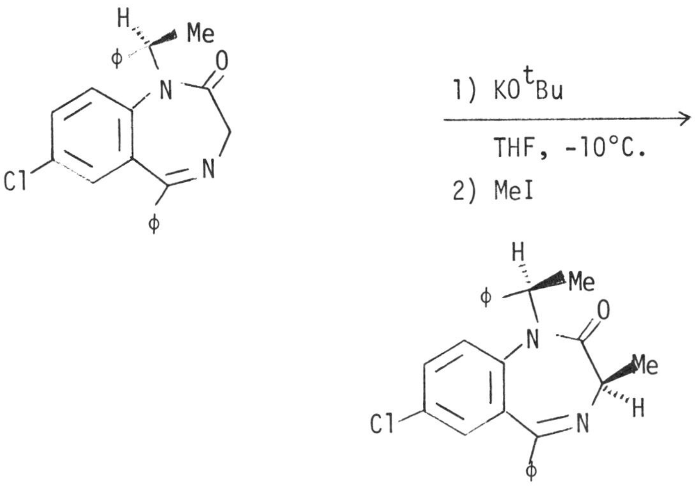

1) KO^{t}Bu

———————————→

THF, -10°C.

2) MeI

69%

31% Diast. Excess.

I.A.3-17 P. A. Wade et al, J. Org. Chem., 46, 765 (1981).

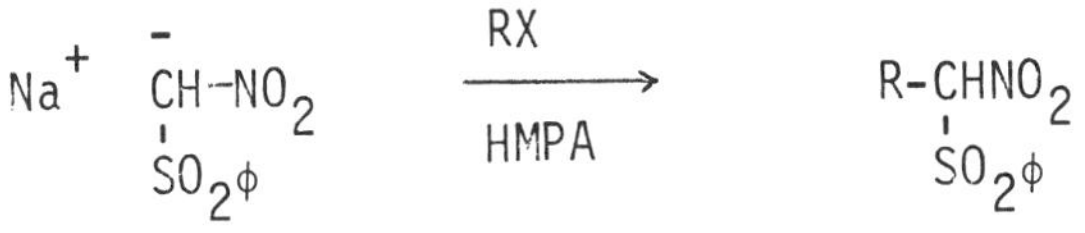

$$Na^+ \quad \overset{-}{\underset{SO_2\phi}{CH-NO_2}} \quad \xrightarrow[HMPA]{RX} \quad \underset{SO_2\phi}{R-CHNO_2}$$

54-75%

I.A.3-18 D. Scholz, Chem. Ber., 114, 909 (1981); K. Sato, S.
Inoue and T. Sakamoto, Synthesis, 796 (1981).

$$
\begin{array}{l}
\text{1) LiH/DMF} \\
\text{2) RX} \\
\text{3) Br}_2\text{/NaOH} \\
\text{4) KOH or KO}^t\text{Bu} \\
\text{(Ramberg-Backlund)}
\end{array}
$$

$$HO_2C-CH_2-CH_2-CH=CH-R$$

Z or E

I.A.4. Alkylation of N-, S- and Se-Stabilized Carbanions

I.A.4-1 A. R. Katritzky, G. DeVille and R. C. Patel, Tetrahedron, 37 (Suppl. 1), 25 (1981); A. R. Katritzky and S. S.
Thind, J. Chem. Soc., Perkin I, 661 (1981).

$$
R^1{-}R^2{>}C{-}NO_2 \quad + \quad \phi\text{-pyridinium} \quad \xrightarrow[\ 50\text{-}100°C.\ ]{DMSO} \quad R^1{-}\underset{R^3}{\overset{R^2}{C}}{-}NO_2
$$

33-78%

I.A.4-2 N. Kornblum and A. S. Erickson, <u>J. Org. Chem.</u>, <u>46</u>, 1037 (1981); S. Hoz, B. A. Feit et al, <u>ibid</u>, 450.

$$\underset{\substack{| \\ CH_3}}{\overset{\substack{CH_3 \\ |}}{R-C-NO_2}} \quad \xrightarrow[\substack{NaH/DMSO \\ 25°C/Light \\ (R = Alkyl\ or\ Aryl)}]{NaCH_2NO_2} \quad \underset{\substack{| \\ CH_3 \\ \\ 60-95\%}}{\overset{\substack{CH_3 \\ |}}{R-C-CH_2NO_2}}$$

Primary nitro product readily converted to aldehyde.

___

I.A.4-3 L. Rene and R. Royer, <u>Synthesis</u>, 878 (1981).

$$R^2CH_2NO_2 \quad \xrightarrow[\substack{ZnCl_2\ or\ Zn\ dust \\ (-R^1OH)}]{(R^1O)_3CH} \quad \underset{\substack{| \\ CH(OR^1)_2 \\ \\ 13-49\%}}{R^2-CH-NO_2}$$

___

I.A.4-4 I. Hoppe and U. Schollkopf, <u>Justus Liebigs Ann. Chem.</u>, 103 (1981); A. H. Schulthess and H. J. Hansen, <u>Helv. Chim. Acta</u>, <u>64</u>, 1322 (1981).

1) 2 BuLi

THF, -78°C.

2) RX

3) H_3O^+

52-64%

I.A.4-5 S. I. Nakatsuka et al, Chem. Lett., 695 (1981).

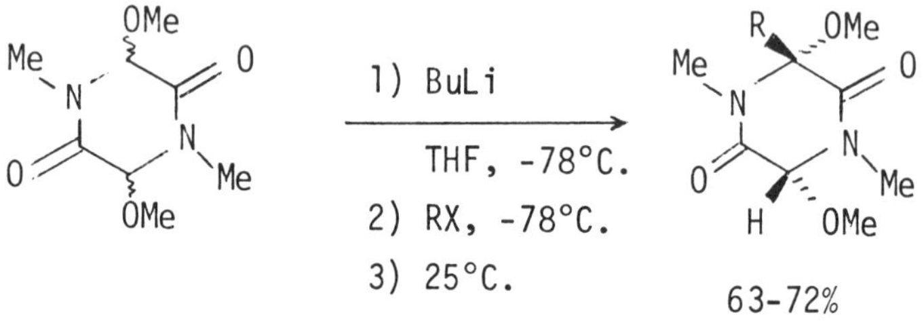
OMe
Me
N
O
N
Me
O
OMe
1) BuLi
THF, -78°C.
2) RX, -78°C.
3) 25°C.
R
OMe
Me
N
O
N
Me
O
H
OMe
63-72%

I.A.4-6 W. S. Johnson, B. Frei and A. S. Gopalan, J. Org. Chem., 46, 1512 (1981); M. Lissel, Synth. Commun., 11, 343 (1981).

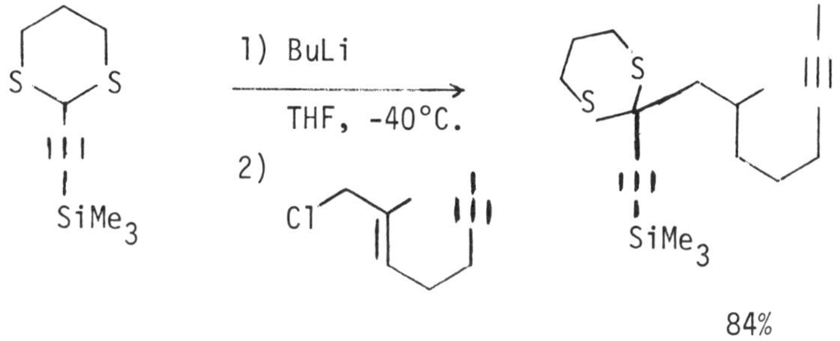
S
S
SiMe3
1) BuLi
THF, -40°C.
2)
Cl
S
S
SiMe3
84%

I.A.4-7 E. Fujita et al., Tetrahedron Lett., 22, 2005 (1981).

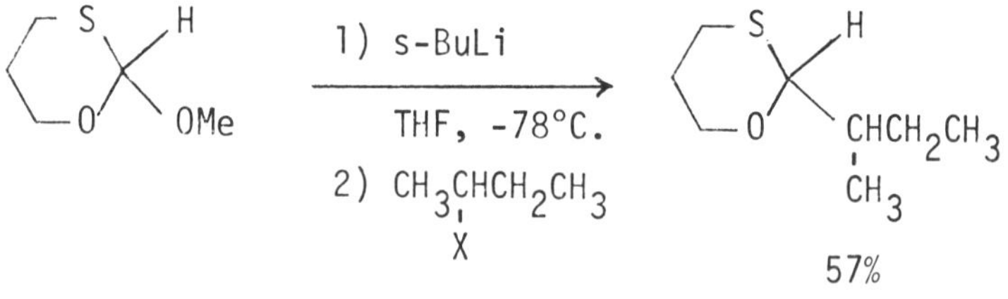
S
H
O
OMe
1) s-BuLi
THF, -78°C.
2) CH3CHCH2CH3
X
S
H
O
CHCH2CH3
CH3
57%

I.A.4-8 W. Oppolzer, R. L. Snowden and P. H. Briner, Helv. Chim. Acta, 64, 2022 (1981); K. Tanaka, M. Terauchi and A. Kaji, Chem. Lett., 315 (1981); K. Sato et al, J. Chem. Soc., Perkin I, 761 (1981).

1) LDA/HMPA

THF, -78°C.

2) RX

3) aq. NH_4Cl

78-91%

I.A.4-9 A. Tomazie and E. Ghera, Tetrahedron Lett., 22, 4349 (1981).

1) 3 eq. LDA/HMPA

THF, -78°C.

2) MeI

3) NH_4Cl

62-76%

I.A.4-10 D. J. Ager, Tetrahedron Lett., 22, 2803 (1981).

$$\phi SCH_2SiMe_3 \xrightarrow[\text{2) } E^+]{\text{1) nBuLi/DME}} \phi SCHSiMe_3$$

$$\underset{E}{|}$$

21-82%

E^+ = RX, RCO_2R^1, $(RCO)_2O$, RCOCl, RCHO, R_2CO and α,β-unsaturated ketones (1,4 addition).

I.A.4-11   T. Cohen et al, <u>Tetrahedron Lett.</u>, <u>22</u>, 3377 (1981);
D. van Leusen, P. H. F. M. Rouwette and A. M. van Leusen, <u>J.</u>
<u>Org. Chem.</u>, <u>46</u>, 5159 (1981);   T. Kato et al, <u>Chem. Lett.</u>, <u>25</u>
(1981).

$(\phi S)_2 CH \qquad CH(S\phi)_2$

1) s-BuLi/TMEDA
   →
   THF, -78°C.
2) H_2O
3) $CuCl_2/TiCl_4$
   aq. HOAc

$\phi S \qquad O$

60%

I.A.4-12   C. Bodeker, E. R. de Waard and H. O. Huisman, <u>Tetra-</u>
<u>hedron</u>, <u>37</u>, 1233 (1981);   T. Durst, M. Lancaster and D. J. H.
Smith, <u>J. Chem. Soc., Perkin I</u>, 1846 (1981).

CH_2OH

ϕSO_2

1) 2 BuLi
   →
2) $Me_2CHCH_2CH_2Br$
3) $Li/H_2NCH_2CH_2NH_2$

CH_2OH

42%

I.A.4-13   H. J. Reich et al, <u>J. Amer. Chem. Soc.</u>, <u>103</u>, 3112
(1981);   E. Negishi, C. L. Rand and K. P. Jadhav, <u>J. Org.</u>
<u>Chem.</u>, <u>46</u>, 5041 (1981).

$\phi SeCH_2C\equiv CH$

1) 2 LDA, -78°C.
   →
2) RX
3) E^+
4) O_3
5) -30°C.

$E^+ = H_2O$, RX, RCHO, R_2CO, Me_3SiCl.

$\phi Se \qquad O$
         $\Vert$ E
R   H

42-68%

I.A.5. Alkylations of Organometallic Reagents
    (see also: I.F, I.G).

I.A.5-1 Y. H. Lai, Synthesis, 585 (1981).

Review: "Grignard Reagents from Chemically Activated Magnesium."

I.A.5-2 D. A. Holt, Tetrahedron Lett., 22, 2243 (1981).

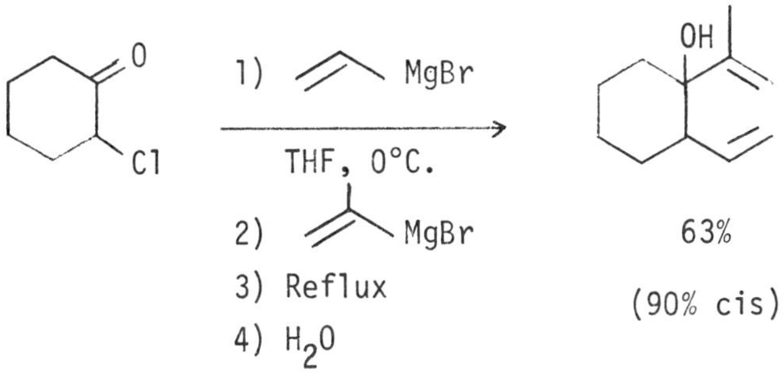

1) ⤳ MgBr

THF, 0°C.

2) ⤳ MgBr
3) Reflux
4) H$_2$O

63%

(90% cis)

I.A.5-3 H. Ishikawa, T. Mukaiyama and S. Ikeda, Bull. Chem. Soc. Jpn., 54, 776 (1981); W. F. Bailey and A. A. Croteau, Tetrahedron Lett., 22, 545 (1981); G. P. Axiotis, ibid, 1509; F. Barbot and P. Miginiac, J. Organometal. Chem., 222, 1 (1981).

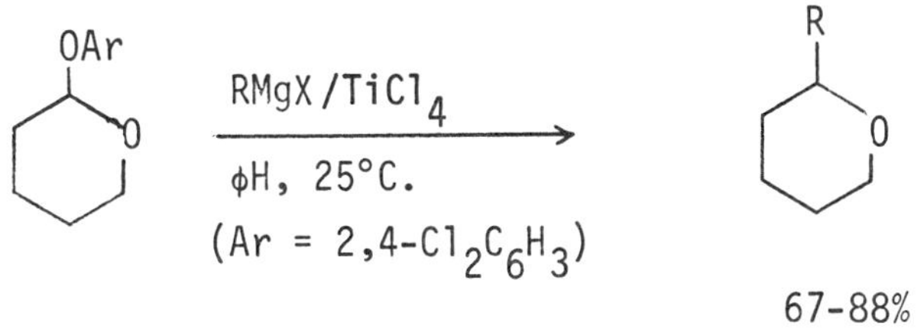

RMgX /TiCl$_4$

φH, 25°C.
(Ar = 2,4-Cl$_2$C$_6$H$_3$)

67-88%

I.A.5-4 H. G. Richey, Jr. et al, $\underline{J.\ Org.\ Chem.}$, $\underline{46}$, 3773, 3780 (1981).

$$\phi CH=CHCH_2NMe_2 \xrightarrow[\substack{2)\ H_2O}]{\substack{1)\ CH_2=CHCH_2MgCl \\ \phi CH_3/THF/\Delta}} \begin{array}{l} \phi CH_2CHCH_2NMe_2 \\ \quad\ \ \, \underset{\displaystyle CH_2CH=CH_2}{|} \end{array}$$

$$44\%$$

I.A.5-5 H. Felkin, S. G. Davies et al, $\underline{Chem.\ Commun.}$, 681 (1981).

$$Me\diagup\diagdown\diagup\diagdown OH \xrightarrow[\substack{[(-)\ DIOP]NiCl_2 \\ Et_2O,\ 20°C.}]{MeMgBr} Me\diagup\diagdown\overset{*}{\diagup\diagdown} \underset{Me}{|}$$

$$15\%\ ee$$

DIOP = 0,0'-Isopropylidene-2,3-dihydroxy-1,4-
bis(diphenylphosphino)butane.

I.A.5-6 I. Degani, R. Fochi and V. Regondi, $\underline{Tetrahedron\ Lett.}$, $\underline{22}$, 1821 (1981); T. Umemoto and Y. Kuriu, $\underline{ibid}$, 5197.

$$\underset{S}{\overset{S}{\diagdown}}\diagup\underset{D}{\overset{D}{\diagdown}} \xrightarrow[\substack{2)\ RMgX/Et_2O \\ 3)\ HgO/THF \\ aq.\ HBF_4}]{1)\ \phi_3C^+\ ClO_4^-} \underset{69-82\%}{\overset{\displaystyle O \atop \displaystyle \|}{R-C-D}}$$

NOTE: See $\underline{Tetrahedron\ Lett.}$ $\underline{22}$ (22) i (1981) for a
        report of an explosion with 1,3-benzodithiolium
        perchlorate.

I.A.5-7 T. L. Macdonald, B. A. Narayanan and D. E. O'Dell,
J. Org. Chem., 46, 1504 (1981).

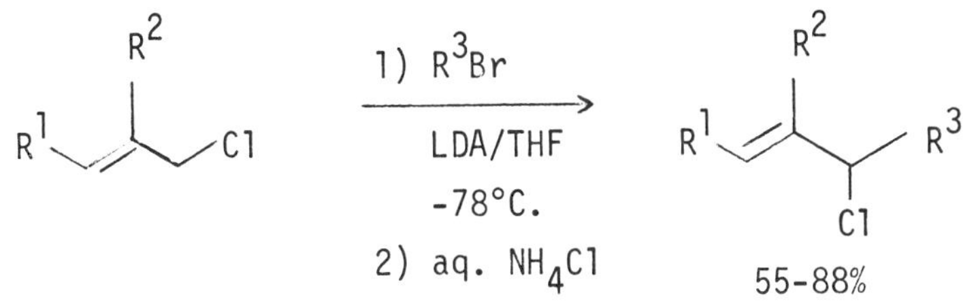

I.A.5-8 R. Hanko and D. Hoppe, Angew Chem., Int. Ed. Engl.,
20, 127 (1981); K. Ohno and M. Machida, Tetrahedron Lett.,
22, 4487 (1981).

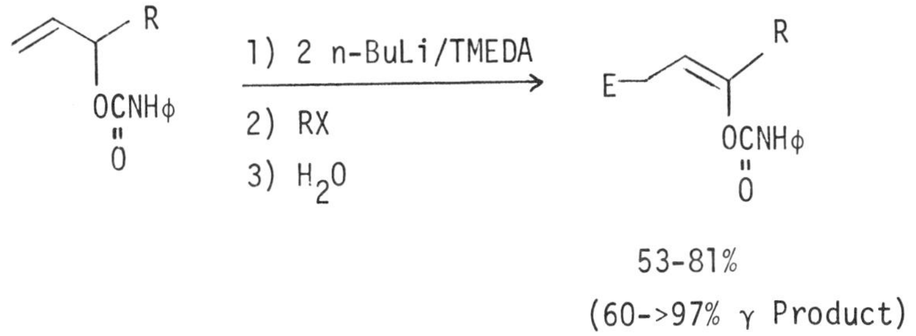

53-81%
(60->97% γ Product)

I.A.5-9 J. J. Fitt and H. W. Gschwend, J. Org. Chem., 46,
3349 (1981).

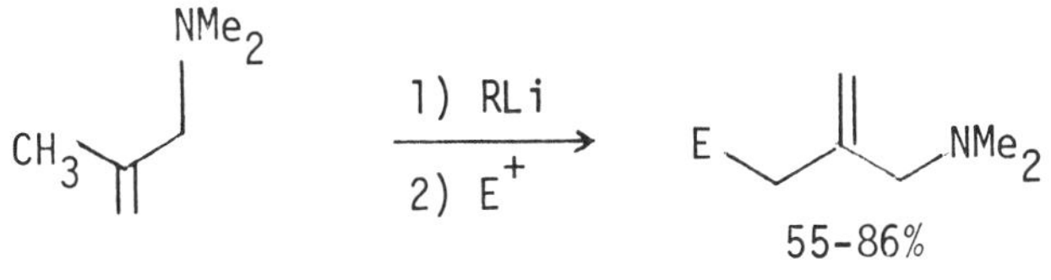

55-86%

E^+ = RX, RCHO, R_2CO, $ArCONMe_2$.

I.A.5-10 D. Seebach et al, Helv. Chim. Acta, 64, 643, 1337 (1981); A. I. Meyers, S. Hellring and W. T. Hoeve, Tetrahedron Lett., 22, 5115 (1981); A. I. Meyers and S. Hellring, ibid, 5119; K. N. Houk, P. Beak, P. v. R. Schleyer et al, J. Org. Chem., 46, 4108 (1981).

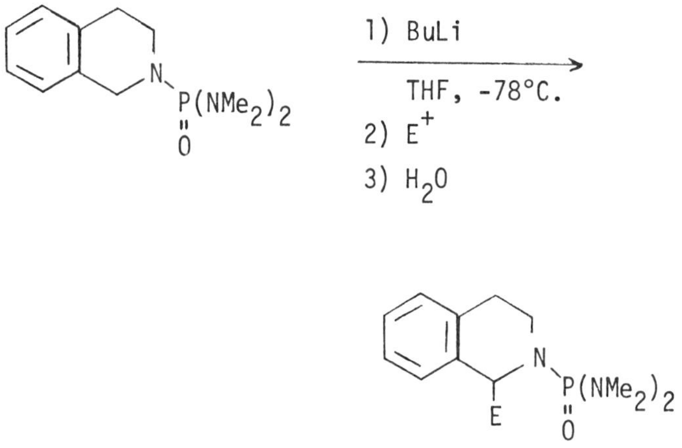

1) BuLi

THF, -78°C.

2) E$^+$

3) H$_2$O

52-95%

E$^+$ = RX, RCHO, R$_2$CO, Epoxides, D$_2$O.

Deprotection of Nitrogen with HCl/MeOH/H$_2$O, Reflux.

I.A.5-11 M. K. Yeh, J. Chem. Soc., Perkin I, 1652 (1981).

1) nBuLi/TMEDA

ϕCH$_2$OCH$_3$

Hexane, -10°C.

2) E$^+$, -10°C.

ϕCHOCH$_3$
  |
  E

22-85%

E$^+$ = BuBr, ϕ_2CO, ϕCH$_2$CO$_2$Et.

I.A.5-12 I. Fleming and C. D. Floyd, J. Chem. Soc., Perkin
I., 969 (1981).

$$Me_3Si-\underset{\underset{Me_3Si}{|}}{\overset{\overset{Me_3Si}{|}}{C}}-Li \quad \xrightarrow{E^+} \quad Me_3Si-\underset{\underset{Me_3Si}{|}}{\overset{\overset{Me_3Si}{|}}{C}}-E$$

E^+ = RX, Some Epoxides and Non-Enolizable RCHO,
R_2CO and RCOCl.

I.A.5-13 J. P. Quintard, B. Elissondo and M. Pereyre, J.
Organometal. Chem., 212, C31 (1981).

$$Bu_3SnCH(OEt)_2 \quad \xrightarrow[\substack{2) \ \phi CH_2Br \\ 3) \ H_3O^+}]{1) \ BuLi} \quad \phi CH_2-CHO$$

Also, reaction with benzaldehyde and cyclohexenone.

I.A.5-14 M. J. Smith and S. E. Wilson, Tetrahedron Lett., 22,
4615 (1981); J. Stapersma and G. W. Klumpp, Tetrahedron,
37, 187 (1981).

$$\xrightarrow[\substack{nBuLi/TMEDA \\ 2) \ H_2O}]{1) \ Li/Et_2O}$$

44%

Without added TMEDA, opposite stereochemistry
observed (33%).

I.A.5-15 B. H. Lipshutz, R. S. Wilhelm and D. M. Floyd, J. Amer. Chem. Soc., 103, 7672 (1981); R. C. Larock and D. R. Leach, Tetrahedron Lett., 22, 3435 (1981).

I.A.5-16 H. L. Goering et al, J. Org. Chem., 46, 5304, 2144 (1981).

I.A.5-17 R. J. P. Corriu, C. Guerin and J. M'Boula, Tetrahedron Lett., 22, 2985 (1981); F. Derguini, Y. Bessiere and G. Linstrumelle, Synth. Commun., 11, 859 (1981).

$$\text{Me}_3\text{SiCH=CHCH}_2\text{-CuCN} \quad \xrightarrow[-78°C.]{E^+} \quad \text{Me}_3\text{SiCH=CHCH}_2\text{-E}$$

Li$^+$ 45-75%

E^+ = RX, RCHO, RCOCl, RCH=CHCHO(1,2 Addn),
         RCH=CHCO$_2$Et (1,4 Addn).

I.A.5-18 M. T. Reetz et al, <u>Synth. Commun.</u>, <u>11</u>, 261 (1981);
<u>Chem. Ind.</u>, 541 (1981).

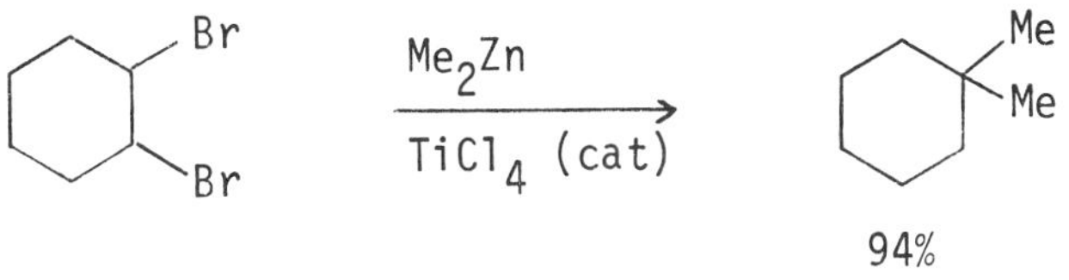

94%

Gem-dihaloalkanes to gem-dimethyl derivatives also.

I.A.5-19 M. Sekine, M. Nakajima and T. Hata, <u>J. Org. Chem.</u>,
<u>46</u>, 4030 (1981); P. Savignac et al, <u>Tetrahedron</u>, <u>37</u>, 1377
(1981).

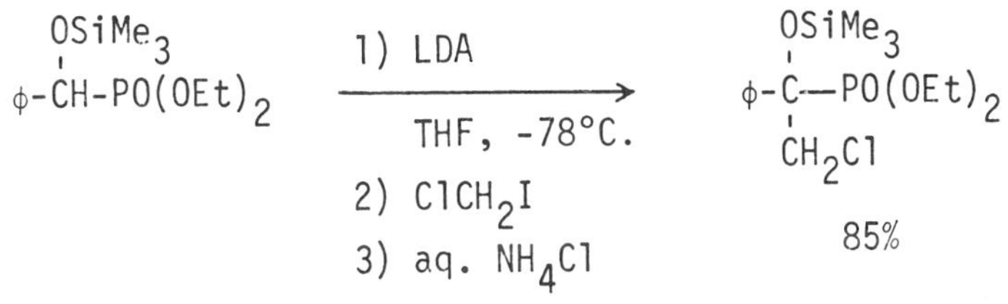

1) LDA

THF, -78°C.

2) ClCH$_2$I

3) aq. NH$_4$Cl

85%

I.A.5-20 I. Fleming and B. W. Au-Yeung, <u>Tetrahedron</u>, <u>37</u>
(Suppl. 1), 13 (1981); H. Sakurai, K. Sasaki and A. Hosomi,
<u>Tetrahedron Lett.</u>, <u>22</u>, 745 (1981).

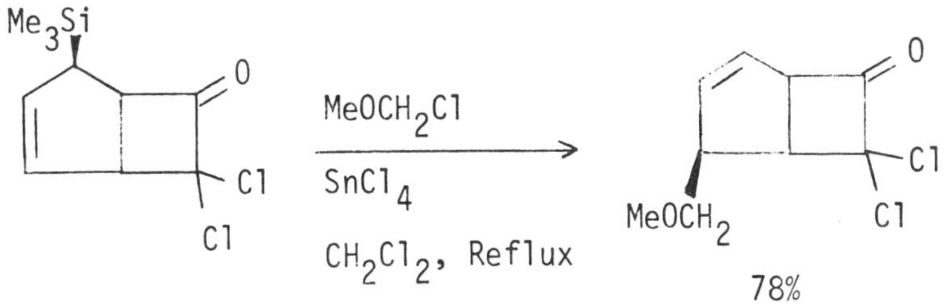

MeOCH$_2$Cl

SnCl$_4$

CH$_2$Cl$_2$, Reflux

78%

I.A.6. Other Alkylation Procedures and Reviews

I.A.6-1   W. Lamanna and M. Brookhart, J. Amer. Chem. Soc.,
$\underline{103}$, 989 (1981).

$$(CO)_3\overset{-}{Mn}\text{—}\hspace{-0.5em}\bigcirc \quad \xrightarrow[\substack{MeOSO_2F \\ 2)\ KH/THF}]{1)\ MeI\ or} \quad (CO)_3\overset{-}{Mn}\text{—}\hspace{-0.5em}\bigcirc^{\blacktriangleleft Me}$$

79%

I.A.6-2   T. Shono, Y. Matsumura and K. Tsubata, Tetrahedron
Lett., $\underline{22}$, 2411 (1981).

$$\underset{\underset{CO_2Me}{|}}{R^1\diagdown N}\diagup\underset{R^2}{\overset{|}{CH}}\diagdown OMe \quad \xrightarrow[\substack{CH_2Cl_2 \\ -70°C.}]{\phi NC/TiCl_4} \quad \underset{\underset{CO_2Me}{|}}{R^1\diagdown N}\diagup\underset{R^2}{\overset{|}{CH}}\diagdown CONH\phi$$

34-82%

Products hydrolyzed to amino acids.

I.A.6-3   T. Saegusa et al, J. Org. Chem., $\underline{46}$, 4980 (1981).

$$\overset{O}{\bigcirc} \quad \xrightarrow[\substack{DMF,\ 50°C \\ 2)\ \diagup\hspace{-0.6em}\diagdown Br,\ 25°C.}]{1)\ NC\text{-}CH_2\text{-}CO_2Cu\cdot nBu_3P} \quad \overset{O}{\bigcirc}\hspace{-1em}\diagup\hspace{-0.5em}CH_2CH=CH_2 \\ CO_2CH_2CH=CH_2$$

87%

I.A.6-4   A. Nakamura et al, Chem. Lett., 719 (1981).

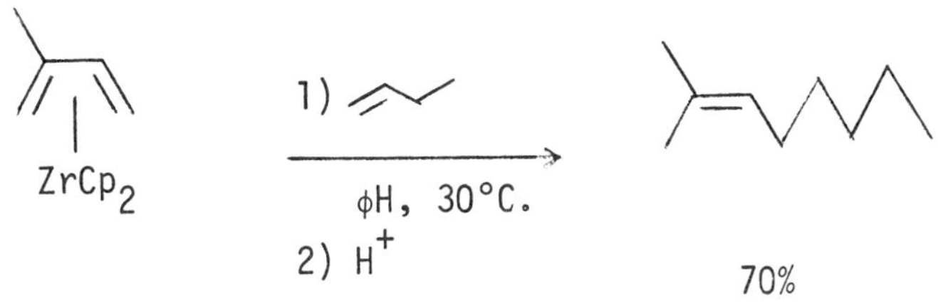

Also reactions with Alkynes and Dienes.

I.A.6-5   H. Lehmkuhl, I. Doring and H. Nehl, J. Organometal.
Chem., 221, 123 (1981);   H. Lehmkuhl and H. Nehl, ibid, 131
(1981).

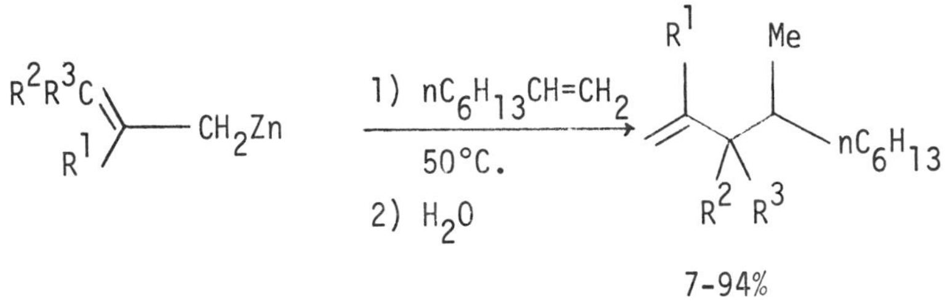

7-94%

I.A.6-6   B. Giese, K. Heuck and U. Luning, Tetrahedron Lett.,
22, 2155 (1981).

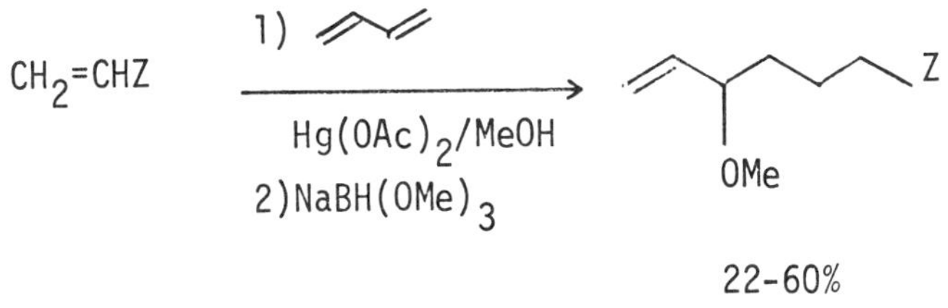

22-60%

$Z = CN, COCH_3, CO_2CH_3$.

I.A.6-7   R. C. Larock et al, Tetrahedron Lett., 22, 5231
(1981).

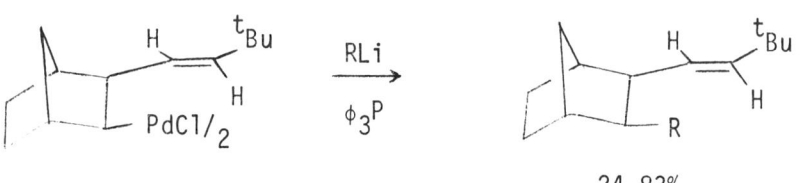

24-83%

Vinylmercurial plus olefin and Li$_2$PdCl$_4$ gives starting
material.

I.A.6-8   E. Hungerbuhler and D. Seebach, Helv. Chim. Acta, 64,
687 (1981);   E. Hungerbuhler, D. Seebach and D. Wasmuth, ibid,
1467.

Chiral Electrophilic Synthetic Building Blocks with
Four Different Functional Groups.

I.A.6-9   M. F. Semmelhack, Pure Appl. Chem., 53, 2379 (1981).

Review: "Nucleophilic Addition to Diene and Arene-
          Metal Complexes."

I.A.6-10   T. A. Hase and J. K. Koskimies, Aldrichimica Acta,
14, 73 (1981).

Review: "A Compilation of References on Formyl and
         Acyl Anion Synthons."

I.A.6-11 R. Noyori, S. Murata and M. Suzuki, Tetrahedron, 37, 3899 (1981).

    Review: "Trimethylsilyl Triflate in Organic Synthesis."

I.A.6-12 I. Fleming, Bull. Soc. Chim. Fr. II, 7 (1981); Chem. Soc. Rev., 10, 83 (1981).

    Review: "Some Uses of Silicon Compounds in Organic
             Synthesis."

I.A.6-13 H. P. Abicht and K. Issleib, Zeit. Chem., 21, 341 (1981).

    Review: "C-Lithiated Phosphines as Synthons."

I.A.6-14 Y. V. Belkin and N. A. Polezhaeva, Russ. Chem. Rev., 50, 481 (1981).

    Review: "The Chemistry of Stabilized Sulphonium
             Ylides."

I.A.6-15 M. A. Winnik, Chem. Rev., 81, 491 (1981).

    Review: "Cyclization and the Conformation of
             Hydrocarbon Chains."

I.A.6-16   G. Illuminati and L. Mandolini, Acct. Chem. Res.,
14, 95 (1981).

Review: "Ring Closure Reactions of Bifunctional
           Chain Molecules."

I.A.6-17   H. Wynberg, Rec. Trav. Chim., 100, 393 (1981).

Review: "Chance, Necessity and Asymmetric Catalysis."

I.A.6-18   T. Mukaiyama, Tetrahedron, 37, 4111 (1981).

Review: "Asymmetric Synthesis based on Chiral
           Diamines having Pyrrolidine Ring."

I.A.6-19   G. Solladie, Synthesis, 185 (1981).

Review: "Asymmetric Synthesis Using Nucleophilic
           Reagents Containing a Chiral Sulfoxide Group."

I.A.6-20   G. H. Posner et al, Pure Appl. Chem., 53, 2307
(1981).

Review: "Asymmetric Synthesis Using Organometallic
           Reagents and Optically Pure Solfoxides."

I.A.6-21 B. Fraser-Reid, K. M. Sun and T. F. Tam, Bull. Soc. Chim. Fr. II, 238 (1981).

Review: "Carbohydrate Derivatives in the Asymmetric Synthesis of Natural Products: Some Applications of Furanose Sugars."

I.A.6-22 S. Rajappa, Tetrahedron, 37, 1453 (1981).

Review: "Nitroenamines: Preparation, Structure and Synthetic Potential."

I.A.6-23 G. Cainelli and G. Cardillo, Acct. Chem. Res., 14, 89 (1981).

Review: "Some Aspects of the Stereospecific Synthesis of Terpenoids by Means of Isoprene Units."

I.A.7. Nucleophilic Addition to Electron Deficient Carbon

I.A.7.a.1a. Intermolecular Aldol-Type 1,2-Additions

I.A.7.a.1a-1 C. H. Brieskorn and W. Schwack, Chem. Ber., 1993 (1981); J. L. Gras, Org. Syn., 60, 88 (1981); S. Serota, W. M. Linfield et al, J. Org. Chem., 46, 4147 (1981); S. Y. Ambekar, T. B. H. McMurry and E. Martin, Chem. Ind., 877 (1981).

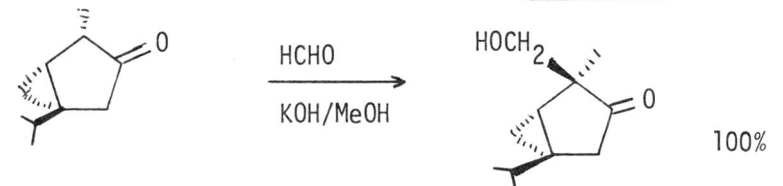

I.A.7.a.1a-2 R. C. Cookson and N. A. Mirza, Synth. Commun.,
11, 299 (1981).

$$+ \quad \xrightarrow[\substack{SnCl_2 \\ CH_2Cl_2, \ 0°C.}]{(HCHO)_n} \quad \text{55\%}$$

I.A.7.a.1a-3 E. V. Dehmlow and A. R. Shamout, J. Chem. Res.
(S), 106 (1981); A. Schulze and H. Oediger, Justus Liebigs
Ann. Chem., 1725 (1981); B. Zupancic and M. Kokalj, Synthesis,
913 (1981).

$$Me \xrightarrow[\substack{(R=Ar \ or \ ^tBu) \\ K_2CO_3/Aliquat \\ \phi CH_3, \ 90°C.}]{RCHO} R \qquad 87\text{-}97\%$$

I.A.7.a.1a-4 L. Engman and M. P. Cava, Tetrahedron Lett., 22,
5251 (1981); C. Fuganti, P. Grasselli et al, ibid, 965.

$$\phi CHO \quad + \quad CH_3 \overset{O}{\overset{\|}{C}} \phi \quad \xrightarrow[\substack{(Ar=pMeOC_6H_4) \\ \phi CH_3, \ reflux}]{\overset{O}{\overset{\|}{Ar\text{-}Te\text{-}Ar}}} \quad \phi CH=CHC\phi \quad 94\%$$

I.A.7.a.1a-5   T. Mukaiyama and M. Murakami, <u>Chem. Lett.</u>, 1129 (1981).

HC≡C-OEt

1) N→O/HgCl$_2$

2) Zn

3) RCHO

4) H$_2$O

OH
|
R⟍⟋⟍CO$_2$Et
|
Cl

67-81%

I.A.7.a.1a-6   A. B. Smith, III and P. A. Levenberg, <u>Synthesis</u>, 567 (1981).

R^1-C-CH-R^2 (with O double bond above C, and R^3 below CH)

1) LDA

THF, -78°C.

2) R^{4}CHO

3) Oxidation

R^1⟍C(=O)⟋C(R^2)(R^3)⟍C(=O)⟋R^4

20-69%

Oxidation Methods: CrO$_3$·2 Pyr or DMSO/CO$_2$Cl$_2$.

I.A.7.a.1a-7   J. A. Findlay, D. N. Desai and J. B. Macaulay, <u>Can. J. Chem.</u>, <u>59</u>, 3303 (1981).

(cyclohexanone)

CH$_3$(CH$_2$)$_8$CHO

150°C.

(cyclohexanone with O and CH(OH)-(CH$_2$)$_8$CH$_3$ substituent)

45%

I.A.7.a.1a-8   K. Oshima et al, Bull. Chem. Soc. Jpn., 54, 274
(1981).

$$\text{structure}\quad\xrightarrow[\substack{\text{THF, 0°C.}\\ \text{2) NaIO}_4 \\ \text{3) Silica Gel}}]{\text{1) Me}_2\text{AlS}\phi}\quad\text{structure}\quad 40\%$$

I.A.7.a.1a-9   D. L. J. Clive and C. G. Russell, Chem. Commun.,
434 (1981);   I. Matsuda and Y. Izumi, Tetrahedron Lett., 22,
1805 (1981);   J. Ojima et al, Ind. J. Chem., 20B, 803 (1981).

$$R^1\overset{OLi}{\diagup}\!\!\!\sim\!\!\sim R^2 \xrightarrow[\substack{\text{Et}_2\text{O, 0°C.}\\ \text{2) }\phi\text{SeCH}_2\text{CHO}\\ \text{3) MeSO}_2\text{Cl/Et}_3\text{N}}]{\text{1) ZnCl}_2} R^1\overset{O}{\diagup}\!\!\sim\!\!\diagdown R^2 \quad 42\text{-}83\%$$

I.A.7.a.1a-10   L. Mayring and T. Severin, Chem. Ber., 114,
3863 (1981);   C. Larsen and D. N. Harpp, J. Org. Chem., 46,
2465 (1981);   U. Quintily, G. Scorrano et al, J. Org. Chem.,
46, 5156 (1981);   E. Vedejs and D. M. Gapinski, Tetrahedron
Lett., 22, 4913 (1981).

$$\text{structure}\quad\xrightarrow[\substack{\text{NaH}\\ \text{2) SiO}_2}]{\text{1) Me}_2\text{NN=C-CHO}\ (CH_3)}\quad\text{structure}\quad 40\%$$

I.A.7.a.1a-11   A. Barco et al, Synthesis, 199 (1981).

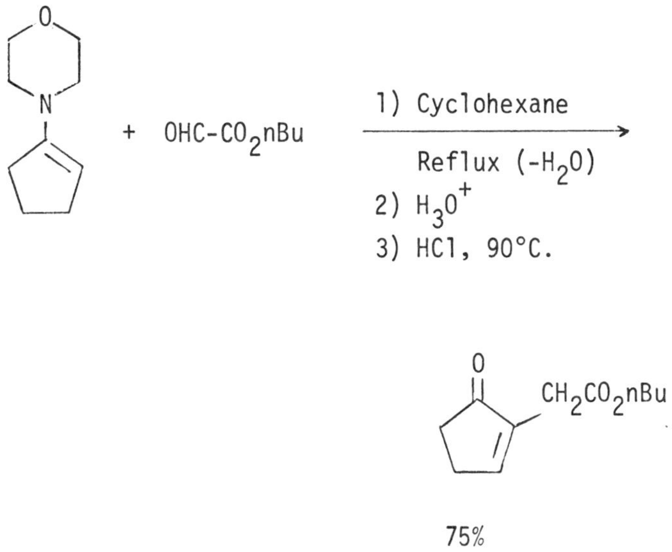

+   OHC-CO$_2$nBu

1) Cyclohexane
   ――――――――――→
   Reflux (-H$_2$O)
2) H$_3$O$^+$
3) HCl, 90°C.

O
CH$_2$CO$_2$nBu

75%

―――――――――――――

I.A.7.a.1a-12   D. Courtheyn et al, J. Org. Chem., 46, 3226
(1981);   J. M. Muchowski and M. C. Venuti, ibid, 459;   J.
Becher, L. Finsen and I. Winckelmann, Tetrahedron, 37, 2375
(1981).

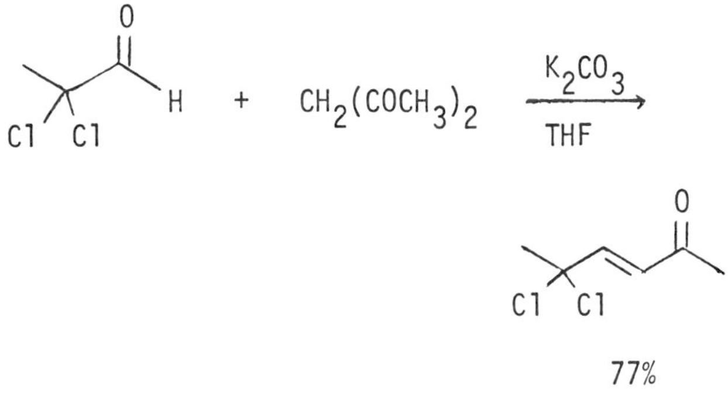

+   CH$_2$(COCH$_3$)$_2$ $\xrightarrow[\text{THF}]{\text{K}_2\text{CO}_3}$

77%

I.A.7.a.1a-13 G. Cardillo et al, J. Org. Chem., 46, 2439
(1981); M. Maleki, J. A. Miller and O. W. Lever, Jr.,
Tetrahedron Lett., 22, 3789 (1981); M. Gaudemar et al, J.
Organometal. Chem., 219, C1 (1981).

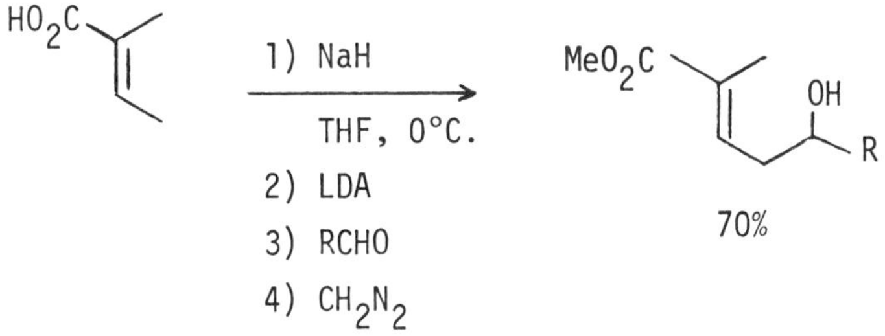

1) NaH

THF, 0°C.

2) LDA

3) RCHO

4) CH$_2$N$_2$

70%

I.A.7.a.1a-14 K. E. Harding, L. N. Moreno and V. M. Nace,
J. Org. Chem., 46, 2809 (1981); W. Adam et al, ibid, 3359.

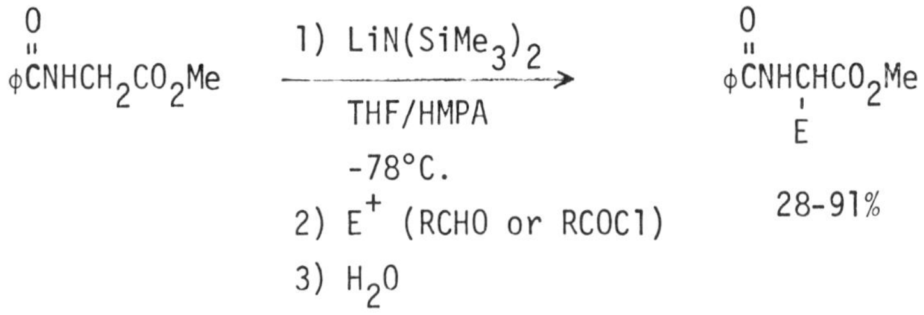

ϕCNHCH$_2$CO$_2$Me

1) LiN(SiMe$_3$)$_2$

THF/HMPA

-78°C.

2) E$^+$ (RCHO or RCOCl)

3) H$_2$O

ϕCNHCHCO$_2$Me
        |
        E

28-91%

I.A.7.a.1a-15 D. A. Evans et al, Pure Appl. Chem., 53, 1109
(1981).

Review: "Chiral Enolate Design."

I.A.7.a.1a-16 W. Choy, P. Ma and S. Masamune, Tetrahedron
Lett., 22, 3555 (1981).

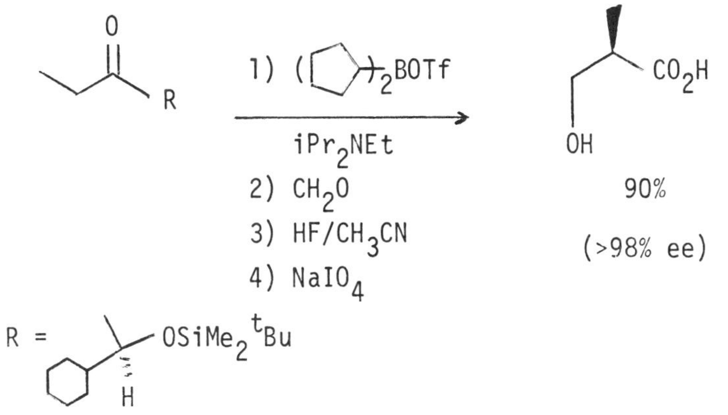

I.A.7.a.1a-17 C. H. Heathcock et al, J. Amer. Chem. Soc., 103,
4972 (1981); C. H. Heathcock et al, Tetrahedron, 37, 4087
(1981); C. T. White and C. H. Heathcock, J. Org. Chem., 46,
191, 1296, 2290 (1981).

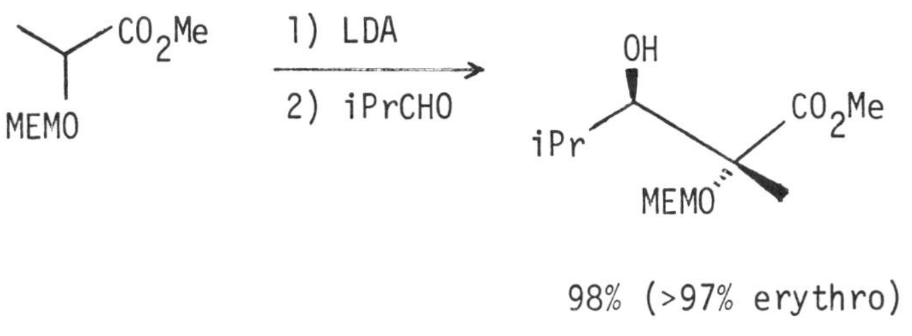

98% (>97% erythro)

I.A.7.a.1a-18 D. A. Evans and L. R. McGee, J. Amer. Chem.
Soc., 103, 2876 (1981).

Erythro Selective Aldols via Zirconium Enolates.

I.A.7.a.1a-19 S. Masamune et al, J. Amer. Chem. Soc., 103,
1566, 1568 (1981); D. A. Evans, J. Bartoli and T. L. Shih,
ibid, 2127; D. A. Evans et al, ibid, 3099; Y. Yamamoto, H.
Yatagai and K. Maruyama, ibid, 3229; A. I. Meyers and Y.
Yamamoto, ibid, 4278; M. Wada, Chem. Lett., 153 (1981); T.
Mukaiyama et al, ibid, 1193.

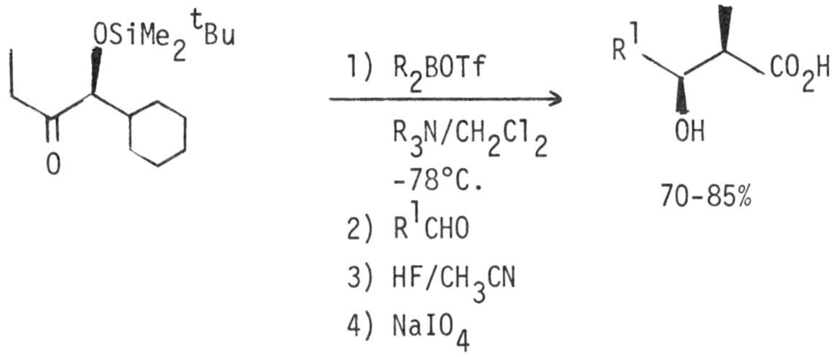

I.A.7.a.1a-20 R. Noyori, I. Nishida and J. Sakata, J. Amer.
Chem. Soc., 103, 2106 (1981); Y. Yamamoto, H. Yatagai and K.
Maruyama, Chem. Commun., 162 (1981); R. A. J. Smith et al,
J. Org. Chem., 46, 2932 (1981).

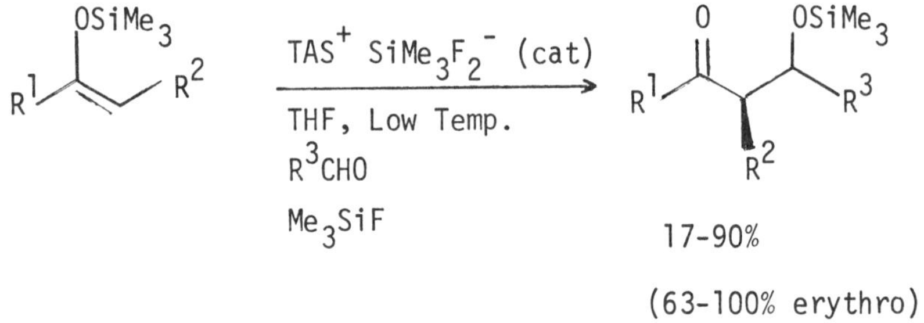

I.A.7.a.1a-21 M. T. Reetz and R. Peter, Tetrahedron Lett.,
22, 4691 (1981).

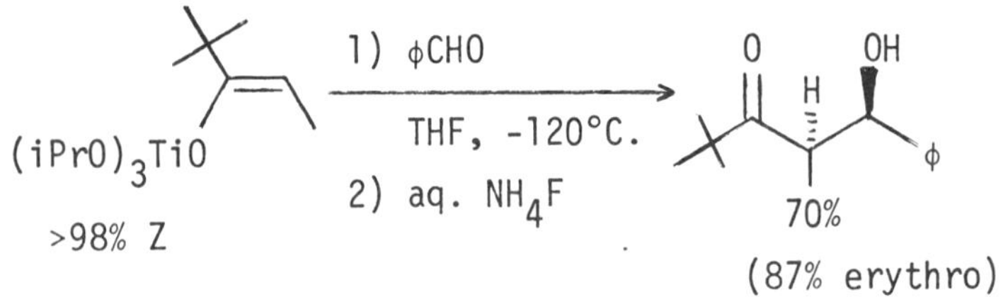

I.A.7.a.la-22 I. Ojima, S. I. Inaba and M. Nagai, _Synthesis_, 545 (1981); G. A. Kraus and M. Shimagaki, _Tetrahedron Lett._, 22, 1171 (1981); M. Ohno et al, _ibid_, 2109.

$$R^2\!\!\diagdown\!\!C\!=\!C\!\diagup\!\!OMe \qquad \begin{array}{c} 1)\ R^1CH=N\text{-}\phi \\ \xrightarrow{\hspace{2cm}} \\ TiCl_4/CH_2Cl_2 \\ 25°C. \\ 2)\ H_2O \end{array} \qquad \begin{array}{c} R^2 \\ R^3\text{-}\overset{|}{C}\text{-}CO_2Me \\ | \\ R^{1}\diagup^{CH\text{-}NH\phi} \\ 54\text{-}98\% \end{array}$$

Products readily converted to β-lactams.

I.A.7.a.la-23 P. Maitte et al, _Synthesis_, 130 (1981); D. G. Desai and R. B. Mane, _Ind. J. Chem._, 20B, 504 (1981).

$$RCN \qquad \begin{array}{c} 1)\ HCl/MeOH \\ \xrightarrow{\hspace{2cm}} \\ 2)\ \text{[structure]}, Et_3N \\ 3)\ NaOEt/EtOH \\ 4)\ H_2O \end{array} \qquad \begin{array}{c} R\diagdown \\ {}_{H_2N}\diagup C=CH\text{-}CO_2Et \\ 20\text{-}66\% \end{array}$$

I.A.7.a.la-24 D. Enders and H. Lotter, _Angew. Chem., Int. Ed. Engl._, 20, 795 (1981); R. Pellicciari et al, _J. Chem. Soc., Perkin I_, 2566 (1981); M. J. Marks and H. M. Walborsky, _J. Org. Chem._, 46, 5405 (1981).

$$R^1\overset{O}{\overset{\|}{C}}R^2 + \text{[pyrrolidine-CHO-OMe]} \qquad \begin{array}{c} 1)\ LiTMP \\ \xrightarrow{\hspace{2cm}} \\ THF,\ \text{-}100°C. \\ 2)\ 0°C.,\ H_3O^+ \end{array} \qquad \begin{array}{c} \text{[product structure]} \\ 71\text{-}80\% \end{array}$$

Separation of diasteriomers then reaction with MeLi gives enantiomerically pure α-hydroxyketones and vicinal diols.

I.A.7.a.1a-25 S. Hunig and M. Oller, Chem. Ber., 114, 959
(1981); L. Streinz and M. Romanuk, Coll. Czech. Chem. Commun.
46, 1682 (1981); R. M. Sandifer, A. K. Bhattacharya and T. M.
Harris, J. Org. Chem., 46, 2260 (1981).

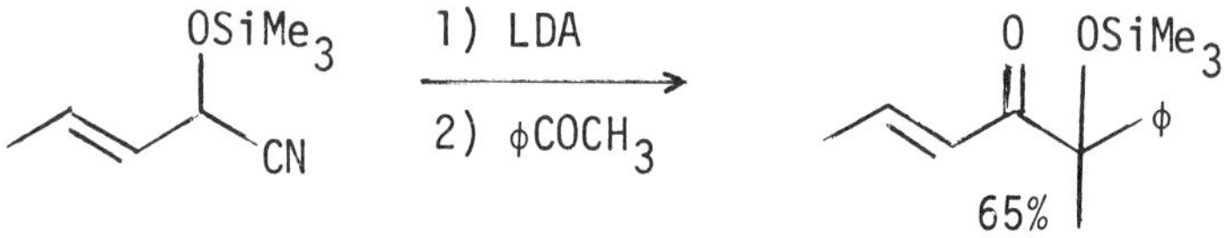

I.A.7.a.1a-26 W. T. Brady and R. D. Watts, J. Org. Chem., 46,
4047 (1981).

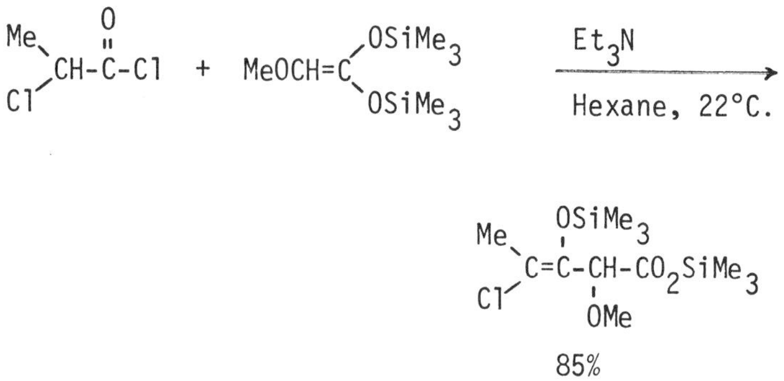

I.A.7.a.1a-27 I. Kopka and M. W. Rathke, J. Org. Chem., 46,
3771 (1981).

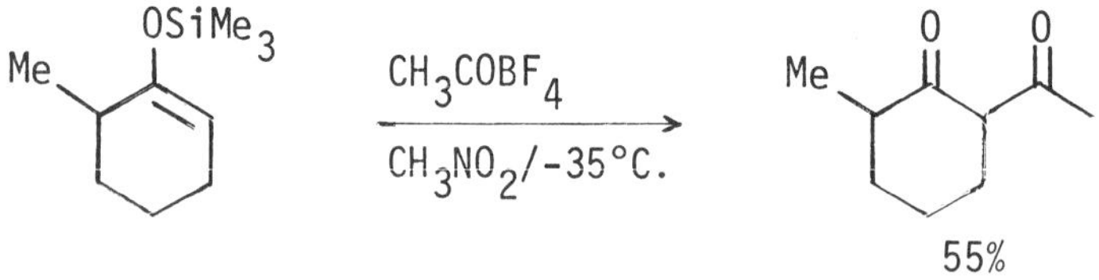

I.A.7.a.1a-28   T. Hata et al, Chem. Lett., 1087 (1981); G.
Doleschall, Synthesis, 478 (1981).

$$R^1-\overset{\overset{O}{\|}}{C}-\overset{\overset{R^2}{\diagup}}{C}H\diagdown R^3 \quad \xrightarrow[\substack{THF, -78°C. \\ O\ O \\ 2)\ \phi\overset{\|}{C}-\overset{\|}{P}(OEt)_2\ (2\ eq)}]{1)\ LiN(SiMe_3)_2} \quad R^1-\overset{\overset{O}{\|}}{C}-\overset{\overset{R^2}{|}}{\underset{R^3}{C}}-\overset{\overset{O}{\|}}{C}-\phi$$

Enolates from carbonyl compounds with pK_a ~ 11-25.

I.A.7.a.1a-29   D. Seebach et al, Helv. Chim. Acta, 64, 716,
736 (1981); M. E. Mueller and E. Leete, J. Org. Chem., 46,
3151 (1981); S. Ohta, A. Tsujimura and M. Okamoto, Chem.
Pharm. Bull., 29, 2762 (1981).

Products can by cyclized to hydroxy-nitroketones with
NaHCO$_3$.

I.A.7.a.1a-30   M. Uher et al, Coll. Czech. Chem. Commun., 46,
3128 (1981).

I.A.7.a.1a-31 J. L. Moreau, R. Couffignal and R. Arous-Chtara
Tetrahedron, 37, 307 (1981); M. Gaudemar et al, J. Organo-
metal Chem., 208, 279 (1981); S. Brandange et al, Acta Chem.
Scand. B, 35, 273 (1981); D. G. Hangauer, Jr., Tetrahedron
Lett., 22, 2439 (1981).

$$RCH_2CO_2SiMe_3 \xrightarrow{\text{1) LDA/Et}_2O, -60°C.}$$

2)

3) H_3O^+

4) Δ

49-70%

I.A.7.a.1b.   Intramolecular Aldol-Type 1,2-Additions

I.A.7.a.1b-1 Y. Inubushi et al, Chem. Pharm. Bull., 29, 766
(1981); J. C. Barriere et al, Helv. Chim. Acta, 64, 1140
(1981); R. J. Stoodley et al, J. Chem. Soc., Perkin I., 1782
(1981).

Morpholine

Camphoric Acid

Et_2O/HMPA

0°C.

16-58%

I.A.7.a.1b-2 J. C. Stowell and H. F. Hauck, Jr., J. Org.
Chem., 46, 2428 (1981); T. Hiyama et al, Bull. Chem. Soc.
Jpn., 54, 2747 (1981).

R-276 Rexyn 300*

CH_3OH, Reflux

48-56%

*Mixture of Sulfonic Acid and Quaternary Ammonium
Hydroxide Resins.

I.A.7.a.1b-3 H. J. Liu and H. K. Lai, Synth. Commun., 11, 65
(1981); Y. Yamada et al, Tetrahedron Lett., 22, 1353 (1981);
J. M. Cook et al, ibid, 3475.

$$EtSC\text{-}CH(CH_2)_3CSEt \quad \xrightarrow[\substack{EtSH \ (cat)/DME \\ 25°C.}]{NaH} \quad \phi CH_2 \text{—} \text{SEt} \quad 74\%$$

I.A.7.a.1b-4 D. R. Williams, A. Abbaspour and R. M. Jacobson,
Tetrahedron Lett., 22, 3565 (1981); D. K. Klipa and H. Hart,
J. Org. Chem., 46, 2815 (1981).

$$\xrightarrow[\substack{2) \ MsCl, \ Et_3N \\ 3) \ CF_3CO_2H}]{1) \ LiC≡C\text{-}CH(OEt)_2} \quad 51\%$$

I.A.7.a.1b-5 W. Kreiser and P. Below, Tetrahedron Lett., 22,
429 (1981); J. Ficini, G. Revial and J. P. Genet, ibid, 629,
633.

$$\xrightarrow[25°C.]{HCl/Me_3SiCl} \quad 70\text{-}93\%$$

$$\xrightarrow[Reflux]{/HOAc/\phi H} \quad 65\text{-}94\%$$

I.A.7.a.1b-6 S. Takano, C. Kasahara and K. Ogasawara, Chem. Commun., 635, 637 (1981); J. Pecher et al, Bull Soc. Chim. Belg., 90, 481 (1981).

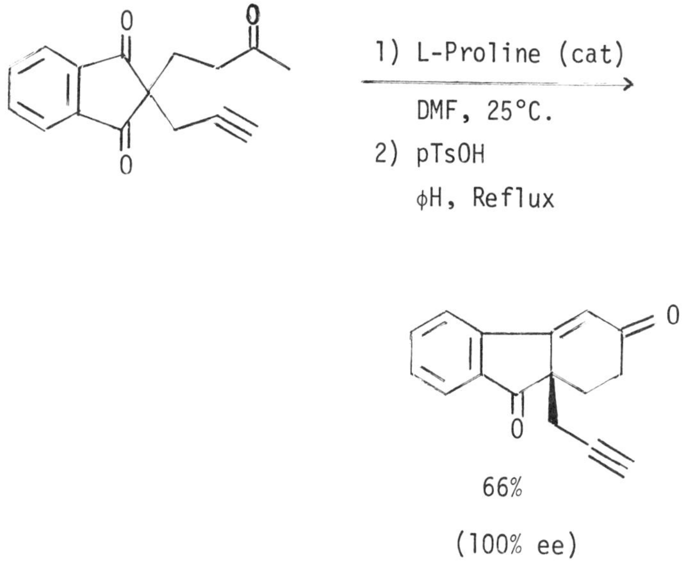

0
0
0
1) L-Proline (cat)
DMF, 25°C.
2) pTsOH
ϕH, Reflux
0
0
66%
(100% ee)

I.A.7.a.2. 1,2-Additions of N-, S- or Se- Stabilized Carbanions

I.A.7.a.2-1 D. Seebach et al, Angew. Chem., Int. Ed. Engl., 20, 397 (1981); Helv. Chim. Acta, 64, 2264 (1981).

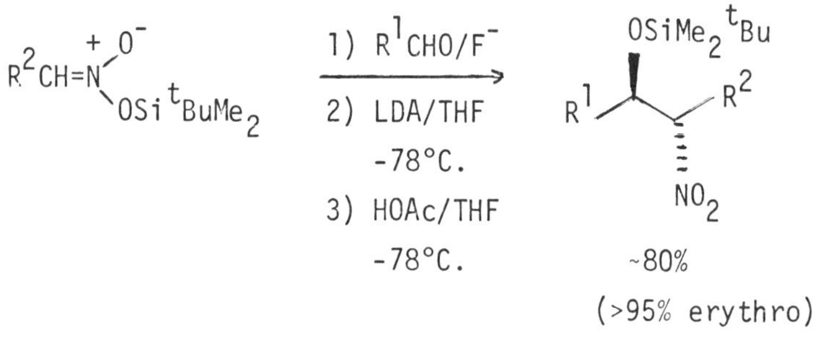

R²CH=N⁺ O⁻ OSi ᵗBuMe₂
1) R¹CHO/F⁻
2) LDA/THF
-78°C.
3) HOAc/THF
-78°C.
OSiMe₂ᵗBu
R¹
R²
NO₂
~80%
(>95% erythro)

I.A.7.a.2-2 Y. Hamada, K. Ando and T. Shioiri, Chem. Pharm. Bull., 29, 259 (1981); G. V. Nekrasova et al, J. Org. Chem. (USSR), 17, 619 (1981).

$$ArCO_2H \xrightarrow[\substack{(EtO)_2POCN \\ Et_3N}]{CH_3NO_2} \underset{28-81\%}{Ar\overset{O}{\overset{\|}{C}}CH_2NO_2}$$

I.A.7.a.2-3 T. Nakatsuka, T. Miwa and T. Mukaiyama, Chem. Lett., 279 (1981); U. Schollkopf and U. Groth, Angew. Chem., Int. Ed. Engl., 20, 977 (1981); U. Schollkopf, U. Groth and W. Hartwig, Justus Liebigs Ann. Chem., 2407 (1981); A. Calcagni, D. Rossi and G. Lucente, Synthesis, 445 (1981); D. Armesto, M. J. Ortiz and R. Perez-Ossorio, Tetrahedron Lett., 22, 2203 (1981).

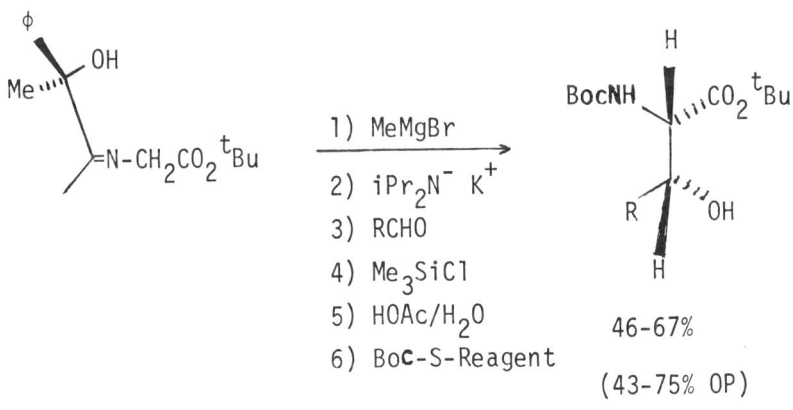

I.A.7.a.2-4 F. Heinzer and D. Bellus, Helv. Chim. Acta, 64, 2279 (1981); J. Rachon and U. Schollkopf, Justus Liebigs Ann. Chem., 1693 (1981).

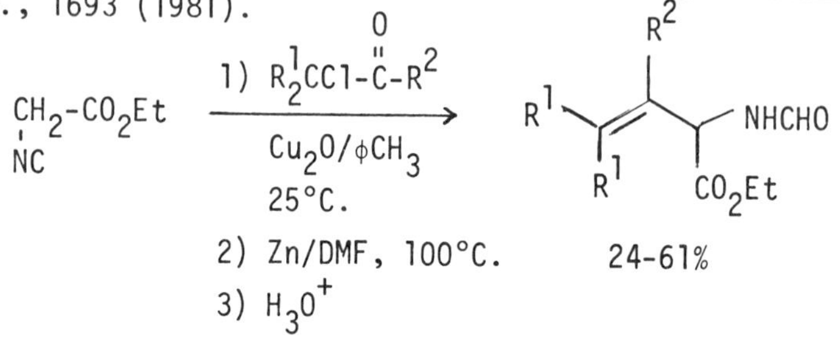

CH_2-CO_2Et
NC
1) R^1_2CCl-C(=O)-R^2
Cu_2O/ϕCH_3
25°C.
2) Zn/DMF, 100°C.
3) H_3O^+
R^1
R^1
R^2
NHCHO
CO_2Et
24-61%

I.A.7.a.2-5 V. Singh, V. V. Kane and A. R. Martin, Synth. Commun., 11, 429 (1981).

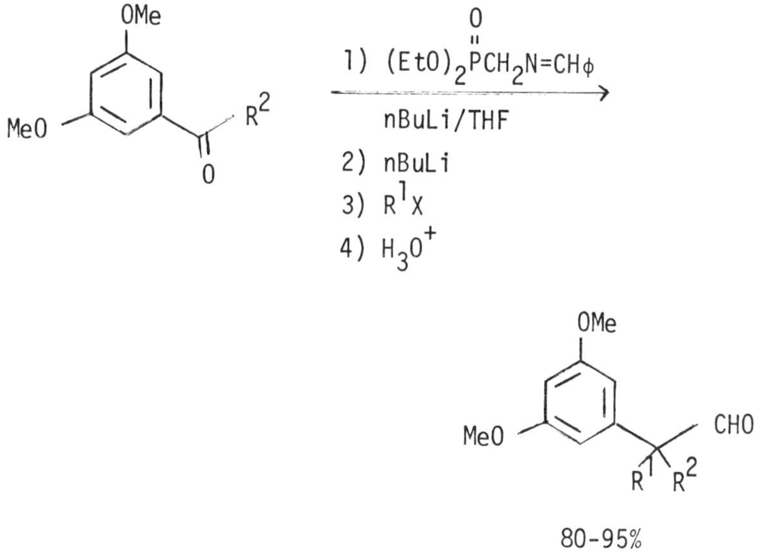

OMe
MeO
R^2
O
1) $(EtO)_2$P(=O)CH_2N=CHϕ
nBuLi/THF
2) nBuLi
3) R^1X
4) H_3O^+
OMe
MeO
CHO
R^1
R^2
80-95%

I.A.7.a.2-6 K. S. Kyler and D. S. Watt, J. Org. Chem., 46, 5182 (1981); K. Hiroi and L. M. Chen, Chem. Commun., 377 (1981).

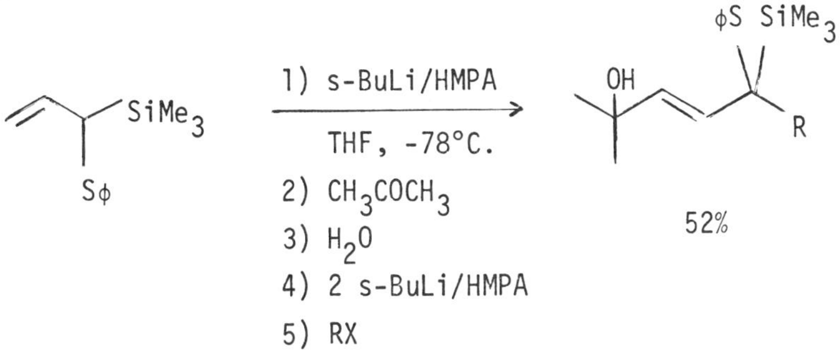

SiMe3
Sϕ
1) s-BuLi/HMPA
THF, -78°C.
2) CH3COCH3
3) H2O
4) 2 s-BuLi/HMPA
5) RX
ϕS SiMe3
OH
R
52%

I.A.7.a.2-7 A. de Groot and B. J. M. Jansen, Tetrahedron Lett., 22, 887 (1981).

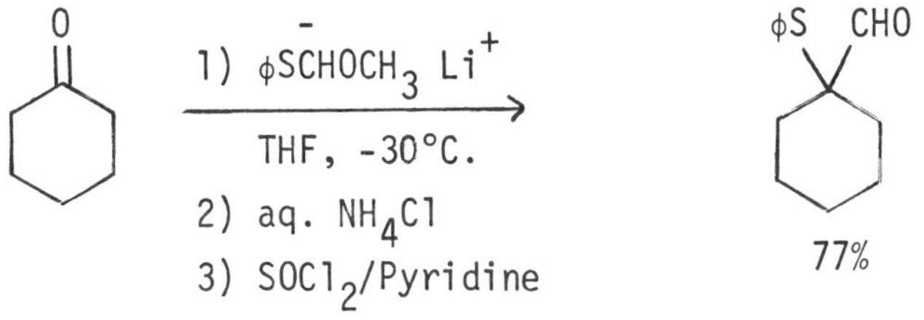

O
1) ϕSCHOCH3 Li+
THF, -30°C.
2) aq. NH4Cl
3) SOCl2/Pyridine
ϕS CHO
77%

I.A.7.a.2-8 Q. B. Cass, A. A. Jaxa-Chamiec and P. G. Sammes, Chem. Commun., 1248 (1981); T. Mandai, M. Kawada et al, Tetrahedron Lett., 22, 4489 (1981).

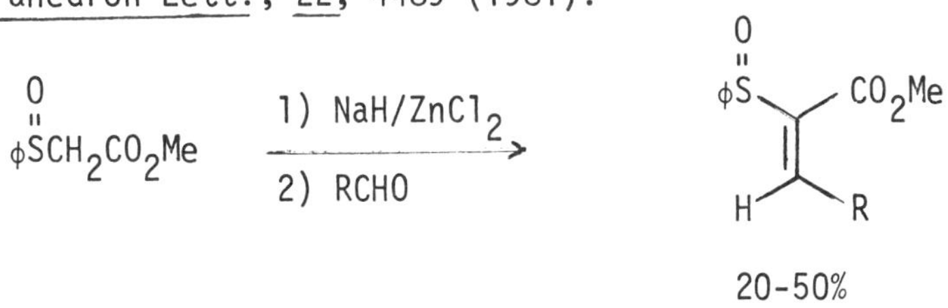

O
ϕSCH2CO2Me
1) NaH/ZnCl2
2) RCHO
O
ϕS
CO2Me
H
R
20-50%

Further reactions of conjugated sulfoxides studied.

I.A.7.a.2-9 D. R. Williams, J. G. Phillips and J. C. Huffman,
J. Org. Chem., 46, 4101 (1981); D. R. Williams and J. G.
Phillips, ibid, 5452.

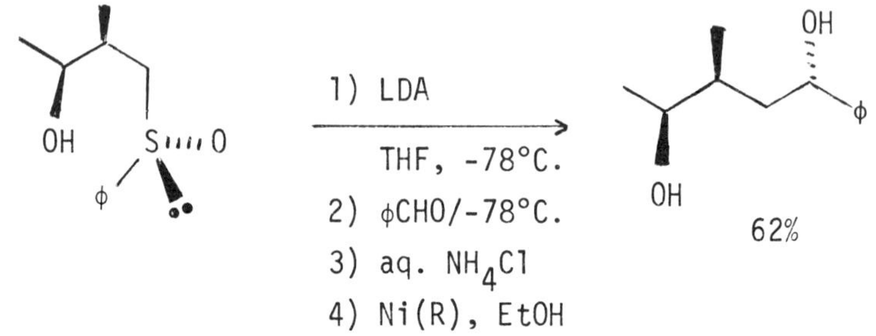

1) LDA

THF, -78°C.

2) φCHO/-78°C.

3) aq. NH₄Cl

4) Ni(R), EtOH

62%

I.A.7.a.2-10 C. Scolastico et al, J. Chem. Soc., Perkin I,
1278, 1284 (1981); N. Kunieda, A. Suzuki and M. Kinoshita,
Bull. Chem. Soc. Jpn., 54, 1143 (1981).

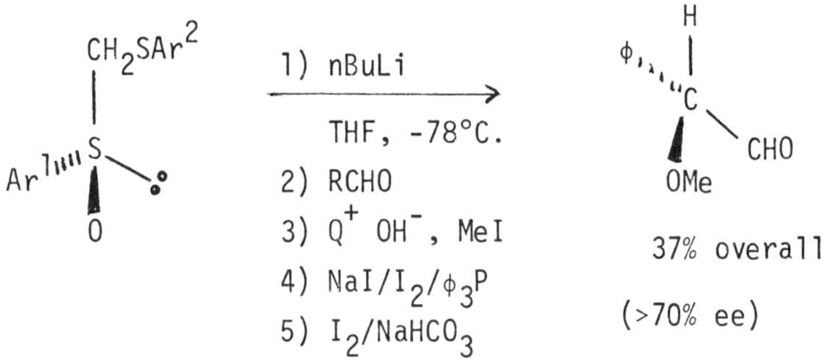

1) nBuLi

THF, -78°C.

2) RCHO

3) Q⁺ OH⁻, MeI

4) NaI/I₂/φ₃P

5) I₂/NaHCO₃

37% overall

(>70% ee)

I.A.7.a.2-11 C. R. Johnson and N. A. Meanwell, J. Amer. Chem.
Soc., 103, 7667 (1981).

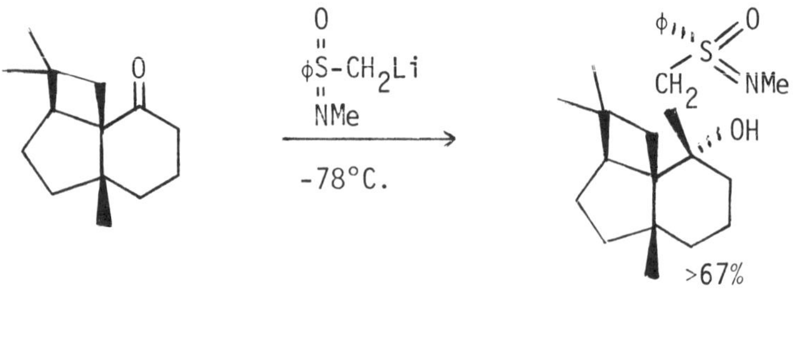

O
‖
φS-CH₂Li
‖
NMe

-78°C.

>67%

I.A.7.a.2-12   J. P. Beny, J. C. Pommelet and J. Chuche, <u>Bull.</u>
<u>Soc. Chim. Fr.</u> II, 369, 377 (1981).

$$RCHO \xrightarrow[\substack{aq.\ NaOH \\ CH_2Cl_2,\ 20°C. \\ (X,\ Y = Halogens)}]{} \quad 23-59\%$$

I.A.7.a.2-13   K. Fuji et al, <u>Synth. Commun.</u>, <u>11</u>, 209 (1981);
N. H. Anderson et al, <u>Tetrahedron</u>, <u>37</u>, 4069 (1981);   M. Braun
and M. Esdar, <u>Chem. Ber.</u>, <u>114</u>, 2924 (1981);   H. Paulsen et al,
<u>Justus Liebigs Ann. Chem.</u>, 2009 (1981).

$$\begin{array}{c} 1)\ \phi\overset{O}{\overset{\|}{C}}CH_3 \\ \xrightarrow{\hspace{2cm}} \\ THF,\ -78°C \\ 2)\ Hg(ClO_4)_2 \cdot 3H_2O \\ MeOH \end{array}$$

$$\phi-\underset{\underset{CO_2Me}{|}}{\overset{\overset{OH}{|}}{C}}-CH_3$$

74%

I.A.7.a.2-14   D. M. Baird and R. D. Bereman, <u>J. Org. Chem.</u>,
<u>46</u>, 458 (1981);   R. K. Dieter, <u>ibid</u>, 5031;   R. Kaya and N. R.
Beller, <u>Synthesis</u>, 814 (1981).

$$\begin{array}{l} 1)\ nBuLi/CS_2 \\ \xrightarrow{\hspace{1.5cm}} \\ 2)\ nBuLi \\ 3)\ BrCH_2CH_2Br \end{array}$$

45-50%

I.A.7.a.2-15 H. J. Reich, W. W. Willis, Jr. and P. D. Clark,
J. Org. Chem., 46, 2775 (1981); L. Hevesi, A. Krief et al,
Tetrahedron Lett., 22, 4009 (1981).

$$ \begin{array}{c} ArSe \\ ArSe \end{array}\!\!\Big\rangle \quad \xrightarrow[\substack{2)\ RCHO \\ 3)\ Et_3N \\ CH_3SO_2Cl}]{1)\ LDA} \quad \begin{array}{c} ArSe \\ \diagdown\!\!=\!\!\diagup \\ R \end{array} $$

62-77%

Also, conversion of products to lithium reagents,
Diels-Alder reactions and comparison with sulfur analogs.

I.A.7.a.2-16 S. W. Rollinson, R. A. Amos and J. A.
Katzenellenbogen, J. Amer. Chem. Soc., 103, 4114 (1981).

$$ \begin{array}{c} R^1CHCO_2Me \\ | \\ Se\phi \end{array} \quad \xrightarrow[\substack{-78°C. \\ 2)\ R^2CHO \\ 3)\ H_2O_2/HOAc/H_2O \\ 0°C.}]{1)\ LDA/THF} \quad R^1\!\!\diagup\!\!=\!\!\underset{HO}{\overset{CO_2Me}{\diagdown}}\!\!R^2 $$

40%

(+10% Z)

I.A.7.a.3. 1,2-Additions of Grignard-type Carbanions

I.A.7.a.3-1 G. A. Olah and M. Arvanaghi, Angew Chem., Int.
Ed. Engl., 20, 878 (1981); D. L. Comins and W. Dernell,
Tetrahedron Lett., 22, 1085 (1981).

$$ \underset{\underset{CHO}{|}}{\overset{\bigcirc}{N}} \quad \xrightarrow[\substack{(M = MgBr\ or\ Li) \\ Et_2O\ or\ THF \\ 2)\ H_3O^+}]{1)\ RM} \quad RCHO $$

72-97%

I.A.7.a.3-2 D. B. Reitz, P. Beak and A. Tse, J. Org. Chem.,
46, 4316 (1981).

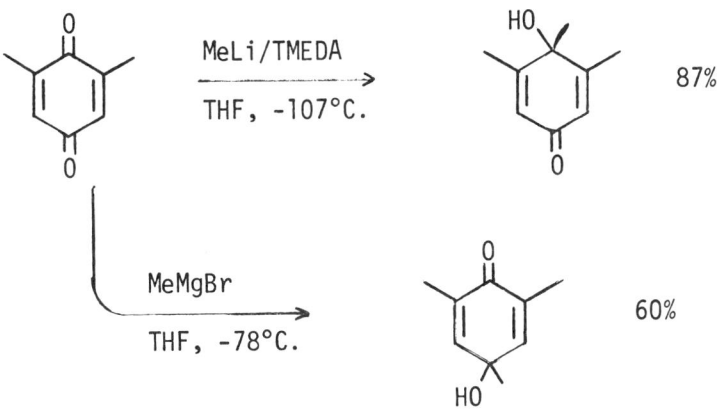

I.A.7.a.3-3 D. Liotta, M. Saindane and C. Barnum, J. Org.
Chem., 46, 3369 (1981); A. J. Guilford and R. W. Turner,
Tetrahedron Lett., 22, 4835 (1981).

I.A.7.a.3-4 R. L. Danheiser, C. Martinez-Davila and H. Sard,
Tetrahedron, 37, 3943 (1981).

Also [4+2] cyclohexenol annulation sequence.

I.A.7.a.3-5 R. Bernardi, C. Fuganti and P. Grasselli,
Tetrahedron Lett., 22, 4021 (1981).

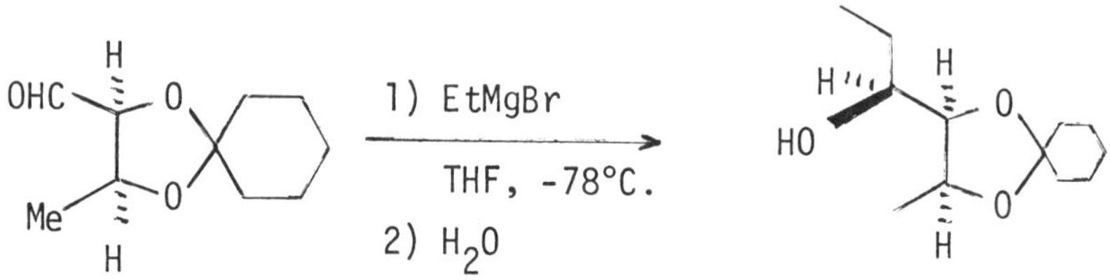

75%

(60% threo)

I.A.7.a.3-6 J. P. Mazaleyrat and D. J. Cram, J. Amer. Chem.
Soc., 103, 4585 (1981).

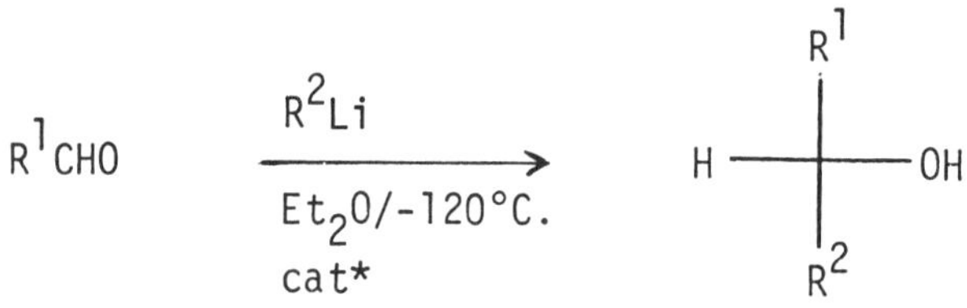

35-75%

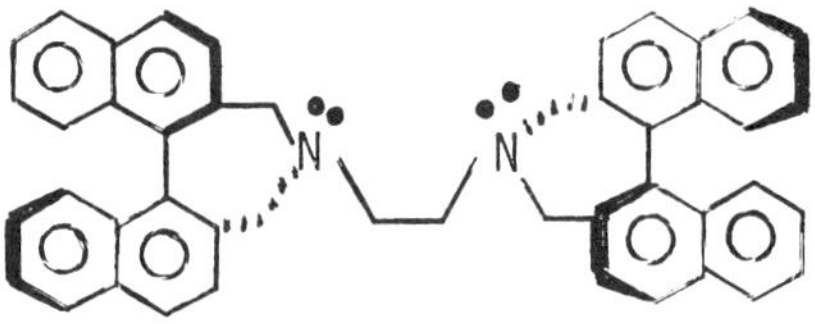

I.A.7.a.3-7 J. K. Whitesell and B. R. Jaw, J. Org. Chem., 46,
2798 (1981).

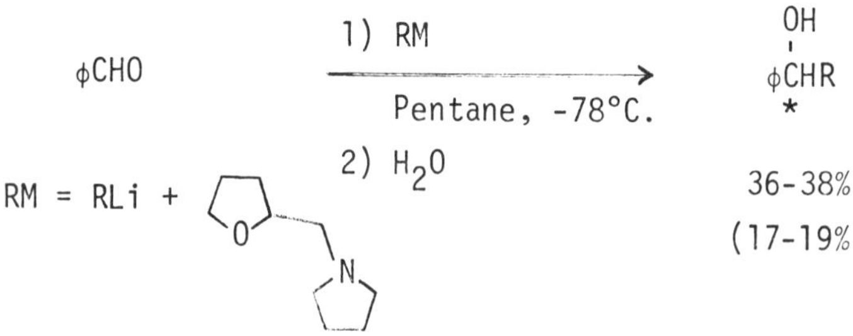

36-38%

(17-19% ee)

I.A.7.a.3-8   K. Krohn, Angew. Chem., Int. Ed. Engl., 20, 576
(1981); D. Avnir et al, Synth. Commun., 11, 241 (1981).

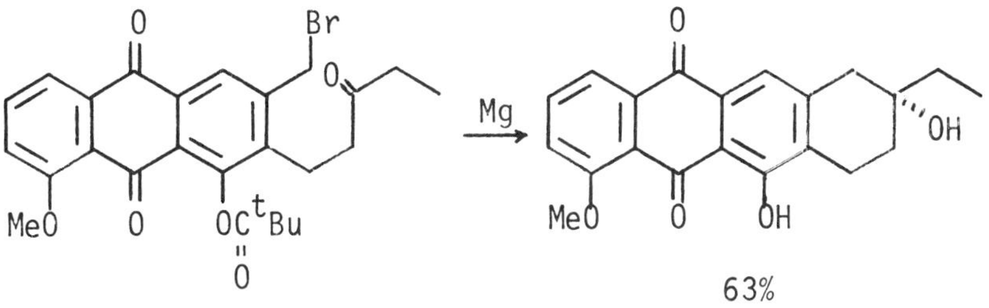

63%

I.A.7.a.3-9   J. Brocard, J. Lebibi and D. Couturier, Chem.
Commun., 1264 (1981); M. Oda, N. Morita and T. Asao, Chem.
Lett., 397 (1981).

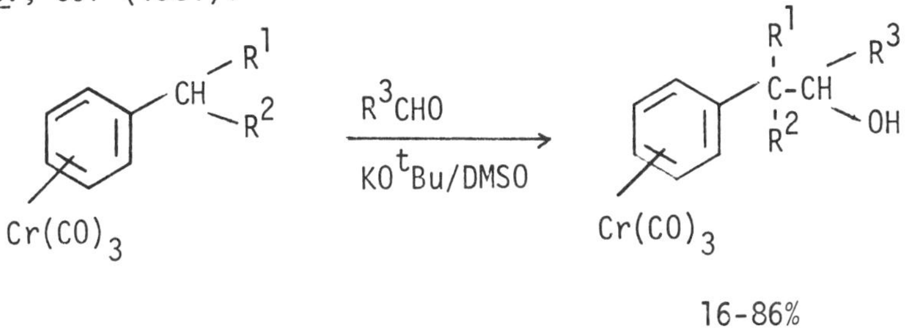

16-86%

I.A.7.a.3-10   T. Cohen and J. R. Matz, Tetrahedron Lett., 22,
2455 (1981).

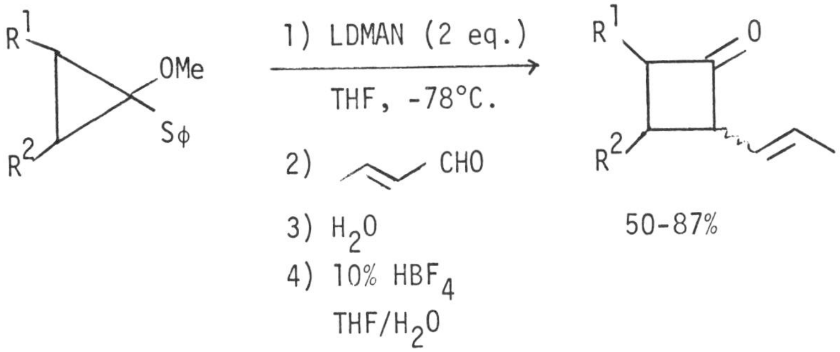

1) LDMAN (2 eq.)

THF, -78°C.

2) $\diagup\diagdown$ CHO

3) H_2O

4) 10% HBF_4

THF/H_2O

50-87%

LDMAN = Lithium 1-dimethylaminonaphthalenide.

I.A.7.a.3-11   D. Seebach and M. Pohmakotr, Tetrahedron, 37, 4047 (1981).

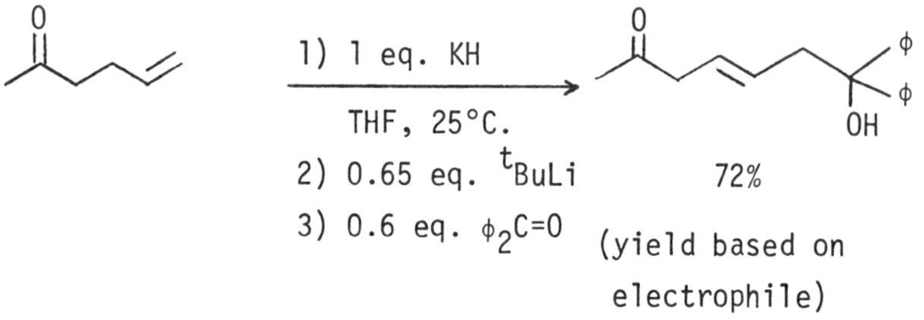

1) 1 eq. KH

THF, 25°C.

2) 0.65 eq. ^{t}BuLi   72%

3) 0.6 eq. ϕ_2C=O   (yield based on electrophile)

I.A.7.a.3-12   W. Oppolzer et al, Helv. Chim. Acta, 64, 2002, 2592 (1981);   D. Hoppe et al, Angew. Chem., Int. Ed. Engl., 20, 1024 (1981).

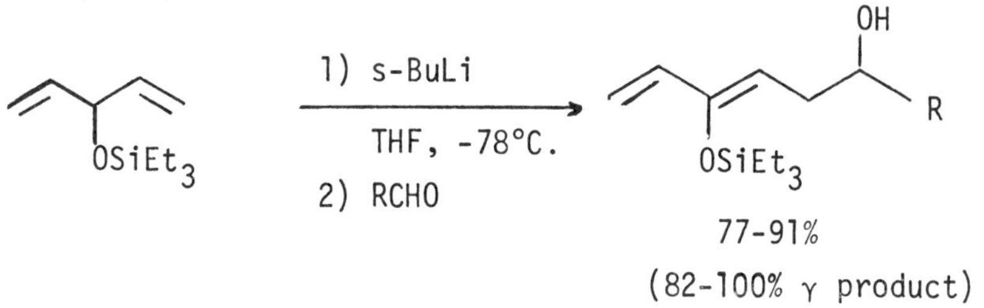

1) s-BuLi

THF, -78°C.

2) RCHO

77-91%

(82-100% γ product)

I.A.7.a.3-13   P. Canonne et al, J. Org. Chem., 46, 3091 (1981); Tetrahedron Lett., 22, 4995 (1981).

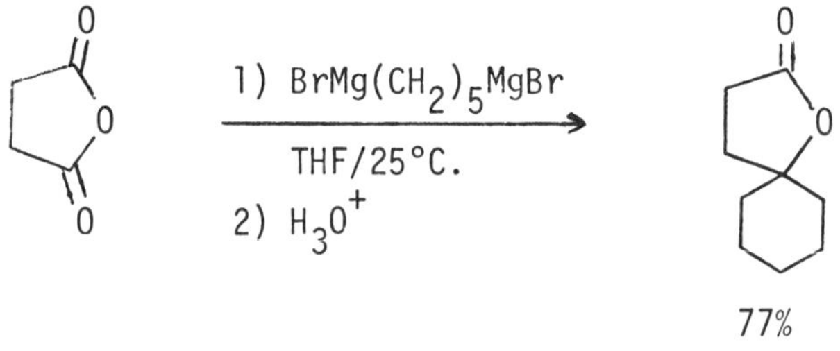

1) BrMg(CH$_2$)$_5$MgBr

THF/25°C.

2) H$_3$O$^+$

77%

I.A.7.a.3-14 J. Pornet, Tetrahedron Lett., 22, 453, 455
(1981); G. Courtois, M. Harama and P. Miginiac, J. Organometal. Chem., 218, 1, 275 (1981); S. F. Karaev, S. O.
Guseinov and E. A. Akhundov, J. Gen. Chem. (USSR), 51, 1163
(1981).

$$HC \equiv C-CH_2-SiMe_3 \xrightarrow[CH_2Cl_2, -60°C.]{RCHO/TiCl_4} CH_2=C-CH=CHR$$
$$\underset{Cl}{|}$$

57-72%

$$\xrightarrow[\underset{THF}{Q^+ F^-}]{RCHO} CH_2=C=CH-\underset{\underset{OH}{|}}{C}HR$$

45-80%

I.A.7.a.3-15 R. W. Hoffmann, Chem. Ber., 114, 375, 2802
(1981); J. Org. Chem., 46, 1309 (1981).

$$\xrightarrow[\underset{2)\ N(CH_2CH_2OH)_3}{n-Hexane, -40°C.}]{1)\ RCHO}$$

81-93%
(24-86% ee)

I.A.7.a.3-16 F. Sato et al, Chem. Commun., 1140, 180 (1981);
F. Sato, S. Iijima and M. Sato, Tetrahedron Lett., 22, 243
(1981); D. Seebach et al, Helv. Chim. Acta, 64, 357, 2485
(1981).

$$Cp_2Ti \xrightarrow{RCHO}$$

87-94%

(90-100% threo)

I.A.7.a.3-17 M. T. Reetz, Chem. Commun., 237 (1981); Synth. Commun., 11, 647 (1981); B. Klei, J. H. Teuben and H. J. De Liefde Meijer, Chem. Commun., 342 (1981).

$$\text{(ketone)} \xrightarrow[\text{CH}_2\text{Cl}_2,\ -30°\text{C}.]{2\ \text{Me}_2\text{TiCl}_2} \text{(product)}$$

34%

Also, methylation of tertiary alcohols.

I.A.7.a.3-18 T. Hiyama, K. Kimura and H. Nozaki, Tetrahedron Lett., 22, 1037 (1981).

$$\text{RCHO} \xrightarrow[\substack{\text{or CrCl}_2 \\ \text{THF, 25°C.}}]{\text{CrCl}_3\ -\ 1/2\ \text{LAH}} \text{(product)}$$

55-100%
(Normally > 90% threo)

I.A.7.a.3-19 T. Mukaiyama and T. Harada, Chem. Lett., 1527 (1981); ibid, 1337, 1679; H. B. Kagan, J. L. Namy and P. Girard, Tetrahedron, 37 (Suppl. 1), 175 (1981); T. Imamoto et al, Tetrahedron Lett., 22, 4987 (1981).

$$\xrightarrow[\substack{\text{Sn/THF} \\ 2)\ \text{H}_2\text{O}}]{1)\ R^1R^2\text{CO}}$$

(X = Br or I)

50-88%

I.A.7.a.3-20 B. Weidmann, C. D. Maycock and D. Seebach, Helv. Chim. Acta, 64, 1552 (1981); Y. Yamamoto and K. Maruyama, Tetrahedron Lett., 22, 2895 (1981); A. Nakamura et al, Chem. Lett., 671 (1981).

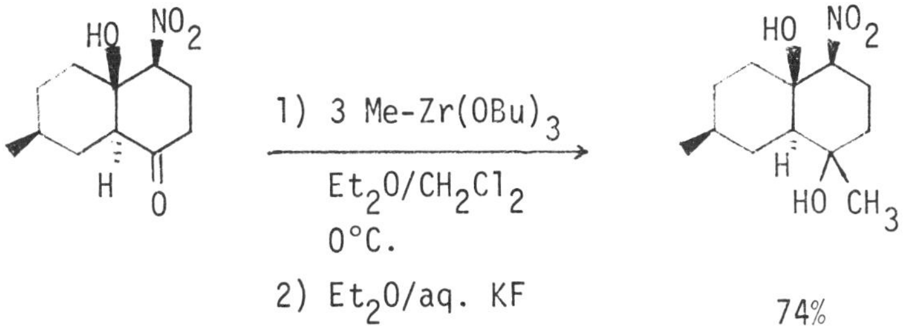

HO
NO_2
H
O
1) 3 $Me-Zr(OBu)_3$
Et_2O/CH_2Cl_2
0°C.
2) Et_2O/aq. KF
HO
NO_2
H
HO CH_3
74%

I.A.7.a.3-21 A. A. Lapidus and Y. Y. Ping, Russian Chem. Rev., 50, 63 (1981).

Review: "Organic Synthesis Based on Carbon Dioxide."

I.A.7.a.3-22 S. Ghosh and U. R. Ghatak, J. Org. Chem., 46, 1486 (1981); G. Cardillo et al, Chem. Commun., 465 (1981).

CH_3 CO_2H
1) Excess LDA
THF, -40°C.
2) CO_2
3) H_3O^+
CO_2H
CH_3 CO_2H
83%

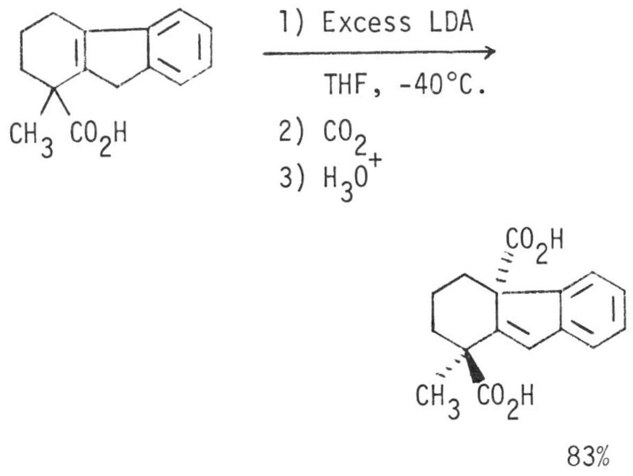

I.A.7.a.3-23   D. E. Seitz and A. Zapata, Synthesis, 557 (1981).

$$Me_3Si-CH_2-SnBu_3 \xrightarrow[\substack{\text{2)} \quad \overset{O}{\underset{\parallel}{}} \\ RCX, -78°C. \\ (X = OMe, Cl, OH) \\ \text{3) aq. } NH_4Cl}]{\substack{\text{1) nBuLi} \\ THF, 0°C.}} R\overset{O}{\underset{\parallel}{C}}-CH_2SiMe_3$$

30-88%

I.A.7.a.3-24   K. Takai, K. Oshima and H. Nozaki, Bull. Chem.
Soc. Jpn., 54, 1281 (1981);   M. Onaka, Y. Matsuoka and T.
Mukaiyama, Chem. Lett., 531 (1981);   T. Fujisawa et al, ibid,
1135.

$$R^1\overset{O}{\underset{\parallel}{C}}X \xrightarrow[\substack{\phi_3P \\ THF, -78°C. \\ (X = Cl \text{ or } S\phi)}]{R^2_3Al/Cu(acac)_2} R^1-\overset{O}{\underset{\parallel}{C}}-R^2$$

55-95%

I.A.7.a.3-25   G. Cahiez, Tetrahedron Lett., 22, 1239 (1981);
S. Nahm and S. M. Weinreb, ibid, 3815;   J. E. Dubois, B. L.
Zhang and C. Lion, Tetrahedron, 37, 4189 (1981);   I. Tabushi,
K. Seto and Y. Kobuke, ibid, 863.

52-93%

I.A.7.a.3-26   J. Villieras et al, Synthesis, 68 (1981);   B.
Mauze, A. Doucoure and L. Miginiac, J. Organometal. Chem., 215,
1 (1981);   E. Elkik, R. Dahan and A. Parlier, Bull. Soc. Chim.
Fr. II, 353 (1981).

$$\underset{\text{(X = Br or Cl)}}{nBuCH\overset{X}{\underset{Br}{\diagup}}} \quad \xrightarrow[\substack{Et_2O/Pentane \\ -113°C. \\ \text{2) } RCO_2CH_3 \\ \text{3) } H_2O}]{\text{1) } nBuLi/THF} \quad \underset{32-82\%}{nBuCH\overset{O}{\underset{X}{-C-R}}}$$

I.A.7.a.3-27   J. S. Nimitz and H. S. Mosher, J. Org. Chem., 46,
211 (1981);   L. M. Weinstock, R. B. Currie and A. V. Lovell,
Synth. Commun., 11, 943 (1981).

$$\underset{}{\overset{N}{\underset{}{\diagdown}}N-\overset{O}{\overset{\|}{C}}-CO_2R^1} \quad \xrightarrow[THF/-50°C.]{R^2MgX} \quad \underset{22-77\%}{R^2-\overset{O}{\overset{\|}{C}}-CO_2R^1}$$

I.A.7.a.3-28   H. Takahashi, K. Tomita and H. Noguchi, Chem.
Pharm. Bull., 29, 3387 (1981);   K. S. Ng and H. Alper, J.
Org. Chem., 46, 1039 (1981);   H. Boshagen, W. Geiger and B.
Junge, Angew. Chem. Int. Ed. Engl., 20, 806 (1981).

$$\underset{\substack{\\ \phi''\text{—}OH \\ H}}{\overset{\substack{H \diagdown \phi \\ \| \\ N \\ | \\ Me_{'',}N}}{}} \quad \xrightarrow[\substack{\text{2) } H_2 \\ Pd/C}]{\text{1) } MeMgBr} \quad \underset{97\% \text{ ee}}{H_2N\overset{H}{\underset{Me}{-\!\!\!\!\!\!\!+\!\!\!\!\!\!\!-}}\phi}$$

I.A.7.a.3-29   J. T. Gupton and C. M. Polaski, Synth. Commun.,
11, 561 (1981);   J. T. Gupton and D. E. Polk, ibid, 571; V.
Nair and C. S. Cooper, J. Org. Chem., 46, 4759 (1981); H. G.
Viehe et al, Angew. Chem., Int. Ed. Engl., 20, 1023 (1981).

$$\begin{array}{c} 1) \text{ RMgX} \\ \xrightarrow{\hspace{1.5cm}} \\ \text{THF} \\ 2) \text{ H}_3\text{O}^+ \end{array} \quad \begin{array}{c} \text{R-CHO} \\ \\ 55\text{-}88\% \end{array}$$

I.A.7.a.3-30   R. Amouroux and G. P. Axiotis, Synthesis, 270
(1981);   G. Rousseau and J. M. Conia, Tetrahedron Lett., 22,
649 (1981).

$$\begin{array}{c} 1) \text{ 2 R}^3\text{Li} \\ \xrightarrow{\hspace{1.5cm}} \\ 2) \text{ H}_2\text{O} \\ 3) \text{ HOAc/H}_2\text{O} \end{array}$$

$$R^1-\underset{\underset{OH}{|}}{\overset{\overset{R^2}{|}}{C}}-\underset{\underset{R^3}{|}}{\overset{\overset{R^3}{|}}{C}}-NH_2$$

55-83%

I.A.7.a.3-31   R. M. Carlson et al, Synth. Commun., 11, 1017
(1981).

$$\begin{array}{c} 1) \text{ 1 eq. nBuLi} \\ \xrightarrow{\hspace{1.5cm}} \\ \text{Hexane/THF} \\ 2) \text{ 25°C, 3 days} \\ 3) \phi_2\text{CO} \\ 4) \text{ H}_2\text{O} \end{array} \qquad 13\text{-}74\%$$

I.A.7.a.3-32 L. N. Pridgen and L. B. Killmer, J. Org. Chem., 46, 5402 (1981); M. M. Gugelchuk, D. J. Hart and Y. M. Tsai, ibid, 3671.

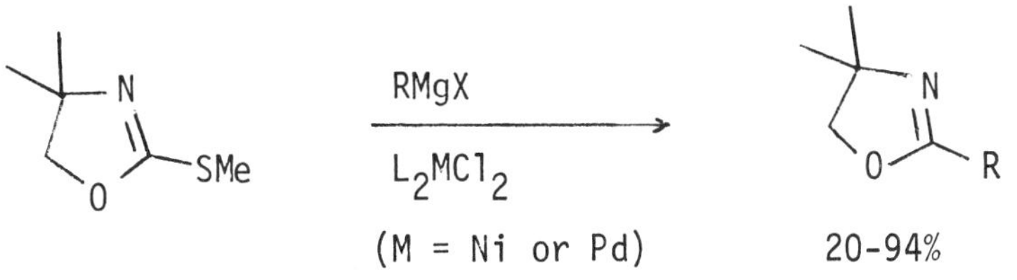

N
SMe
O
RMgX
L_2MCl_2
(M = Ni or Pd)
N
O
R
20-94%

I.A.7.a.3-33 E. W. Colvin, D. Seebach et al, Chem. Commun., 952 (1981).

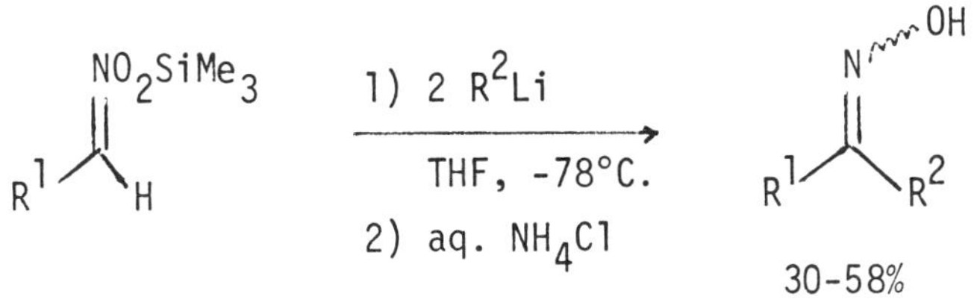

NO_2SiMe_3
R^1
H
1) 2 R^2Li
THF, -78°C.
2) aq. NH_4Cl
N
OH
R^1
R^2
30-58%

I.A.7.b.1. Conjugate Additions of Enolate-type Carbanions

I.A.7.b.1-1 J. Novak and C. A. Salemink, Tetrahedron Lett., 22, 1063 (1981).

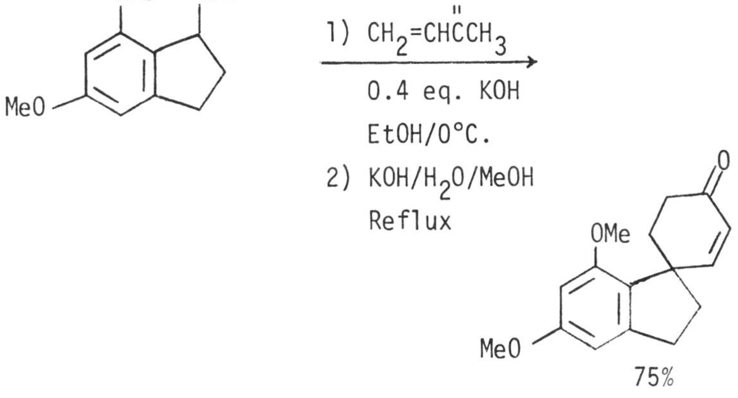

OMe
CHO
MeO
1) CH_2=CHCCH$_3$
O
0.4 eq. KOH
EtOH/0°C.
2) KOH/H_2O/MeOH
Reflux
O
OMe
MeO
75%

I.A.7.b.1-2   J. P. Corriu et al, Chem. Commun., 122 (1981).

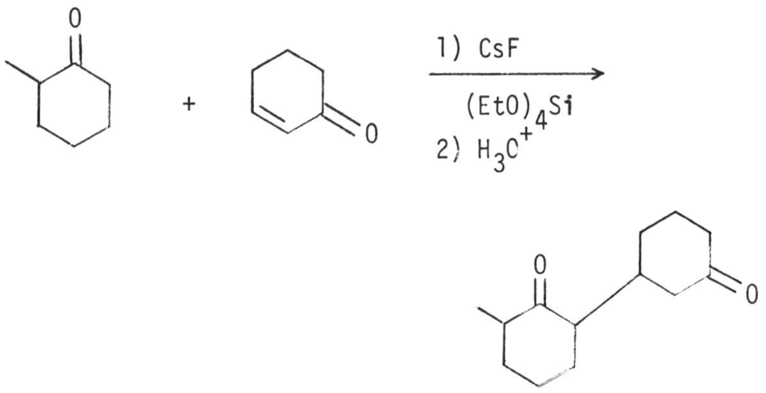

1) CsF
   (EtO)$_4$Si
2) H$_3$C$^+$

65%

I.A.7.b.1-3   G. H. Posner, J. P. Mallamo and A. Y. Black,
Tetrahedron, 37, 3921 (1981).

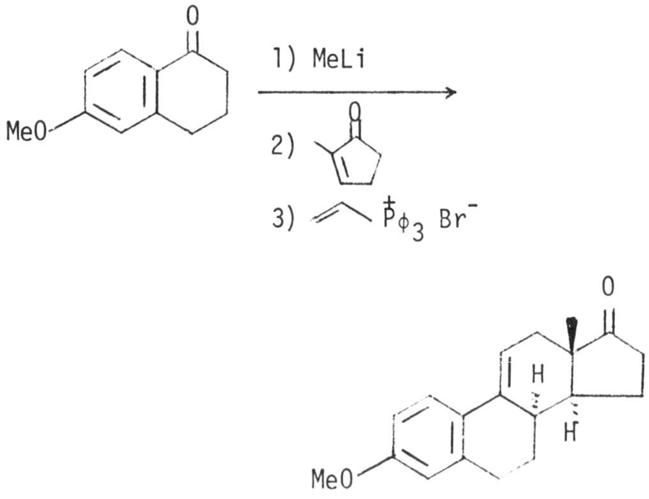

1) MeLi

2)

3) $\overset{+}{P}\phi_3$ Br$^-$

8.4%

I.A.7.b.1-4 M. Miyashita, T. Yanami and A. Yoshikoshi, Org. Syn., 60, 117, 101 (1981); R. M. Cory et al, Chem. Commun., 73 (1981).

61-70%

I.A.7.b.1-5 K. T. Potts et al, J. Amer. Chem. Soc., 103, 3584, 3585 (1981); Y. N. Stefanovsky and L. Z. Viteva, Monat. Chem., 112, 125 (1981); C. Ivanov and T. Tcholakova, Synthesis, 392 (1981); J. Mulzer et al, Chem. Ber., 114, 3701 (1981).

42-100%

Products used to prepare heterocycles.

I.A.7.b.1-6 D. Seebach and J. Golinski, Helv. Chim. Acta, 64, 1413 (1981); S. Fabrissin, S. Fatutta and A. Risaliti, J. Chem. Soc., Perkin I, 109 (1981); Y. Watanabe et al, Bull. Chem. Soc. Jpn., 54, 3875 (1981); J. Ficini, G. Revial and S. Jeannin, Tetrahedron Lett., 22, 2367 (1981).

81%

(>90% Diasterioselective)

I.A.7.b.1-7   K. E. Harding et al, J. Org. Chem., 46, 940
(1981).

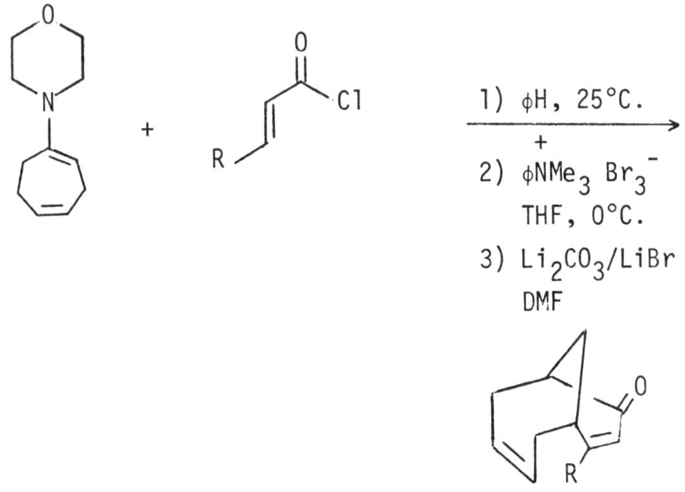

37-95%

I.A.7.b.1-8   K. Popandova-Yambolieva, A. Dobrev and C. Ivanov,
Synth. Commun., 11, 335 (1981);   M. C. Roux-Schmitt, L.
Wartski and J. Seyden-Penne, ibid, 85;   M. C. Roux-Schmitt and
J. Seyden-Penne, Tetrahedron Lett., 22, 2171, 2175 (1981).

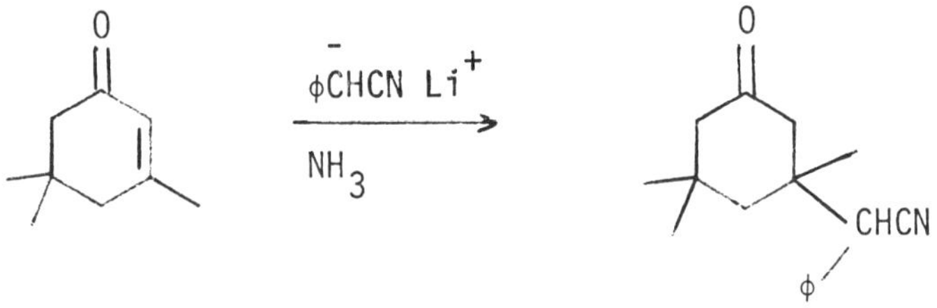

51%

I.A.7.b.1-9 K. Takabe, S. Ohkawa and T. Katagiri, <u>Synthesis</u>, 358 (1981); H. Tucker, G. Golding and S. R. Purvis, <u>Tetrahedron Lett</u>., <u>22</u>, 1373 (1981); M. R. Britten-Kelly, B. J. Willis and D. H. R. Barton, J. Org. Chem., <u>46</u>, 5027 (1981).

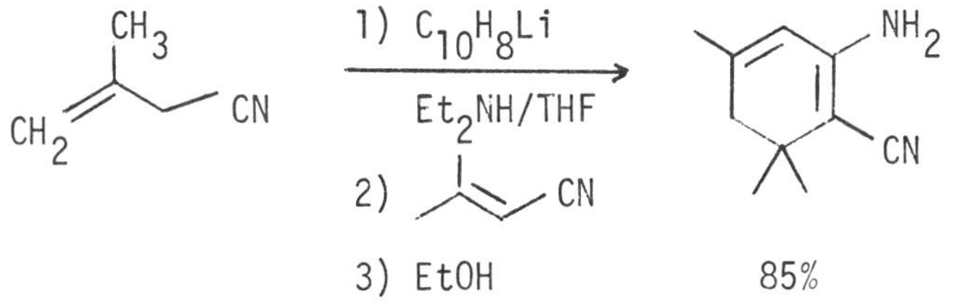

I.A.7.b.1-10 H. Schick et al, <u>Zeit. Chem</u>., <u>21</u>, 68 (1981); E. Guittet and S. Julia, <u>Synth. Commun</u>., <u>11</u>, 697, 709, 723 (1981).

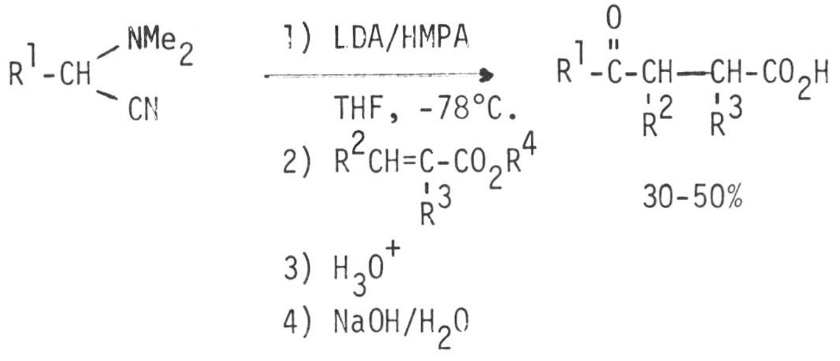

I.A.7.b.1-11 S. Danishefsky and M. Kahn, <u>Tetrahedron Lett</u>., <u>22</u>, 485 (1981).

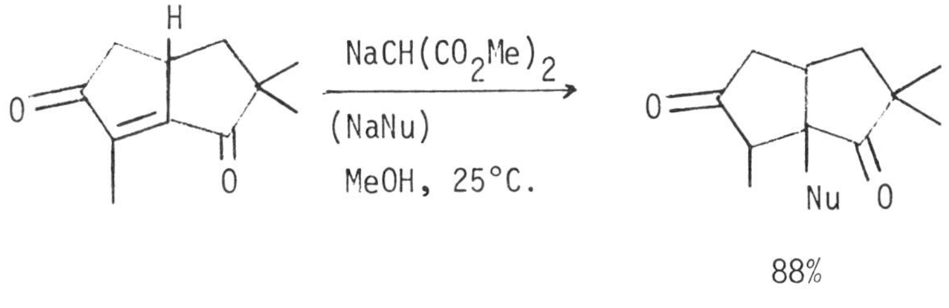

I.A.7.b.1-12 Z. Ahmed and M. P. Cava, Tetrahedron Lett., 22, 5239 (1981).

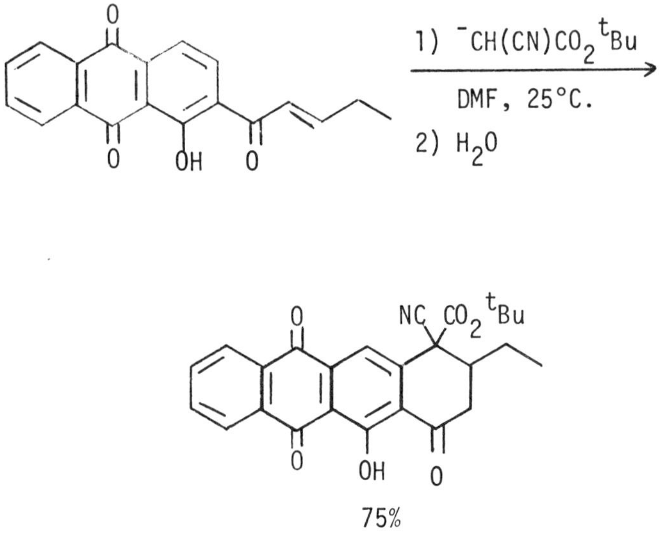

1) $^-CH(CN)CO_2{}^tBu$

DMF, 25°C.

2) H_2O

75%

I.A.7.b.1-13 D. J. Cram and G. D. Y. Sogah, Chem. Commun., 625 (1981).

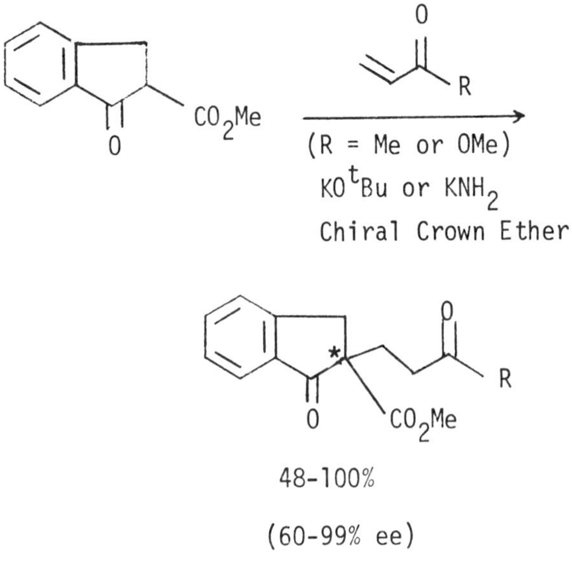

(R = Me or OMe)

KO^tBu or KNH_2

Chiral Crown Ether

48-100%

(60-99% ee)

I.A.7.b.1-14 J. Hajicek and J. Trojanek, Coll. Czech. Chem.
Commun., 46, 1262 (1981); M. Yamato and Y. Kusunoki, Chem.
Pharm. Bull., 29, 1214, 2832 (1981).

$$\begin{array}{ccc} \text{OH} & & \text{O} \\ \text{CN} & \xrightarrow[\text{EtOH, 10°C.}]{\text{1) NaCH(CN)CO}_2\text{Et}} & \\ & \text{2) H}_3\text{O}^+ & \text{NC} \quad \text{CO}_2\text{Et} \end{array}$$

46%

I.A.7.b.1-15 M. Mikolajczyk, S. Grzejszczak and K. Korbacz,
Tetrahedron Lett., 22, 3097 (1981); K. Peseke, J. Prakt.
Chem., 323, 499 (1981); H. J. Liu, L. K. Ho and H. K. Lai,
Can. J. Chem., 59, 1685 (1981).

$$\underset{\underset{\text{SMe}}{|}}{(\text{EtO})_2\overset{\overset{\text{O}}{\|}}{P}\text{C=CH}_2} \xrightarrow{\text{NaCH(CO}_2\text{Me})_2}$$

$$\underset{\underset{\text{SMe}}{|}}{(\text{EtO})_2\overset{\overset{\text{O}}{\|}}{P}\text{-CH-CH}_2\text{-CH(CO}_2\text{Me})_2}$$

81%

I.A.7.b.1-16 J. Tsuji, H. Kataoka and Y. Kobayashi, Tetrahedron Lett., 22, 2575 (1981); B. M. Trost and G. A. Molander,
J. Amer. Chem. Soc., 103, 5969 (1981).

$$\text{R}\diagup\!\!\diagdown\!\!\diagup\!\!\diagdown \xrightarrow[\underset{\text{THF}}{(\phi_3\text{P})_4\text{Pd (cat)}}]{\text{CH}_2(\text{CO}_2\text{Me})_2} \underset{\underset{\text{OH}}{|}}{\text{R}\diagdown\!\!\diagup\!\!\diagdown\!\!\diagup}\text{CH(CO}_2\text{Me})_2$$

64-92%

(96% E)

I.A.7.b.1-17 B. M. Trost and T. P. Klun, J. Amer. Chem. Soc.,
103, 1864 (1981); T. Hirao et al, Tetrahedron Lett., 22,
3079 (1981); J. Tsuji et al, Tetrahedron Lett., 22, 2573
(1981); M. Moreno-Manas and A. Trius, Tetrahedron, 37, 3009
(1981); N. H. Rama, E. Turner and W. B. Whalley, J. Chem.
Res. (S), 149 (1981).

I.A.7.b.1-18 H. Wegmann and W. Steglich, Chem. Ber., 114,
2580 (1981); T. Takeda, T. Hoshiko and T. Mukaiyama, Chem.
Lett., 797 (1981).

I.A.7.b.2. Conjugate Additions of Organometallic Reagents

I.A.7.b.2-1 D. L. J. Clive, V. Farina and P. Beaulieu, Chem.
Commun., 643 (1981); C. Chuit, J. P. Foulon and J. F.
Normant, Tetrahedron, 37, 1385 (1981).

I.A.7.b.2-2 R. T. Taylor and J. G. Galloway, J. Organometal.
Chem., 220, 295 (1981); T. Saegusa et al, J. Org. Chem., 46,
192 (1981).

$$\xrightarrow[\text{Cu}_2\text{Br}_2 \ (\text{cat.})]{\text{Me}_3\text{SiCH}_2\text{MgCl}}$$

77%

I.A.7.b.2-3 G. H. Posner, J. P. Mallamo and K. Miura, J.
Amer. Chem. Soc., 103, 2886 (1981); Y. Tamura, M. Kagotani
and Z. Yoshida, Tetrahedron Lett., 22, 3409 (1981); S.
Hanessian, P. C. Tyler and Y. Chapleur, ibid, 4583.

1) MeMgI

THF, -78°C.

2) H$_3$O$^+$

3) Al(Hg)

77% (79% ee)

I.A.7.b.2-4 T. Takahashi, K. Hori and J. Tsuji, Tetrahedron
Lett., 22, 119 (1981); Chem. Lett., 1189 (1981).

1) R$^1_2$CuLi

Et$_2$O, -78°C.

2) R$^2_2$CuLi

Et$_2$O, -78°C.

55%

I.A.7.b.2-5 I. Fleming and D. A. Perry, Tetrahedron, 37, 4027
(1981).

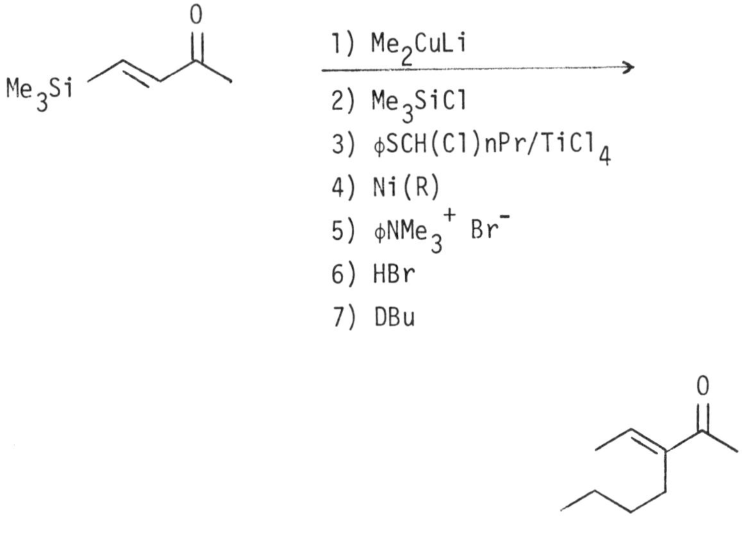

Me₃Si — CH=CH — C(=O)CH₃

1) Me₂CuLi

2) Me₃SiCl
3) φSCH(Cl)nPr/TiCl₄
4) Ni(R)
5) φNMe₃⁺ Br⁻
6) HBr
7) DBu

45%

I.A.7.b.2-6 D. J. Ager, Tetrahedron Lett., 22, 587 (1981).

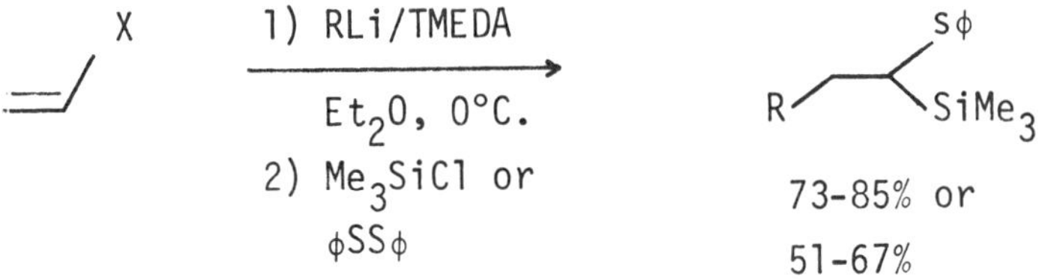

1) RLi/TMEDA

Et₂O, 0°C.

2) Me₃SiCl or
   φSSφ

73-85% or
51-67%

X = φS or Me₃Si

I.A.7.b.2-7 T. Fujisawa et al, Chem. Lett., 1159 (1981).

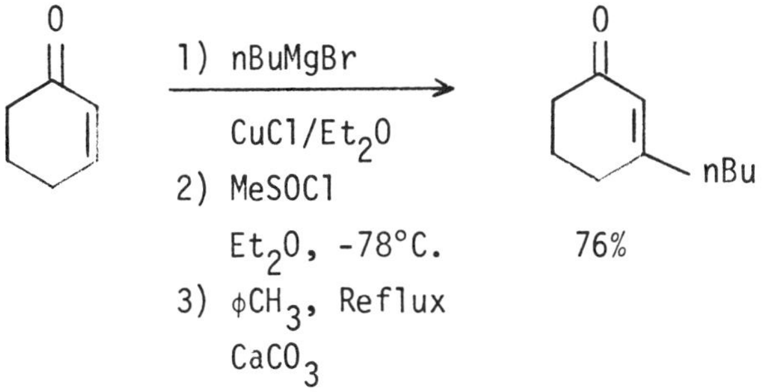
1) nBuMgBr
CuCl/Et$_2$O
2) MeSOCl
Et$_2$O, -78°C.
3) ϕCH$_3$, Reflux
CaCO$_3$
nBu
76%

I.A.7.b.2-8 C. J. Kowalski and K. W. Fields, J. Org. Chem., 46, 197 (1981).

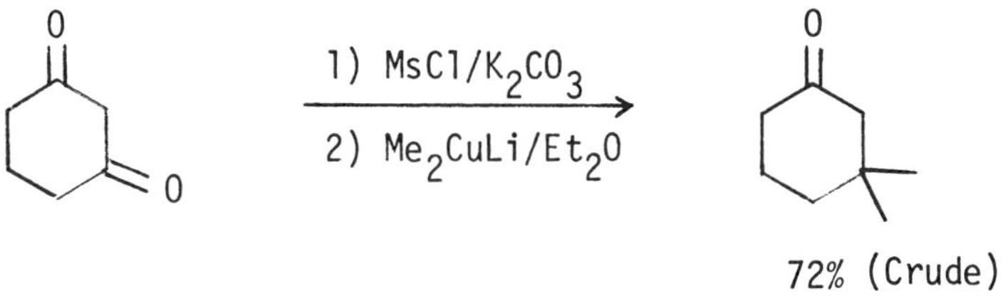
1) MsCl/K$_2$CO$_3$
2) Me$_2$CuLi/Et$_2$O
72% (Crude)

Other nucleophiles (X^-, EtO^-, Me_2NH, ϕCH_2S^-) give β-substituted cyclohexenones.

I.A.7.b.2-9 N. J. P. Broom and P. G. Sammes, J. Chem. Soc., Perkin I, 465 (1981).

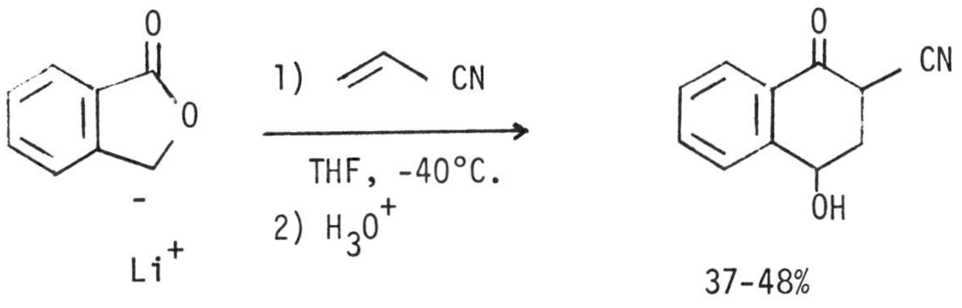
Li$^+$
1) CN
THF, -40°C.
2) H$_3$O$^+$
CN
OH
37-48%

I.A.7.b.2-10   P. Knochel and D. Seebach, <u>Tetrahedron Lett.</u>, <u>22</u>, 3223 (1981); M. Ochiai, M. Arimoto and E. Fujita, <u>ibid</u>, 1115.

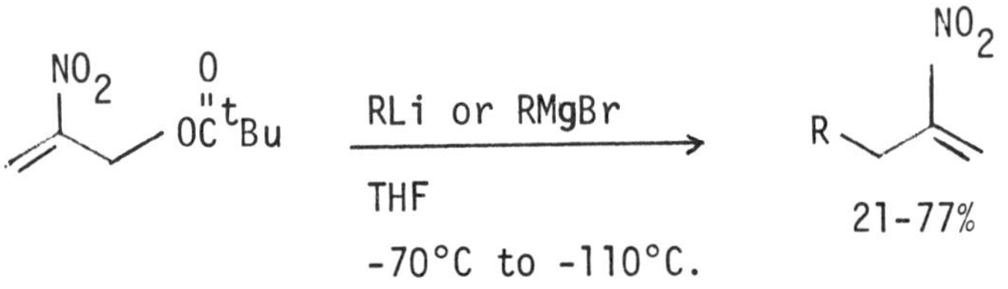

R = 1°, 2°, 3° Alkyl, Vinylic, Acetylenic and Aromatic.

I.A.7.b.2-11   G. Bartoli, M. Bosco and A. C. Boicelli, <u>Synthesis</u>, 570 (1981).

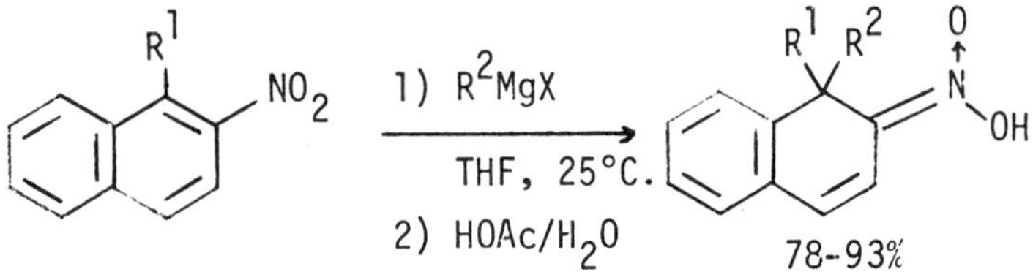

I.A.7.b.2-12   T. Fujisawa et al, <u>Chem. Lett.</u>, 1307 (1981); <u>Tetrahedron Lett.</u>, <u>22</u>, 1817, 2375 (1981).

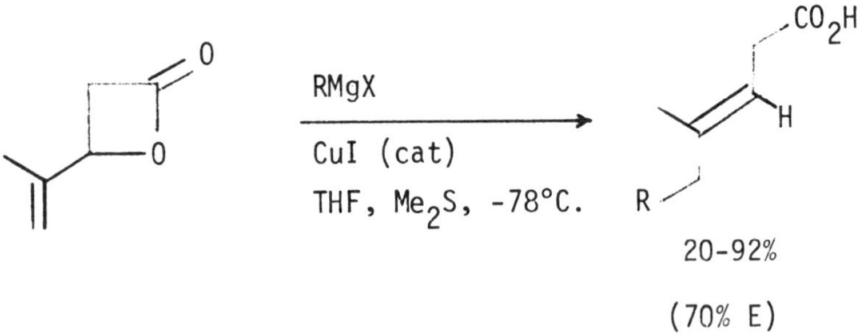

I.A.7.b.2-13 J. P. Marino and H. Abe , J. Amer. Chem. Soc.,
103, 2907 (1981); J. Org. Chem., 46, 5379 (1981); F. E.
Ziegler and M. A. Cady, ibid, 122; M. Tamura and G. Suzukamo,
Tetrahedron Lett., 22, 577 (1981).

$$2(RCuCN)Li$$
$$Et_2O, -78°C.$$
(R = isohexyl)

82%

I.A.7.b.2-14 P. A. Grieco and C. V. Srinivasan, J. Org. Chem.,
46, 2591 (1981); F. E. Ziegler and P. J. Gilligan, ibid, 3874;
S. M. Roberts, R. F. Newton et al, J. Chem. Soc. Perkin I,
1729, 1725 (1981); M. Isobe, M. Kitamura and T. Goto, Tetrahedron Lett., 22, 239 (1981).

1) $LiMe_2Cu$
Et_2O
-20°C.
2) H_3O^+

84%

I.A.7.b.2-15 T. Mukaiyama and N. Iwasawa, Chem. Lett., 913
(1981); M. Huche et al, Tetrahedron Lett., 22, 1329 (1981);
H. Malmberg, M. Nilsson and C. Ullenius, Acta Chem. Scand. B,
35, 625 (1981).

1) 2 R^2MgBr
2) H^+
3) $H_2SO_4/HOAc$

44-63%

I.A.7.b.2-16   P. G. M. Wuts, Synth. Commun., 11, 139 (1981);
A. B. Theis and C. A. Townsend, ibid, 157.

CuBr·Me$_2$S   Complex Preparation

I.A.7.b.3.   Other Conjugate Additions

I.A.7.b.3-1   K. Matsumoto and T. Uchida, Chem. Lett., 1673
(1981); S. Colonna, A. Re and H. Wynberg, J. Chem. Soc.,
Perkin I, 547 (1981); S. Banfi, M. Cinquini and S. Colonna,
Bull. Chem. Soc. Jpn., 54, 1841 (1981).

$$\phi CH=CHC\phi \quad \xrightarrow[\substack{\text{Quinidine} \\ \phi CH_3, \, 14°C. \\ 10 \text{ kbar.}}]{CH_3NO_2} \quad \phi \overset{*}{C}HCH_2 \overset{O}{\overset{\|}{C}}\phi$$

$\overset{O}{\overset{\|}{}}$

CH_2NO_2

100%

(51% ee)

I.A.7.b.3-2   A. Lorenzi-Riatsch, Y. Nakashita and M. Hesse,
Helv. Chim. Acta, 64, 1854 (1981); K. Matsumoto, Angew. Chem.,
Int. Ed. Engl., 20, 770 (1981).

92%

I.A.7.b.3-3 S. K. Mukerji and K. B. G. Torssell, Acta Chem. Scand. B, 35, 643 (1981).

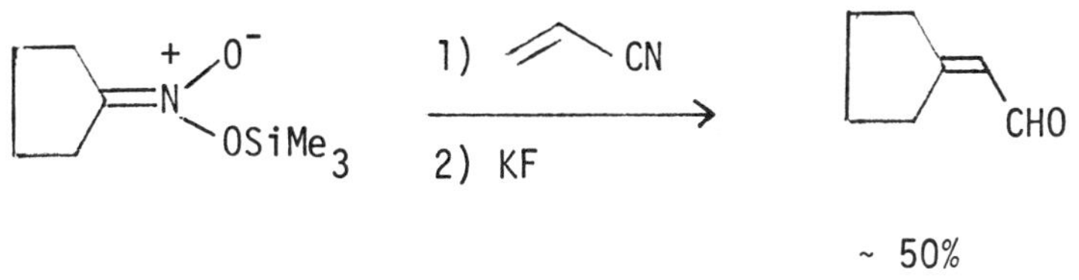

~ 50%

I.A.7.b.3-4 C. Scolastico et al, Synthesis, 74 (1981); G. H. Posner, J. Clardy et al, J. Org. Chem., 46, 5244 (1981).

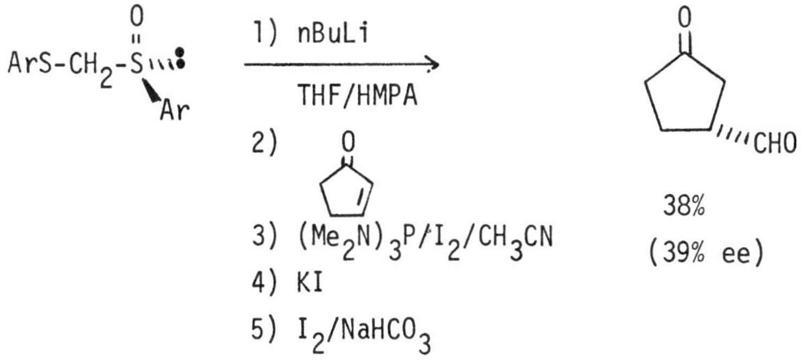

38%
(39% ee)

I.A.7.b.3-5 M. Hirama, Tetrahedron Lett., 22, 1905 (1981); I. H. Sanchez and M. A. Aguilar, Synthesis, 55 (1981).

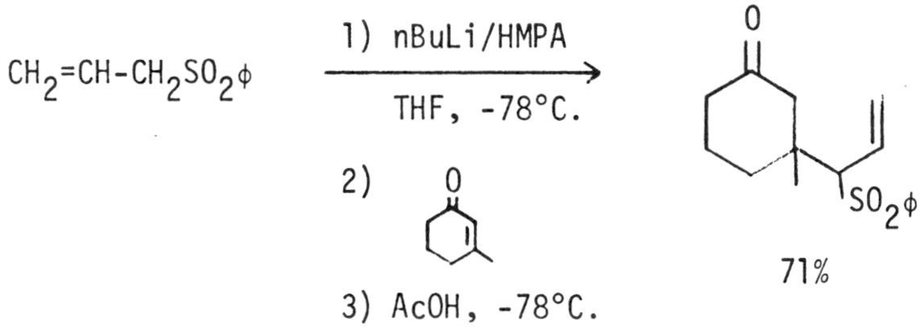

$CH_2=CH-CH_2SO_2\phi$

1) nBuLi/HMPA
THF, -78°C.

2)

3) AcOH, -78°C.

71%

I.A.7.b.3-6 F. E. Ziegler and J. M. Fang, J. Org. Chem., 46, 825 (1981).

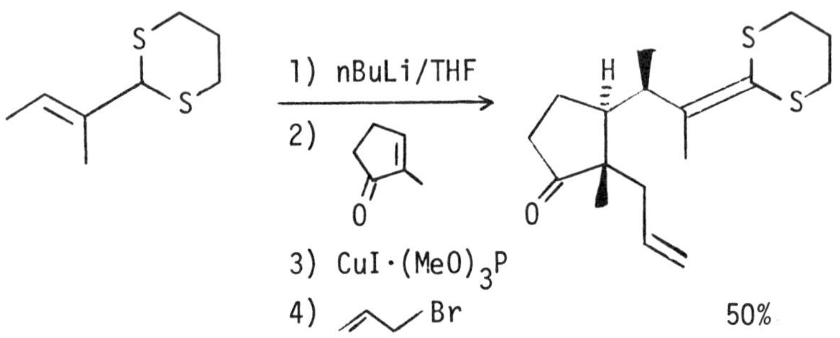
S
S
1) nBuLi/THF
2)
O
3) CuI·(MeO)3P
4)
Br
H
S
S
O
50%

I.A.7.b.3-7 D. B. Grotjahn and N. H. Anderson, Chem. Commun., 306 (1981); J. Lucchetti and A. Krief, Tetrahedron Lett., 22, 1623 (1981).

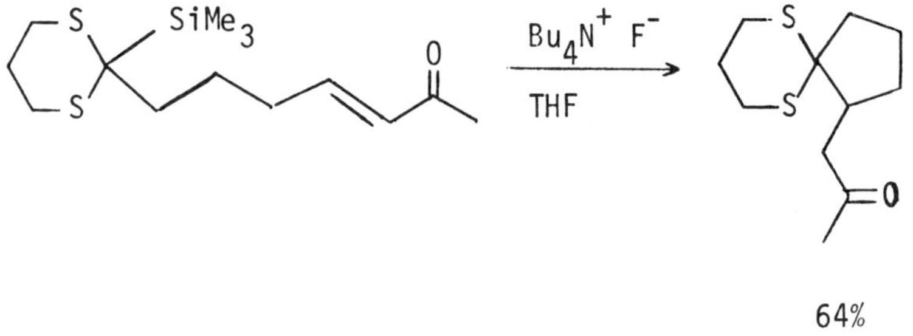
S
S
SiMe3
O
Bu4N+ F-
THF
S
S
O
64%

I.A.7.b.3-8 R. H. B. Galt and Z. S. Matusiak, Tetrahedron Lett., 22, 2913 (1981).

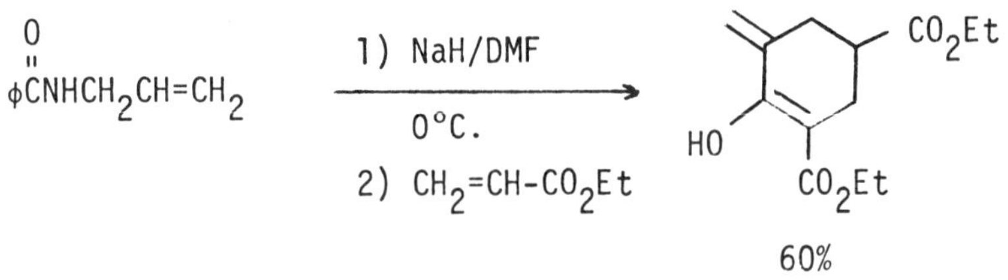
O
ϕCNHCH2CH=CH2
1) NaH/DMF
0°C.
2) CH2=CH-CO2Et
CO2Et
HO
CO2Et
60%

I.A.7.b.3-9 B. B. Snider and G. B. Phillips, J. Org. Chem., 46, 2563 (1981).

R¹ / R² / R³ structure + $CH_2=C$ with CO_2Me and CN → Me_2AlCl	Low to Mod. yields of ring closure product + others

I.A.7.b.3-10 B. Giese and G. Kretzschmar, Angew. Chem., Int. Ed. Engl., 20, 965 (1981); B. Giese and K. Heuck, Chem. Ber., 114, 1572 (1981).

cyclohexene-$CH=CH_2$

1) BH_3
   THF, 0°C.
2) $Hg(OAc)_2$
3) $CH_2=CHCN$
4) $NaBH_4$

→ cyclohexene-$CH_2CH_2-CH_2CH_2CN$

54%

I.A.7.b.3-11 H. Stetter et al, Synthesis, 129, 626 (1981); Chem. Ber., 114, 564, 581, 1226, 2479 (1981); Justus Liebigs Ann. Chem., 1550 (1981).

$$\begin{array}{c} EtO \\ \diagdown \\ CH-CHO \\ \diagup \\ EtO \end{array} \quad \xrightarrow[\substack{\text{Thiazolium Salt (cat)} \\ Et_3N/\text{Dioxane, 80°C.}}]{\overset{\displaystyle O}{\overset{\|}{CH_2=CHCR}}}$$

$$(EtO)_2CH\overset{O}{\overset{\|}{C}}-CH_2CH_2\overset{O}{\overset{\|}{C}}R$$

58-71%

I.A.8. Other Carbon-Carbon Single Bond Forming Reactions

I.A.8-1 H. Mayr, Angew. Chem., Int. Ed. Engl., 20, 184 (1981).

"Lewis Acid Catalyzed Alkylations of CC-Multiple
Bonds: Rules for Selective Enlargements of
Carbon Skeletons."

I.A.8-2 D. Babin, J. D. Fourneron and M. Julia, Tetrahedron,
37 (Suppl. 1), 1 (1981).

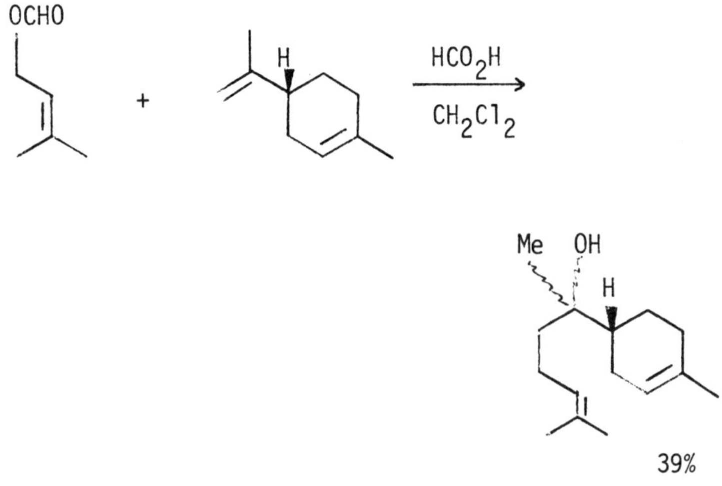

39%

I.A.8-3 M. Grignon-Dubois, J. Dunogues and R. Calas, Tetrahedron Lett., 22, 2883 (1981); Can. J. Chem., 59, 802 (1981);
M. Fetizon et al, Synthesis, 139 (1981).

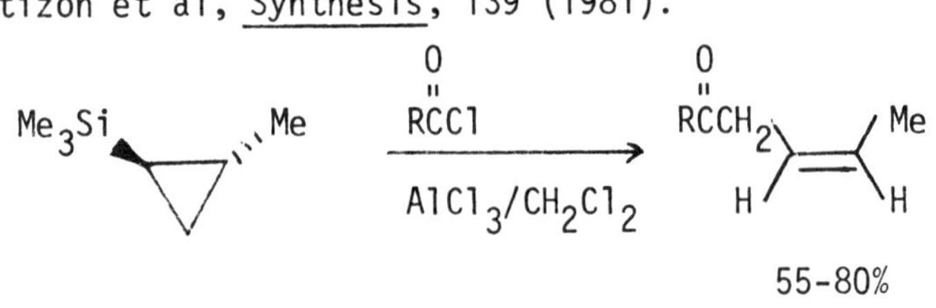

55-80%

I.A.8-4   R. W. Franklin, R. S. Ward and D. W. Roberts, <u>J.</u>
<u>Chem. Res. (S)</u>, 272 (1981);   R. Pardo and M. Santelli,
<u>Tetrahedron Lett.</u>, <u>22</u>, 3843 (1981).

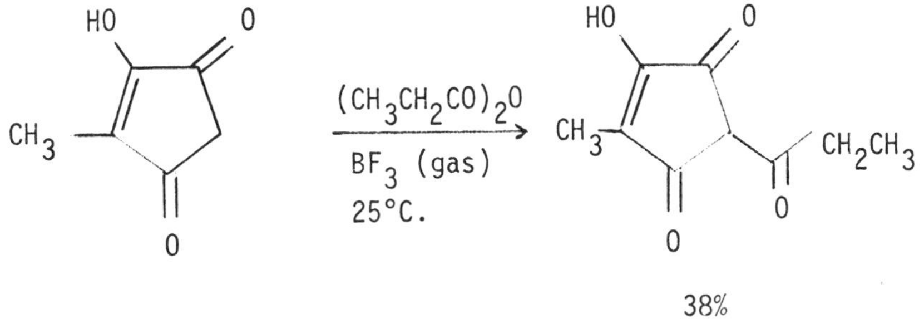

38%

I.A.8-5   R. D. Dawe and B. Fraser-Reid, <u>Chem. Commun.</u>, 1180
(1981).

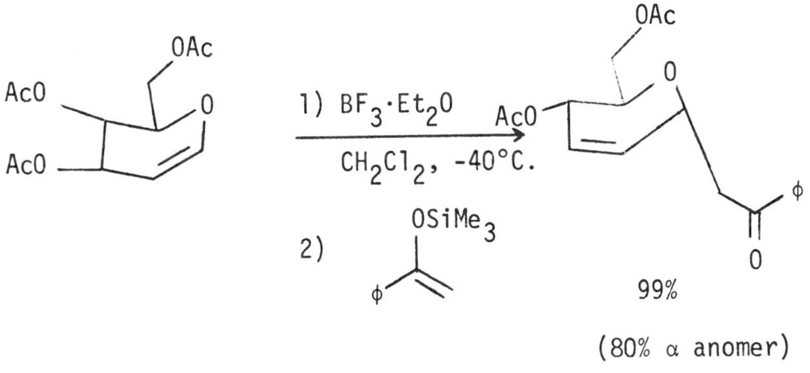

99%

(80% α anomer)

I.A.8-6   N. Ishikawa et al, <u>Bull. Chem. Soc. Jpn.</u>, <u>54</u>, 832
(1981).

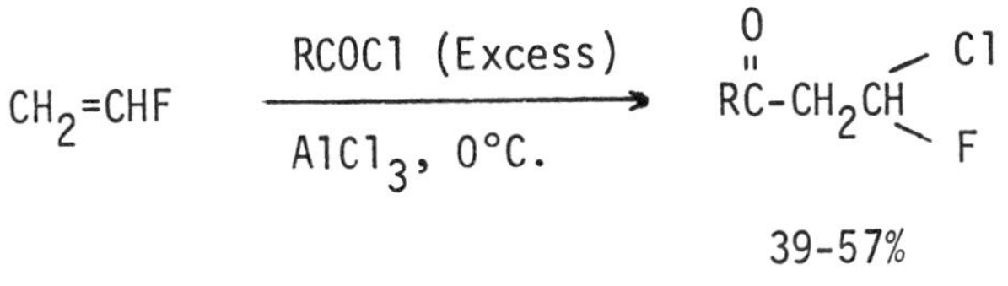

39-57%

I.A.8-7   Z. N. Parnes et al, J. Org. Chem. (USSR), 17, 1203 (1981).

$$\underset{CH_3}{\overset{CH_3}{>}}CH-CH_2-CH_2-Cl \quad \xrightarrow[Me_4Si]{AlCl_3} \quad \underset{CH_3}{\overset{CH_3}{>}}CH-CH\underset{CH_3}{\overset{CH_3}{<}}$$

50-55%

I.A.8-8   H. Mayr and H. Klein, J. Org. Chem., 46, 4097 (1981).

$$\phi C\equiv C-\underset{CH_3}{\overset{CH_3}{\underset{|}{\overset{|}{C}}}}-Cl \quad \xrightarrow[-78°C.]{ZnCl_2, \; Et_2O} \quad \phi C\equiv C-\underset{CH_3}{\overset{CH_3}{\underset{|}{\overset{|}{C}}}}-CH_2CH=CHCH\underset{CH_3}{\overset{}{\underset{|}{}}}Cl$$

67%

I.A.8-9   H. Vathke-Ernst and H. M. R. Hoffmann, Chem. Ber., 114, 1464, 2208, 2898 (1981).

pTsOH/H$_2$O
Pentane, 25°C.

29-38%

I.A.8-10   P. A. Reddy and G. S. K. Rao, Ind. J. Chem., 20B, 100, 487 (1981);   N. S. Narasimhan, T. Mukhopadhyay and S. S. Kusurkar, ibid, 546.

DMF/POCl$_3$

64%

I.A.8-11 E. E. van Tamelen, Pure Appl. Chem., 53, 1259 (1981).

Review: "The Role of Organic Synthesis in Bioorganic
         Chemistry."

I.A.8-12 T. Kametani et al, Tetrahedron Lett., 22, 3653
(1981); Chem. Pharm. Bull., 29, 105 (1981); J. Chem. Soc.,
Perkin I, 756 (1981); A. Rouessac and F. Rouessac, Tetrahedron, 37, 4165 (1981); F. Rouessac and H. Zamarlik, Tetrahedron Lett., 22, 2643 (1981).

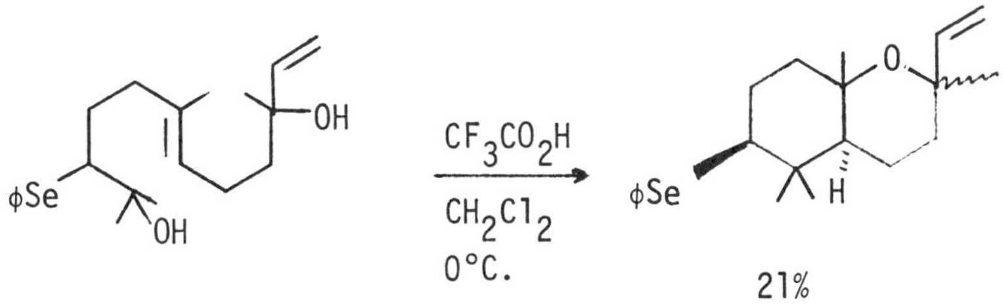

I.A.8-13 T. L. Macdonald and S. Mahalingam, Tetrahedron
Lett., 22, 2077 (1981); A. Murai et al, Chem. Lett., 1125
(1981); K. Oshima et al, Bull. Chem. Soc. Jpn., 54, 1456
(1981).

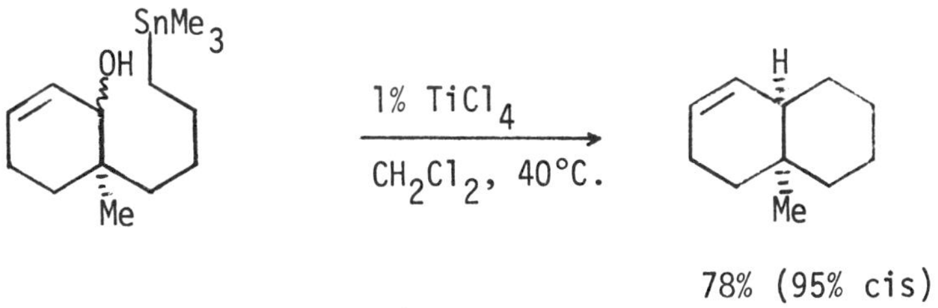

I.A.8-14 W. S. Johnson et al, J. Amer. Chem. Soc., 103, 88
(1981); W. P. Jackson and S. V. Ley, J. Chem. Soc., Perkin
I, 1516 (1981); A. A. Macco and H. M. Buck, J. Org. Chem., 46
2655 (1981); W. Schumacher and M. Hanack, Synthesis, 490
(1981).

$$CF_3CO_2H$$
$$CH_2Cl_2, -78°C.$$

56%

I.A.8-15 F. H. Koster and H. Wolf, Tetrahedron Lett., 22,
3937 (1981); J. R. Williams et al, J. Org. Chem., 46, 2665
(1981); E. J. Brunke, F. J. Hammerschmidt and H. Struwe,
Tetrahedron, 37, 1033 (1981).

$$Ac_2O/HOAc$$
$$HClO_4$$

76%

I.A.8-16 H. Kaneko et al, Chem. Lett., 757 (1981); T. L.
Macdonald, S. Mahalingam and D. E. O'Dell, J. Amer. Chem. Soc.,
103, 6767 (1981); J. D. Fourneron and M. Julia, Bull. Soc.
Chim. Fr. II, 387 (1981).

$$SnCl_4$$
$$CH_2Cl_2, -78°C.$$

57%

I.A.8-17 I. Fleming and A. Pearce, J. Chem. Soc., Perkin I, 251 (1981); G. Demailly and G. Solladie, J. Org. Chem., 46, 3102 (1981).

72%

Cyclization without the TMS Group Gives Five Products.

I.A.8-18 M. T. Reetz, I. Chatziiosifidis and K. Schwellnus, Angew. Chem., Int. Ed. Engl., 20, 687 (1981); M. Alderdice and L. Weiler, Can. J. Chem., 59, 2239 (1981); W. P. Jackson, S. V. Ley and J. A. Morton, Tetrahedron Lett., 22, 2601 (1981).

1) $SnCl_4$ (cat)

$CH_2Cl_2/0°C.$

2) 25°C.

3) H_2O

96%

Intramolecular α-tert Alkylation.

I.A.8-19 J. R. Matz and T. Cohen, Tetrahedron Lett., 22, 2459 (1981); R. R. Rao and S. Bhattacharya, Ind. J. Chem., 20B, 207 (1981).

MeSO$_3$H

P_2O_5

65%

I.A.8-20 L. N. Mander and P. H. C. Mundill, Synthesis, 620
(1981); T. R. Kasturi and S. M. Reddy, Ind. J. Chem., 20B,
64 (1981).

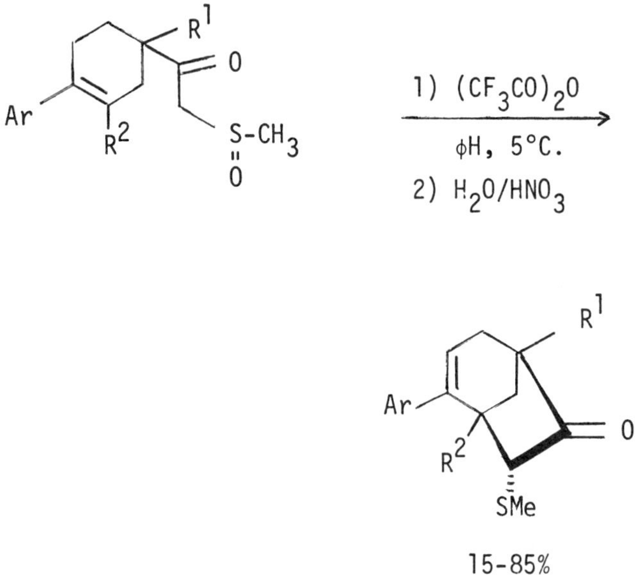

1) $(CF_3CO)_2O$

ϕH, 5°C.

2) H_2O/HNO_3

15-85%

I.A.8-21 J. M. Cook et al, Tetrahedron Lett., 22, 211 (1981);
A. Murai, S. Sato and T. Masamune, ibid, 1033.

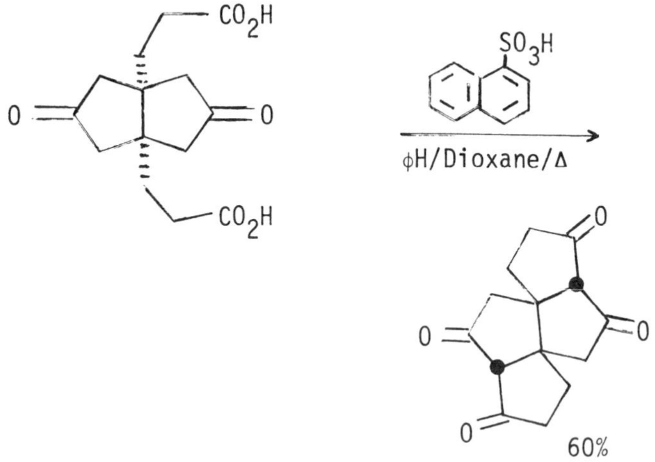

$\phi H/Dioxane/\Delta$

60%

I.A.8-22 M. T. Reetz and I. Chatziiosifidis, <u>Angew. Chem.</u>,
<u>Int. Ed. Engl.</u>, <u>20</u>, 1017 (1981); P. G. Gassman and J. J.
Tally, <u>Org. Syn.</u>, <u>60</u>, 14 (1981); T. Livinghouse, <u>ibid</u>, 126;
M. J. O. Anteunis et al, <u>Tetrahedron Lett.</u>, <u>22</u>, 141 (1981);
K. Utimoto et al, <u>ibid</u>, 4279.

$$\text{Cl} \xrightarrow[\substack{\text{SnCl}_4 \\ \text{CH}_2\text{Cl}_2, \ 25°\text{C.}}]{\text{Me}_3\text{SiCN}} \qquad \underset{\text{CN}}{\text{Cl}} \qquad 72\%$$

I.A.8-23 J. Oku and S. Inoue, <u>Chem. Commun.</u>, 229 (1981); J.
E. Backvall and O. S. Andell, <u>ibid</u>, 1098.

$$\phi\text{CHO} \xrightarrow[\substack{\text{cyclo (L-PHE-L-HIS)} \\ \phi\text{H, } 35°\text{C.}}]{\text{HCN}} \overset{*}{\phi}\text{CH} \underset{\text{CN}}{\overset{\text{OH}}{}}$$

40-90% Conversion

(90-12% ee)

I.A.8-24 D. F. Taber and S. A. Saleh, <u>J. Org. Chem.</u>, <u>46</u>,
4817 (1981); M. Yasuda, C. Pac and H. Sakurai, <u>J. Chem. Soc.</u>,
<u>Perkin I</u>, 746 (1981); E. K. Metzner et al, <u>Synthesis</u>, 791
(1981).

$$\underset{\text{ArSO}_2}{} \xrightarrow[\substack{t_\text{BuOH/18-C-6} \\ \Delta}]{\text{KCN}} \underset{\text{NC}}{}$$

60%

I.A.8-25 R. Davis and K. G. Untch, J. Org. Chem., 46, 2985
(1981).

$$ROH \xrightarrow[\substack{2\ eq.\ NaCN \\ NaI\ (cat) \\ DMF/CH_3CN}]{2\ eq.\ Me_3SiCl} RCN$$

16-98%

I.A.8-26 M. Kurauchi, T. Imamoto and M. Yokoyama, Tetrahedron
Lett., 22, 4985 (1981).

$$ArCHO \xrightarrow[Bu_3P,\ 0°C.]{MeSCN} ArCH_2CN + ArČSMe$$

31-45%   16-47%

I.A.8-27 D. Kahne and D. B. Collum, Tetrahedron Lett., 22,
5011 (1981).

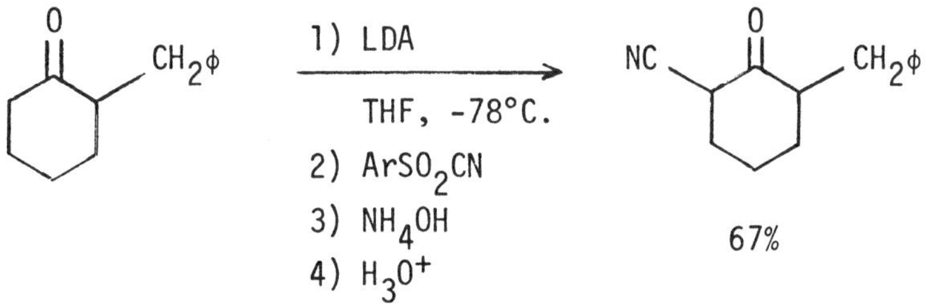

1) LDA

THF, -78°C.

2) ArSO$_2$CN

3) NH$_4$OH

4) H$_3$O$^+$

67%

I.A.8-28 A. L. J. Beckwith, Tetrahedron, 37, 3073 (1981).

Review: "Regioselectivity and Stereoselectivity in
          Radical Reactions."

I.A.8-29 D. H. R. Barton and W. B. Motherwell, Pure Appl. Chem., 53, 15, 1081 (1981).

Review: "New and Selective Reactions and Reagents in Carbohydrate Chemistry."

I.A.8-30 S. Wolff and W. C. Agosta, J. Chem. Res. (S), 78 (1981); D. Brule et al, Bull. Soc. Chim. Fr. II, 57 (1981).

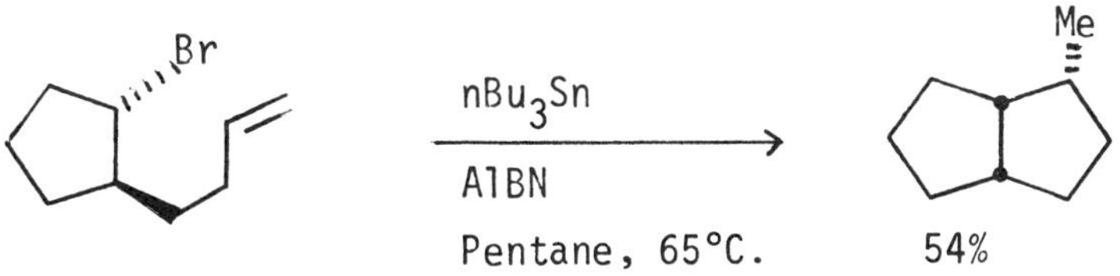
Br
nBu_3Sn
AIBN
Pentane, 65°C.
Me
54%

I.A.8-31 S. Dolezal and J. Jary, Coll. Czech. Chem. Commun., 46, 2709 (1981).

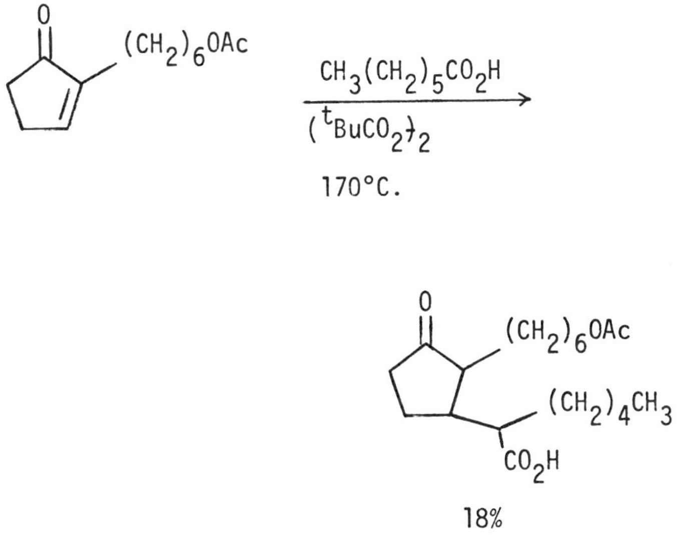
O
$(CH_2)_6OAc$
$CH_3(CH_2)_5CO_2H$
$(^tBuCO_2)_2$
170°C.
O
$(CH_2)_6OAc$
$(CH_2)_4CH_3$
CO_2H
18%

I.A.8-32 H. G. Viehe et al, Tetrahedron, 37 (Suppl. 1), 111
(1981).

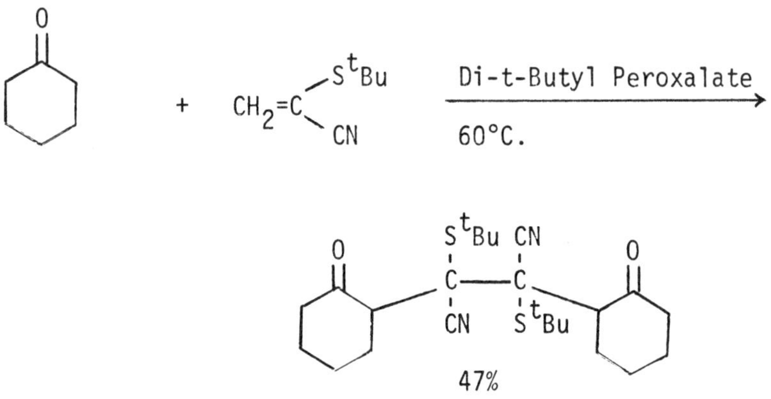

47%

I.A.8-33 R. Sustmann and R. Altevogt,Tetrahedron Lett., 22,
5167 (1981); B. H. Han and P. Boudjouk, ibid, 2759; E. Fritz,
H. Langhals and C. Ruchardt, Justus Liebigs Ann. Chem., 1015
(1981).

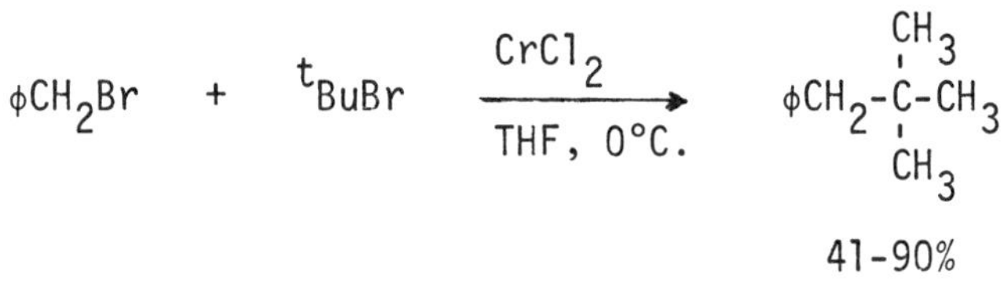

41-90%

I.A.8-34 Y. M. Chang, R. Profetto and J. Warkentin, J. Amer.
Chem. Soc., 103, 7189 (1981).

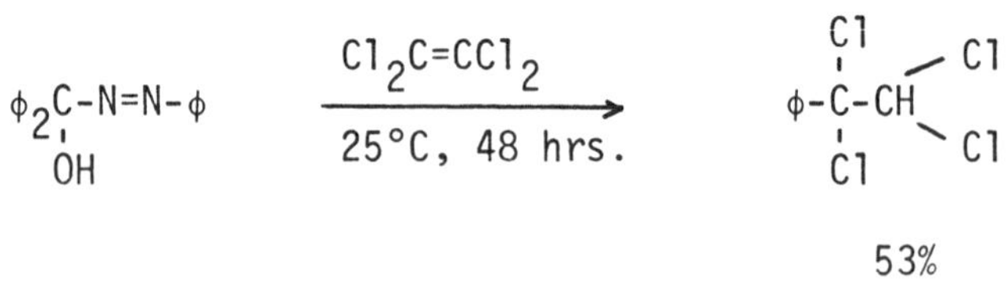

53%

I.A.8-35 A. Albini, Synthesis, 249 (1981).

Review: "Photosensitization in Organic Synthesis."

I.A.8-36 P. A. Wender and J. J. Howbert, J. Amer. Chem. Soc., 103, 688 (1981).

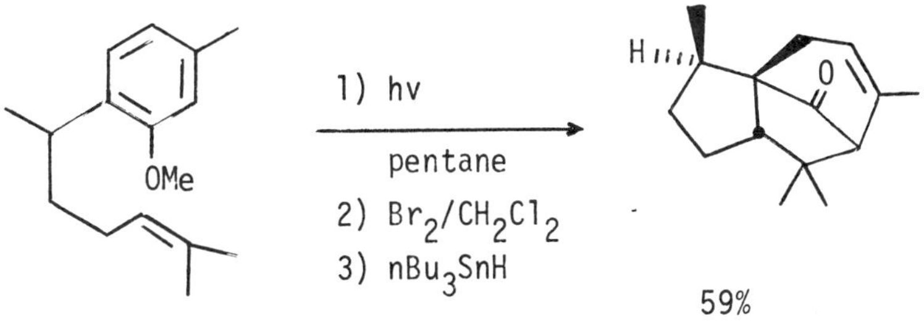

1) hv
———————→
pentane
2) Br$_2$/CH$_2$Cl$_2$
3) nBu$_3$SnH

59%

I.A.8-37 T. Sato, K. Maemoto and A. Kohda, Chem. Commun., 1116 (1981); J. Becher, O. Buchardt et al, Tetrahedron, 37, 789 (1981).

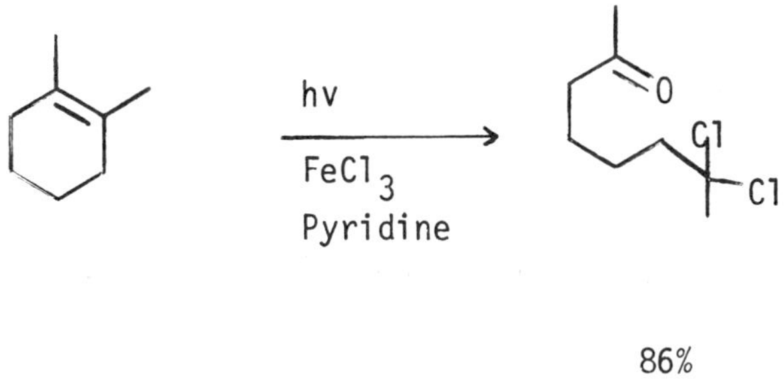

hv
———————→
FeCl$_3$
Pyridine

86%

I.A.8-38 J. W. Bruno, T. J. Marks and F. D. Lewis, J. Amer. Chem. Soc., 103, 3608 (1981).

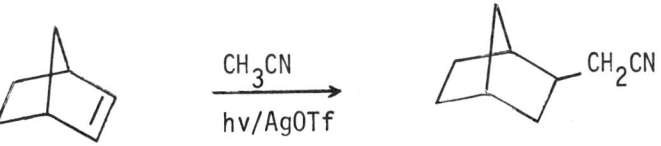

CH$_3$CN
———————→
hv/AgOTf

35-40%

I.A.8-39 T. Sato, H. Kaneko and T. Takahashi, Chem. Lett.,
1469 (1981).

$$HCO_2CH_2CH_3 \xrightarrow[\text{TiCl}_4/hv]{CH_3CH_2CH_2OH} CH_3CH_2\underset{OH}{CHCH} \overset{OC_3H_7}{\underset{OC_3H_7}{\diagup}}$$

52%

I.A.8-40 H. M. Bartels and P. Boldt, Justus Liebigs Ann.
Chem., 40 (1981).

77%

I.A.8-41 H. J. Schafer, Angew. Chem., Int. Ed. Engl., 20,
911 (1981).

Review: "Anodic and Cathodic CC-Bond Formation."

I.A.8-42 L. Mandell, F. J. Heldrich and R. A. Day, Jr.,
Synth. Commun., 11, 55 (1981); G. Mabon, C. Moinet and J.
Simonet, Chem. Commun., 1040 (1981); T. Troll, W. Elbe and
G. W. Ollmann, Tetrahedron Lett., 22, 2961 (1981); D. A.
White, Org. Syn., 60, 58, 78 (1981).

20-40%

I.A.8-43 M. Rabinovitz et al, Angew. Chem., Int. Ed. Engl., 20, 1033 (1981); J. Shabtai, E. Ney-Igner and H. Pines, J. Org. Chem., 46, 3795 (1981).

$$\phi_2C=O \quad \xrightarrow[\substack{\text{Pet. Ether, 25°C.} \\ \text{2) } H_2O}]{\text{1) } K/Al_2O_3} \quad \underset{\substack{| \\ OH}}{\phi-C}\underset{\substack{| \\ OH}}{-C}-\phi$$

with φ groups

72%

I.A.8-44 S. Satoh, H. Suginome and M. Tokuda, Tetrahedron Lett., 22, 1895 (1981); Bull. Chem. Soc. Jpn., 54, 3456 (1981).

$$MeO_2C\diagup\diagdown\diagup Br \quad \xrightarrow[\substack{\text{Electrolysis, Pt-Pt} \\ Q^+ \ OTs^-/DMF}]{EtO_2C\diagdown\diagup\diagup CO_2Et} $$

CO_2Et / CO_2Et / CO_2Et structure

39%

Also addition of allyl halides to acetone.

I.A.8-45 P. G. Gassman et al, J. Org. Chem., 46, 5455 (1981).

$$\xrightarrow[\substack{\text{Electrochemical} \\ \text{Reductive Cycl.}}]{+ \ e^-/DMF}$$

—OSO_2CH_3

82%

I.A.8-46   T. Shono et al, Tetrahedron Lett., 22, 871 (1981).

$$RCHO \quad + \quad CCl_4 \quad \xrightarrow[\text{DMF}]{+2e^-} \quad \overset{\overset{\text{OH}}{|}}{RCH\text{-}CCl_3}$$

$$46\text{-}81\%$$

I.A.8-47   D. A. White, Org. Syn., 60, 1 (1981); M. Chkir, D. Lelandais and C. Bacquet, Can. J. Chem., 59, 945 (1981); U. Jensen and H. J. Schafer, Chem. Ber., 114, 292 (1981).

$$\begin{array}{c} CO_2H \\ | \\ (CH_2)_4 \\ | \\ CO_2Me \end{array} \quad \xrightarrow[\text{NaOMe/MeOH}]{\text{Electrolysis}} \quad \begin{array}{c} CO_2Me \\ | \\ (CH_2)_8 \\ | \\ CO_2Me \end{array}$$

$$70\%$$

I.A.8-48   M, Kawanisi et al, Tetrahedron Lett., 22, 2569 (1981).

$$\xrightarrow[\substack{\text{NaOMe/MeOH} \\ 2)\ H_3O^+}]{1)\ -2e^-}$$

$$62\%$$

I.A.8-49   A. B. Smith, III, Tetrahedron, 37, 2407 (1981).

Review: "The Acid Promoted Decomposition of α-Diazo Ketones."

I.A.8-50 D. W. Johnson, L. N. Mander and T. J. Masters, Austr. J. Chem., 34, 1243 (1981); B. Basu, S.K. Maity and D. Mukherjee, Synth. Commun., 11, 803 (1981); K. C. Nicolaou and R. E. Zipkin, Angew. Chem., Int. Ed. Engl., 20, 785 (1981).

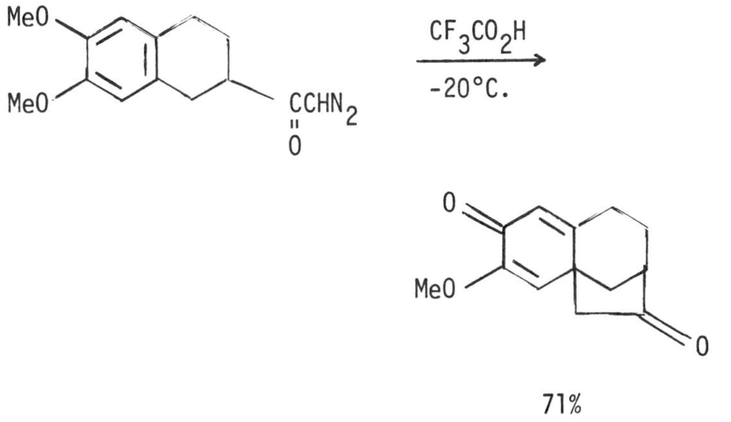

$$CF_3CO_2H \xrightarrow{} -20°C.$$

71%

I.A.8-51 U. R. Ghatak et al, J. Chem. Soc., Perkin I, 1203 (1981); J. Chem. Res. (S), 5 (1981); Chem. Commun., 746 (1981).

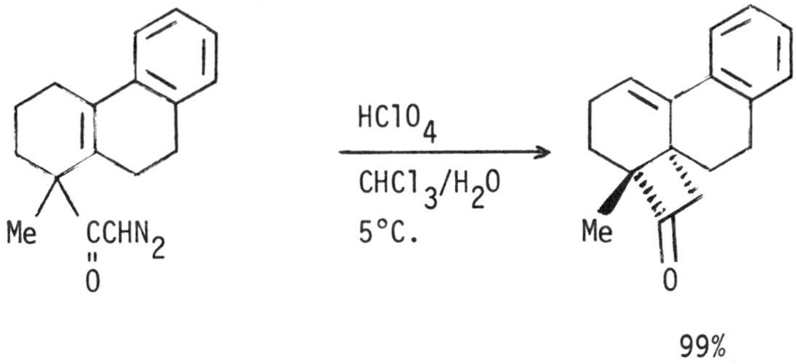

$$\xrightarrow[\text{CHCl}_3/\text{H}_2\text{O} \atop 5°C.]{\text{HClO}_4}$$

99%

I.A.8-52 A. B. Smith, III et al, J. Amer. Chem. Soc., 103, 1996, 2009, 2017 (1981); T. Hudlicky et al, J. Org. Chem., 46, 2911 (1981).

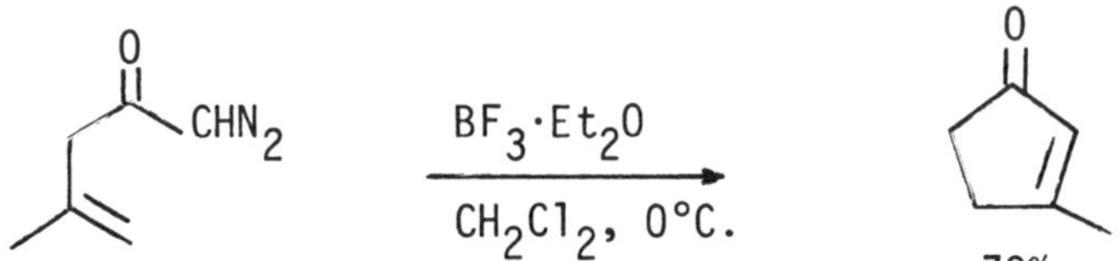

Also used in Polyolefin Cyclizations.

73%

I.A.8-53 M. P. Doyle, W. H. Tamblyn and V. Bagheri, J. Org. Chem., 46, 5094 (1981); K. Saigo, S. Okagawa and H. Nohira, Bull. Chem. Soc. Jpn., 54, 3603 (1981); E. Wenkert et al, Synth. Commun., 11, 533 (1981).

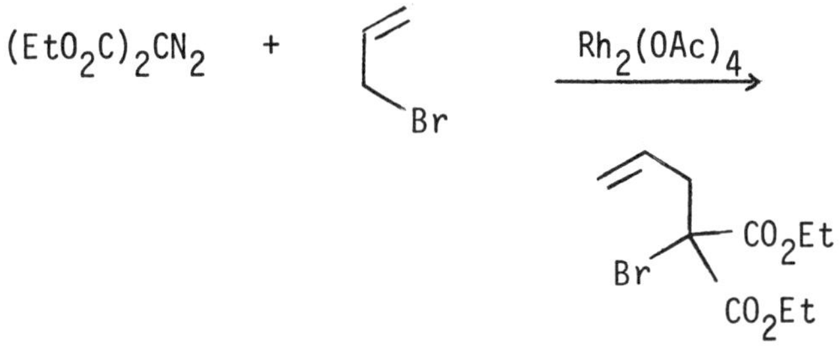

$(EtO_2C)_2CN_2$ +

$Rh_2(OAc)_4$

86%

I.A.8-54 C. A. Loeschorn, M. Nakajima and J. P. Anselme, Bull. Soc. Chim. Belg., 90, 985 (1981).

ϕCHN_2 + ArCHO $\xrightarrow[Et_2O, \ 0°C.]{LiBr}$ $\phi CH_2\overset{O}{\overset{\|}{C}}Ar$

53-100%

I.A.8-55 T. Aoyama and T. Shioiri, Chem. Pharm. Bull., 29, 3249 (1981); I. G. Vasi and R. H. Acharya, Ind. J. Chem., 20B, 509 (1981).

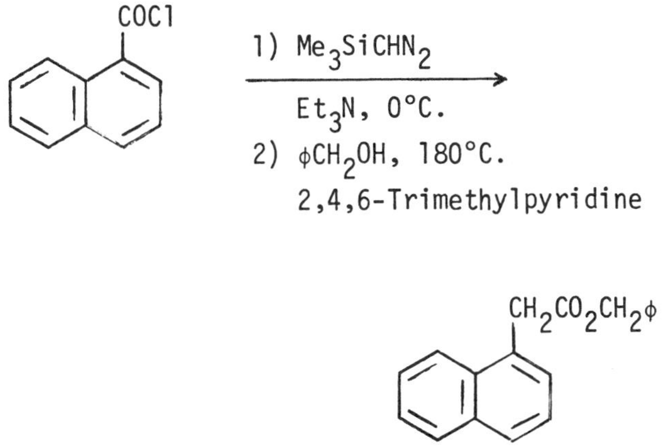

$COCl$

1) Me_3SiCHN_2

$\xrightarrow{\text{Et}_3\text{N, 0°C.}}$

2) ϕCH_2OH, 180°C.
2,4,6-Trimethylpyridine

$CH_2CO_2CH_2\phi$

Safe Arndt-Eistert
Synthesis.

68-82%

I.A.8-56 R. L. Danheiser et al, J. Amer. Chem. Soc., 103, 2443 (1981); M. A. O'Leary, G. W. Richardson and D. Wege, Tetrahedron, 37, 813 (1981).

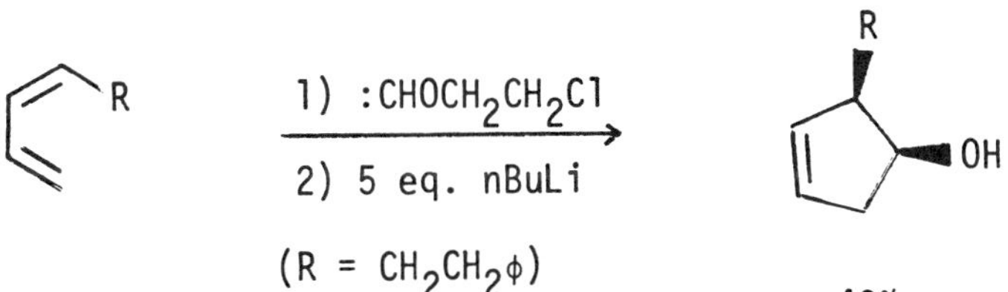

R

1) $:CHOCH_2CH_2Cl$

$\xrightarrow{}$

2) 5 eq. nBuLi

(R = $CH_2CH_2\phi$)

R

OH

40%

I.A.8-57 M. R. DeCamp and L. A. Viscogliosi, <u>J. Org. Chem.</u>,
<u>46</u>, 3918 (1981); A. F. Noels, A. J. Hubert et al, <u>ibid</u>, 873;
N. G. Connelly et al, <u>Chem. Commun.</u>, <u>17</u>, 19 (1981).

$$\text{(cycloheptatriene)} \quad + \quad CHBr_3 \quad \xrightarrow[\substack{18\text{-}C\text{-}6 \\ 140°C.}]{K_2CO_3} \quad \text{(benzocyclobutene-Br)}$$

18-45%

I.A.8-58 T. Mukaiyama, M. Yamaguchi and J. I. Kato, <u>Chem.</u>
<u>Lett.</u>, 1505 (1981); H. Nagashima, K. Sato and J. Tsuji, <u>ibid</u>,
<u>1605</u>.

$$RCHO \quad \xrightarrow[\text{DMSO, } 25°C.]{CBr_4/SnF_2} \quad \underset{\underset{OH}{|}}{RCH\text{-}CBr_3}$$

46-80%

I.A.8-59 T. Harada and A. Oku, <u>J. Amer. Chem. Soc.</u>, <u>103</u>, 5965
(1981); R. Carrie et al, <u>Tetrahedron Lett.</u>, <u>22</u>, 441 (1981).

$$\text{(allyl-ONa)} \quad \xrightarrow[\substack{KO^tBu/THF \\ 0°C.}]{ClCH_2S\phi} \quad \text{(product with OH, CH}_2\text{S}\phi) \quad +$$

29%

$$\text{(allyl-OCH}_2\text{S}\phi)$$

64%

I.A.8-60 I. Crossland, <u>Org. Syn.</u>, <u>60</u>, 6 (1981).

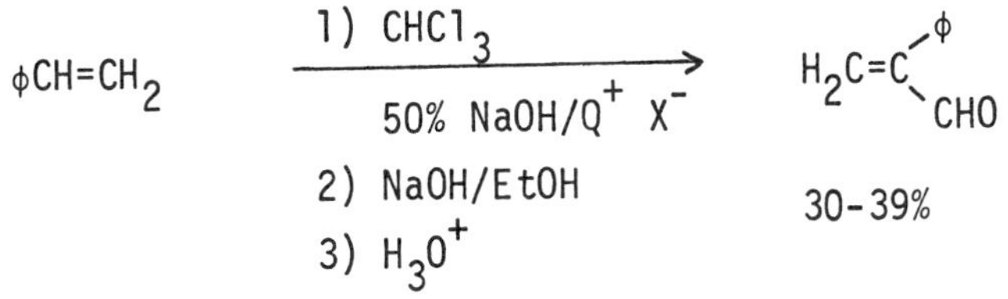
ϕCH=CH2
1) CHCl3
50% NaOH/Q+ X-
2) NaOH/EtOH
3) H3O+
H2C=C
ϕ
CHO
30-39%

I.A.8-61 J. M. Conia et al, <u>Tetrahedron Lett.</u>, <u>22</u>, 645 (1981).

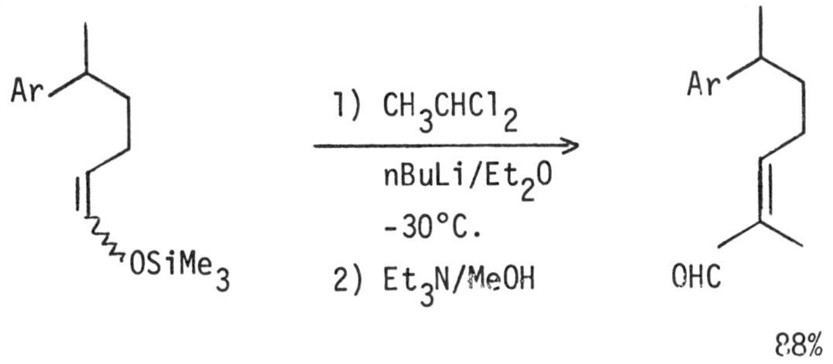
Ar
OSiMe3
1) CH3CHCl2
nBuLi/Et2O
-30°C.
2) Et3N/MeOH
Ar
OHC
88%

I.A.8-62 J. Tsuji, K. Sato and H. Nagashima, <u>Chem. Lett.</u>, 1169 (1981).

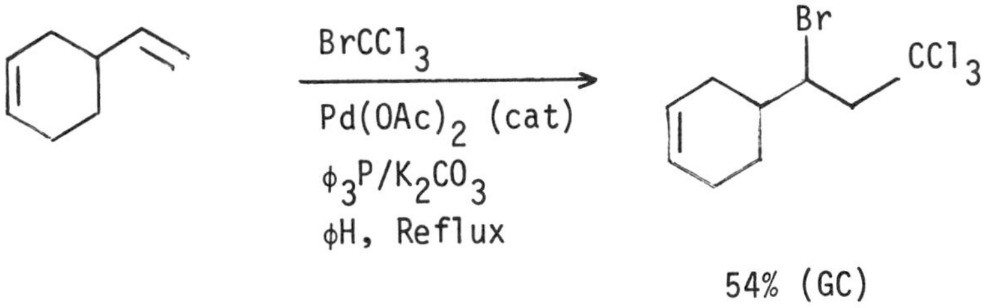
BrCCl3
Pd(OAc)2 (cat)
ϕ3P/K2CO3
ϕH, Reflux
Br
CCl3
54% (GC)

I.A.8-63 B. M. Trost and E. Murayama, J. Amer. Chem. Soc.,
103, 6529 (1981); H. Stetter and R. Y. Ramsch. Synthesis, 477
(1981).

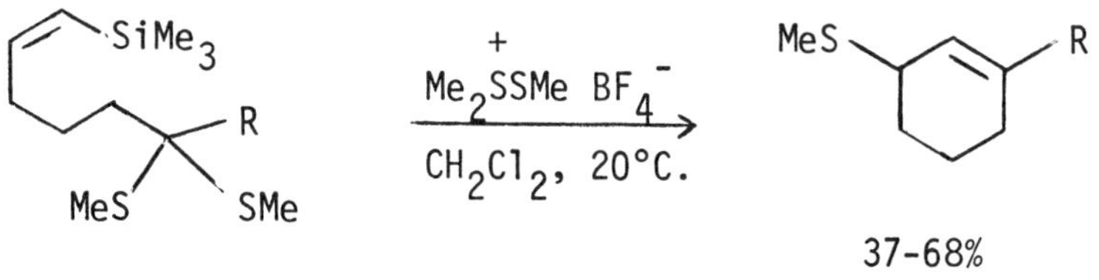

37-68%

I.A.8-64 K. Shiosaki, G. Fels and H. Rapoport, J. Org. Chem.,
46, 3230 (1981).

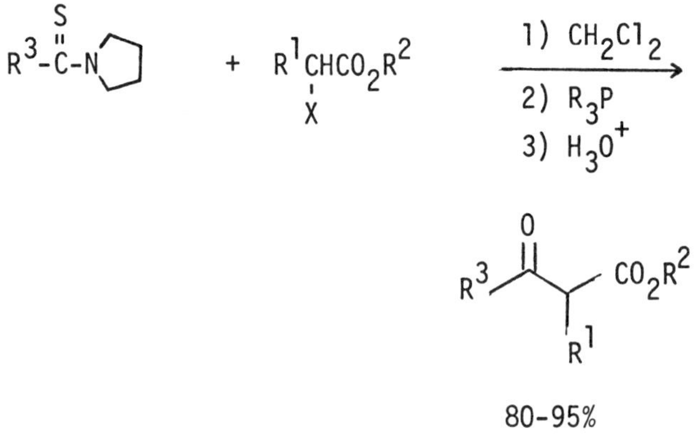

80-95%

X = Br or OSO$_2$CF$_3$

I.A.8-65 H. M. Bartels, P. Boldt and D. Schomburg, Chem. Ber.,
114, 3997 (1981).

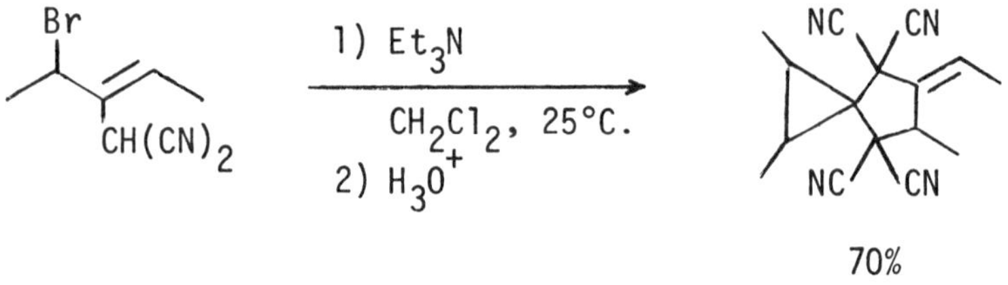

70%

I.A.8-66 A. R. Katritzky and N. K. Ponkshe, <u>Tetrahedron Lett.</u>,
<u>22</u>, 1215 (1981); M.Sekiya et al, <u>ibid</u>, 123.

$$RCH_2O-C\underset{\phi}{\overset{NAr}{\Big\langle}} \xrightarrow[-5°C.]{LDA/THF} \underset{\phi}{\overset{O}{RC-CH}} \underset{\phi}{\overset{NHAr}{\diagdown}}$$

42-46%

I.A.8-67 G. Balme, M. Malacria and J. Gore, <u>J. Chem. Res. (S)</u>,
244 (1981).

$$\xrightarrow{O_2N-\langle\underline{}\rangle-CO_2H}$$

73%

I.A.8-68 H. Stetter and F. Jonas, <u>Tetrahedron Lett.</u>, <u>22</u>,
4945 (1981).

$$\underset{EtO_2C}{\overset{O}{\underset{}{R-C}}}\diagdown_{CH-CH}\diagup\overset{\overset{O}{\parallel}}{\underset{\underset{O}{\parallel}}{\underset{C-CH_3}{C-CH_3}}} \xrightarrow[140°C.]{DMSO/H_2O} \overset{O}{\underset{}{RCCH_2-CH}}\diagup\overset{\overset{O}{\parallel}}{\underset{\underset{O}{\parallel}}{\underset{CCH_3}{CCH_3}}}$$

I.A.8-69 W. C. Lumma, Jr., <u>J. Org. Chem.</u>, <u>46</u>, 3668 (1981).

$$ArCHO + {}^tBuN=C: \xrightarrow[\substack{Pyridine \\ -5°C. \\ 2)\ H_2O,\ NaHSO_3}]{1)\ CF_3CO_2H} Ar\overset{OH}{\underset{\underset{O}{\parallel}}{\diagup\diagdown}}NH^tBu$$

60%

I.A.8-70   S. Dev, Acct. Chem. Res., 14, 82 (1981).

Review: "Aspects of Longifolene Chemistry. An Example
of Another Facet of Natural Products
Chemistry."

I.A.8-71   A. A. Shchegolev and M. I. Kanishchev, Russ. Chem.
Rev., 50, 553 (1981).

Review: "Rearrangements in Vinyl Cations."

I.A.8-72   M. Christl, Angew. Chem., Int. Ed. Engl., 20, 529
(1981).

Review: "Benzvalene - Properties and Synthetic
Potential."

I.A.8-73   A. Akelah and D. C. Sherrington, Chem. Rev., 81,
557 (1981).

Review: "Application of Functionalized Polymers in
Organic Synthesis."

I.A.8-74   J. M. J. Frechet, Tetrahedron, 37, 663 (1981).

Review: "Synthesis and Applications of Organic Polymers
as Supporting and Protecting Groups."

I.A.8-75 A. Akelah, Synthesis, 413 (1981).

Review: "Heterogeneous Organic Synthesis using
         Functionalized Polymers."

I.A.8-76 B. Reuben and K. Sjoberg, CHEMTECH, 315 (1981).

Review: "Phase-Transfer Catalysis in Industry."

I.A.8-77 J. F. Bunnett, Pure Appl. Chem. , 53, 305 (1981).

"Nomenclature for Straightforward Transformations."
(A General System of Nomenclature for Transformations
of One Organic Compound into Another.)

I.B. Carbon-Carbon Double Bonds
       (see also: I.E.1, III.G, VI.A.16).

I.B.1. Wittig Type Olefination Reactions

I.B.1-1 H. J. Bestmann et al, Angew. Chem., Int. Ed. Engl.,
20, 575 (1981); Justus Liebigs Ann. Chem., 1705, 2117 (1981);
R. K. Muller et al, Helv. Chim. Acta, 64, 2419, 2436, 2447,
2463, 2469 (1981); G. Kossmehl and R. Nuck, Chem. Ber., 114,
2914 (1981).

$$R^1CH=P\phi_3 \quad \xrightarrow{\begin{array}{l} 1)\ R^2CH_2\overset{\overset{\displaystyle O}{\|}}{C}X \\ 2)\ EtBr \\ 3)\ NaNH_2 \\ 4)\ R^3CHO \end{array}} \quad R^3CH=C\overset{\displaystyle R^1}{\underset{\displaystyle OEt}{|}}-C=CHR^2$$

29-52%

I.B.1-2 A. D. Buss and S. Warren, Chem. Commun., 100 (1981);
R. Baker and R. J. Sims, Synthesis, 117 (1981); J. M. Clough
and G. Pattenden, Tetrahedron, 37, 3911 (1981); M. Suda,
Chem. Lett., 967 (1981); H. J. C. Jacobs et al, Org. Prep.
Proc. Int., 13, 9 (1981).

$$\phi_2\overset{O}{\overset{\|}{P}}CH_2R^1 \xrightarrow[\substack{\text{THF, } -78°C. \\ 2) R^2CHO \\ 3) NaH/DMF \\ 50°C.}]{1) nBuLi} \quad \overset{R^1}{\underset{H}{}}C=C\overset{R^2}{\underset{H}{}}$$

63%

I.B.1-3 M. Delmas et al, Synth. Commun., 11, 125 (1981); E.
V. Dehmlow and S. Barahona-Naranjo, J. Chem. Res. (S), 142,
143 (1981).

$$\phi_3\overset{+}{P}CH_2R \; Br^- \xrightarrow[\substack{\text{NaOH/Dioxane} \\ 70°C. \\ (R = nPr)}]{\phi CHO} \phi CH=CHR$$

95%

(84% Z)

I.B.1-4 M. P. Cooke, Jr. and K. P. Biciunas, Synthesis, 283
(1981).

$$(EtO)_2\overset{O}{\overset{\|}{P}}CH_2\underset{\underset{P\phi_3}{\|}}{\overset{O}{\overset{\|}{C}}}-C-CO_2Et \xrightarrow[\substack{2) R_2C=O}]{1) NaH/THF}$$

$$\underset{R}{\overset{R}{}}C=CH-\underset{\underset{P\phi_3}{\|}}{C}-\overset{O}{\overset{\|}{C}}-CO_2Et$$

71-96%

I.B.1-5 G. L. Larson, J. A. Prieto and A. Hernandez, Tetrahedron Lett., 22, 1575 (1981).

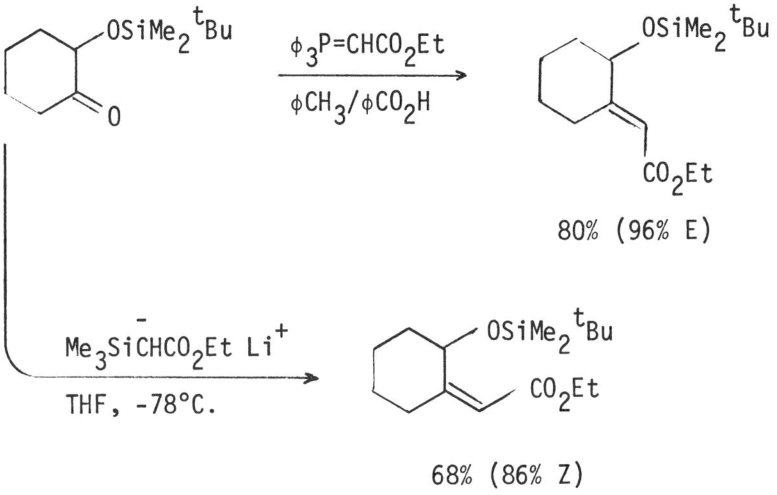

OSiMe2 tBu
O
ϕ3P=CHCO2Et
ϕCH3/ϕCO2H
OSiMe2 tBu
CO2Et
80% (96% E)
Me3SiCHCO2Et Li+
THF, -78°C.
OSiMe2 tBu
CO2Et
68% (86% Z)

I.B.1-6 M. Larcheveque et al, Chem. Commun., 877 (1981); Tetrahedron Lett., 22, 1595 (1981).

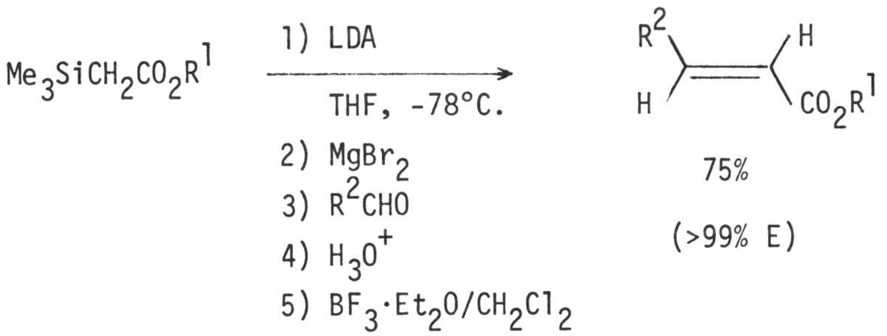

Me3SiCH2CO2R1
1) LDA
THF, -78°C.
2) MgBr2
3) R2CHO
4) H3O+
5) BF3·Et2O/CH2Cl2
R2
H
H
CO2R1
75%
(>99% E)

I.B.1-7 J. Brugidou et al, Tetrahedron Lett., 22, 4709 (1981); S. Tanimoto, S. Jo and T. Sugimoto, Synthesis, 53 (1981).

ϕ3P-CH2
S
S
Br-
1) KOtBu
THF
2) RCHO
RCH=CH
S
S
70%

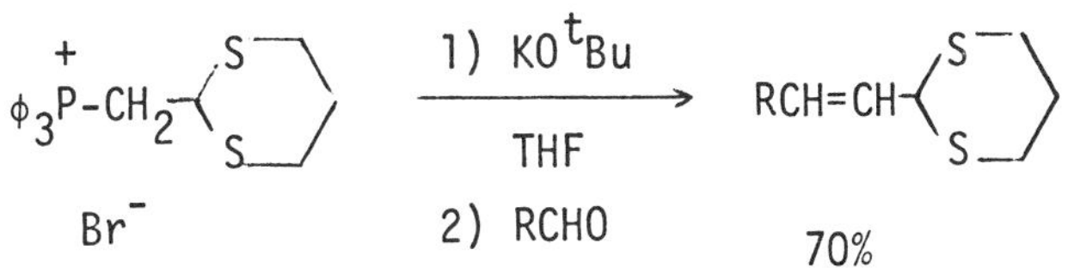

I.B.1-8 M.Slopianka and A. Gossauer, Justus Liebigs Ann.
Chem., 2258 (1981).

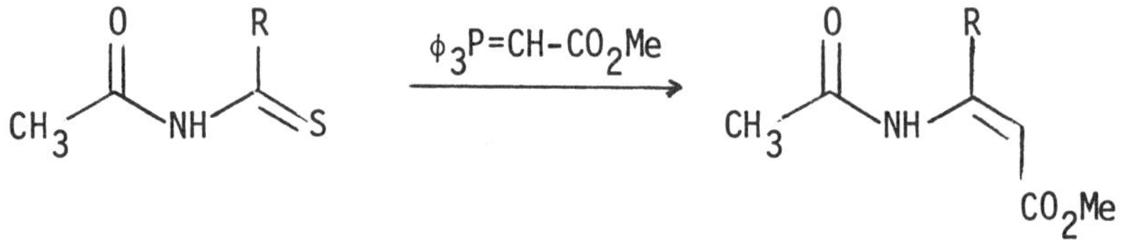

34-81%

Reduction and hydrolysis gives β-Amino Acids.

I.B.1-9 A. I. Meyers, J. P. Lawson and D. R. Carver, J. Org.
Chem., 46, 3119 (1981).

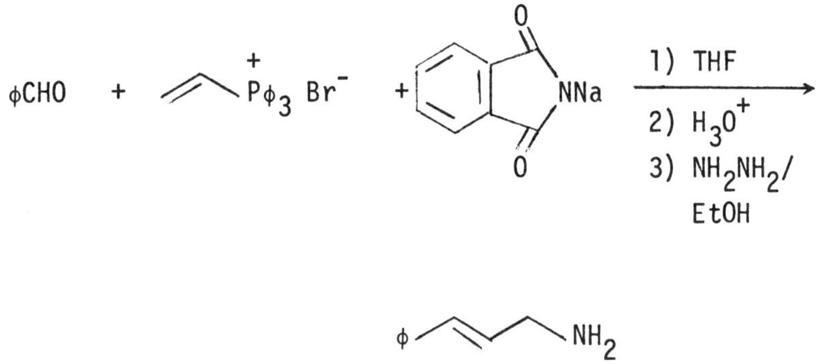

69%

I.B.1-10 N. L. J. M. Broehof and A. van der Gen, Tetrahedron
Lett., 22, 2799 (1981); B. Cortisella, I. Keitel and H. Gross,
Tetrahedron, 37, 1227 (1981).

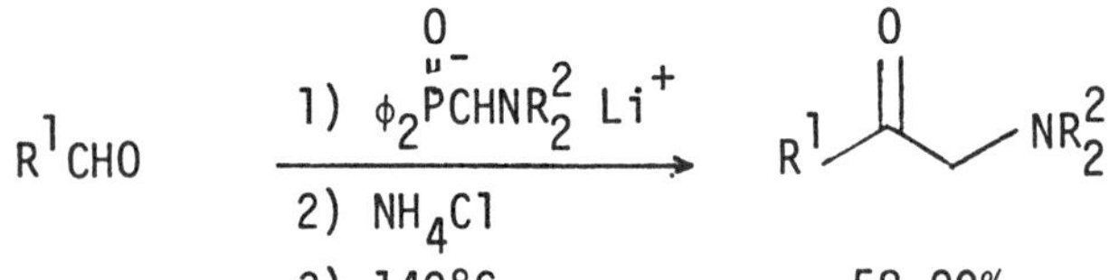

58-90%

I.B.1-11 H. Yamamoto et al, J. Amer. Chem. Soc., 103, 5568
(1981); F. Plenat, Tetrahedron Lett., 22, 4705 (1981); K.
Sato et al, Chem. Lett., 1711 (1981).

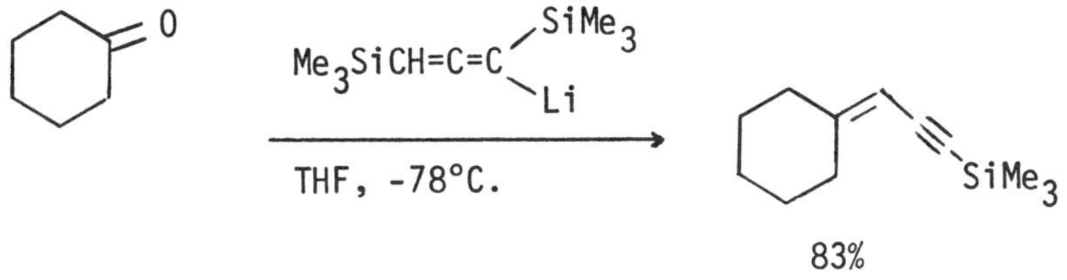

83%

I.B.1-12 H. Nagaoka and Y. Kishi, Tetrahedron, 37, 3873 (1981).

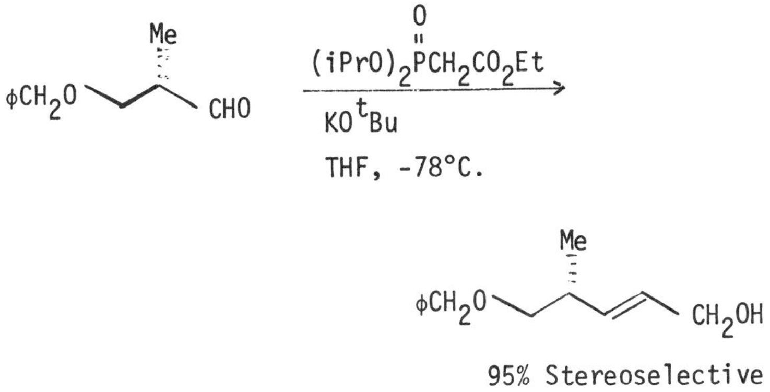

95% Stereoselective

Asymmetric epoxidation then cuprate addition to
above product.

I.B.1-13 B. E. Maryanoff and B. A. Duhl-Emswiler, Tetrahedron
Lett., 22, 4185 (1981); B. E. de Jong, H. de Koning and H. O.
Huisman, Rec. Trav. Chim., 100, 410 (1981); R. Baker and R. J.
Sims, Tetrahedron Lett., 22, 161 (1981).

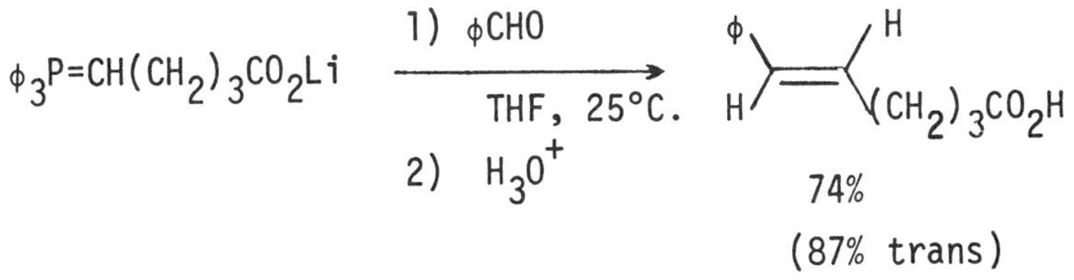

74%
(87% trans)

I.B.1-14   J. Font et al, <u>Tetrahedron</u>, <u>37</u>, 2493, 2391 (1981);
D. B. Tulshian and B. Fraser-Reid, <u>J. Amer. Chem. Soc.</u>, <u>103</u>,
474 (1981).

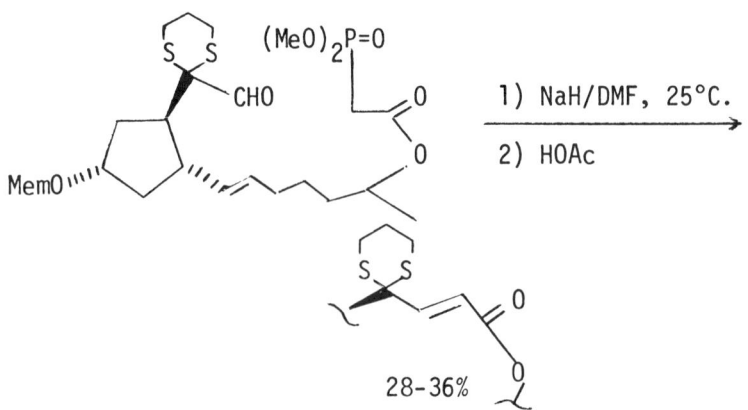

59%

(90% EE)

I.B.1-15   P. Raddatz and E. Winterfeldt, <u>Angew. Chem.</u>, <u>Int.</u>
<u>Ed. Engl.</u>, <u>20</u>, 286 (1981);   W. Kemp and C. D. Tulloch, <u>J.</u>
<u>Chem. Res.</u> (S), 28 (1981);   M. Marky, <u>Helv. Chim. Acta</u>, <u>64</u>,
957 (1981);   B. Bogdanovic et al, <u>Chem. Ber.</u>, <u>114</u>, 2261 (1981).

28-36%

I.B.1-16   E. Nakamura, Tetrahedron Lett., 22, 663 (1981).

$$(MeO)_2\overset{\overset{O}{\|}}{P}CH\begin{cases}OSiMe_2{}^tBu\\CO_2Me\end{cases} \xrightarrow[\text{2) RCHO}]{\begin{array}{c}\text{1) LDA}\\ \text{THF, -40°C.}\end{array}} RCH=\begin{cases}CO_2Me\\OSiMe_2{}^tBu\end{cases}$$

$$63\text{-}78\%$$

I.B.1-17   J. Rachon and U. Schollkopf, Justus Liebigs Ann. Chem., 99 (1981); L. Nedelec, V. Torelli and M. Hardy, Chem. Commun., 775 (1981); D. H. R. Barton, W. B. Motherwell and S. Z. Zard, Chem. Commun., 774 (1981).

$$\phi_2\overset{\overset{O}{\|}}{P}-CH\begin{cases}NC\\CO_2{}^tBu\end{cases} \xrightarrow[\substack{\text{NaH/CH}_2Cl_2\\ \text{or}\\ \text{50\% NaOH/CH}_2Cl_2}]{\text{RCHO}} \overset{R}{\underset{H}{\diagdown}}{=}\overset{NC}{\underset{CO_2{}^tBu}{\diagup}}$$

$$70\text{-}82\%$$

I.B.1-18   D. R. Williams et al, Tetrahedron Lett., 22, 3745 (1981); M. Suda and A. Fukushima, Tetrahedron Lett., 22, 759, 1421 (1981).

Also works with aldehydes.

I.B.1-19   D. J. Ager, Tetrahedron Lett., 22, 2923 (1981).

$$\phi S \diagdown \overset{\overset{\displaystyle R^1}{|}}{C} \diagup SiMe_3 \quad \xrightarrow[\substack{THF, -78°C. \\ 2)\ R^2R^3C=O}]{1)\ Li\ Naphthalenide} \quad \overset{R^2}{\underset{R^3}{\diagup}} C=CHR^1$$

59-86%

I.B.1-20   D. R. Brittelli, J. Org. Chem., 46, 2514 (1981).

$$BrCH_2CO_2H \quad \xrightarrow[\substack{3\ NaOEt \\ Glyme \\ 2)\ ArCHO}]{1)\ (EtO)_2\overset{\overset{\displaystyle O}{\|}}{P}H} \quad \overset{Ar}{\diagdown}\diagup\diagup\diagdown CO_2H$$

54-89%

I.B.1-21   Y. I. Baukov et al, J. Gen. Chem. (USSR), 51, 1105 (1981).

$$EtOC≡CSiMe_3 \quad \xrightarrow[\substack{(Aldehydes\ or\ Ketones) \\ BF_3·Et_2O/Et_2O}]{R^1R^2CO} \quad R^1R^2C=C \diagup \overset{CO_2Et}{\underset{SiMe_3}{}}$$

36-73%

I.B.1-22   W. C. Still and V. J. Novack, J. Amer. Chem. Soc., 103, 1283 (1981).

$$\xrightarrow[\substack{THF/HMPA \\ -78°C.}]{\phi_3As=\diagup}$$

84% (> 99% E)

I.B.1-23 T. Kauffmann et al, Tetrahedron Lett., 22, 5031
(1981); T. Otsubo, F. Ogura and H. Yamaguchi, Chem. Lett.,
447 (1981).

$$Me_3SiCH_2TiCl_3 \quad \xrightarrow[\quad Et_2O, -15°C. \quad]{1)\ RCHO} \quad RCH=CH_2$$

2) H_2O 59-65%

I.B.1-24 H. Bohme and P. Sutoyo, Tetrahedron Lett., 22, 1671
(1981).

$$\underset{Cl^-}{\overset{+}{Me_2N=CHCl}} \quad \xrightarrow[THF]{iPr_2NEt} \quad \underset{Cl}{\overset{Me_2N}{>}}C=C(Cl)NMe_2$$

63%

I.B.2.a. Eliminations of Alcohols and Derivatives to Form
        Double Bonds

I.B.2.a-1 I. Fleming et al, Tetrahedron Lett., 22, 2321
(1981); J. Chem. Soc., Perkin I, 256 (1981); I. Cutting and
P. J. Parsons, Tetrahedron Lett., 22, 2021 (1981).

$$Me_3Si \overset{R^2 \quad R^3}{\diagup\diagdown} OH \quad \xrightarrow[\substack{CH_2Cl_2,\ 0°C. \\ (R^1 = H\ or\ \phi)}]{BF_3 2AcOH} \quad \overset{R^2 \quad R^3}{\diagup} R^1$$

72-85%

I.B.2.b.   Eliminations of Halides to Form Doubld Bonds.

I.B.2.b-1   J. Zavada, R. A. Bartsch et al, Coll. Czech. Chem. Commun., 46, 850 (1981);   Z. Machkova and J. Zavada, ibid, 833.

"The Effect of Base Concentration on Orientiation in E2 Reactions."

I.B.2.b-2   D. Boschelli, R. M. Scarborough, Jr. and A. B. Smith, III, Tetrahedron Lett., 22, 19 (1981).

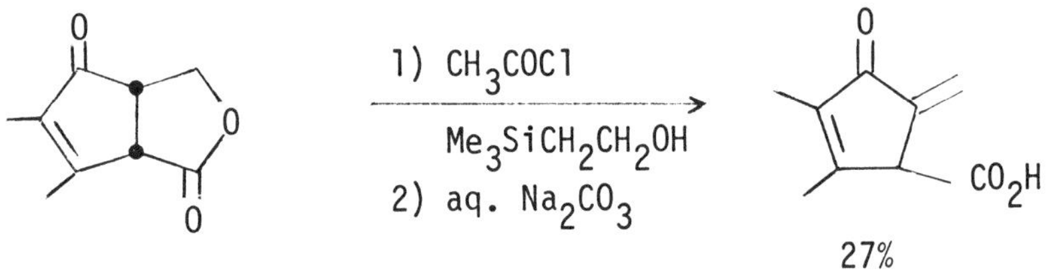

I.B.2.c.   Other Eliminations to Form Double Bonds.

I.B.2.c-1   R. S. Ward and K. S. Thatcher, Tetrahedron Lett., 22, 4831 (1981);   D. L. J. Clive and V. N. Kale, J. Org. Chem., 46, 231 (1981);   J. Mattay, W. Thunker and H. D. Scharf, Justus Liebigs Ann. Chem., 1105 (1981).

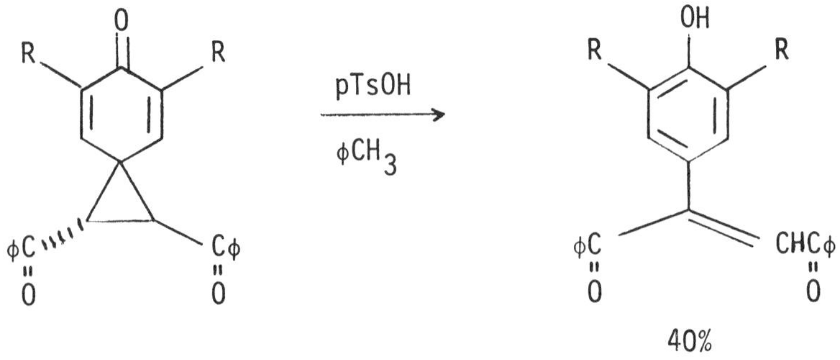

I.B.2.c-2 I. Fleming and D. A. Perry, <u>Tetrahedron Lett.</u>, <u>22</u>,
5095 (1981); A. de Groot et al, <u>ibid</u>, 4137.

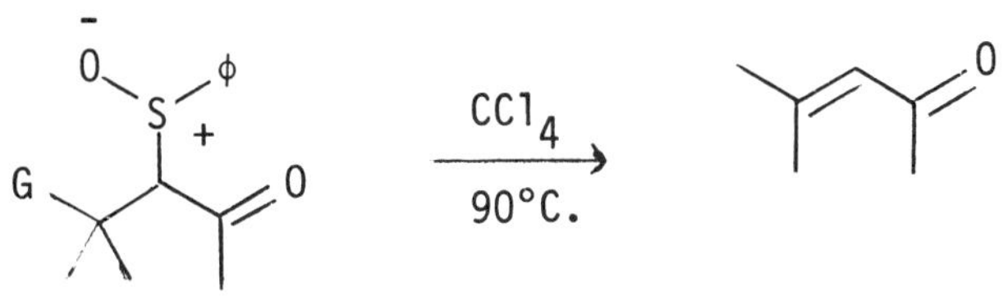

$\underline{G}$	$\underline{t_{1/2}}$
H	4.6 hr.
$SiMe_3$	1.4 hr.

I.B.2.c-3 H. G. Viehe et al, <u>Angew. Chem., Int. Ed. Engl.</u>, <u>20</u>,
585 (1981); R. F. Atkinson et al, <u>J. Org. Chem.</u>, <u>46</u>, 2804
(1981).

53%

I.B.2.c-4 A. R. Katritzky and A. M. El-Mowafy, <u>Chem. Commun.</u>,
96 (1981); J. T. Roberts, B. R. Rittberg and P. Kovacic, <u>J.
Org. Chem.</u>, <u>46</u>, 4111 (1981).

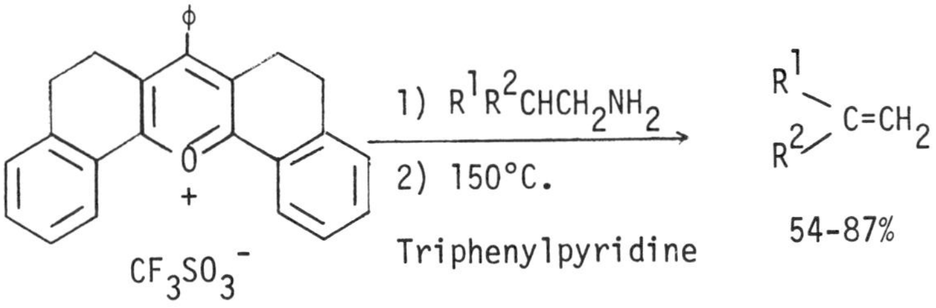

54-87%

I.B.2.c-5 Y. Ueno, M. Okawara et al, Tetrahedron Lett., 22, 2675 (1981).

Tos—CH=CH₂ (structure)
R

1) BuLi
2) $(CH_2O)_n$
3) Bu_3SnH
   AIBN, 80°C.
4) Distillation

R (diene structure)

36-88%

I.B.2.c-6 D. Liotta et al, J. Org. Chem., 46, 2920 (1981); Tetrahedron Lett., 22, 3043 (1981).

(cyclohexanone with CHO)

1) φSeCl
   ──────────
   Pyridine
2) 30% H_2O_2
   CH_2Cl_2

(cyclohexenone with CHO)

85%

I.B.2.c-7 M. Suzuki, T. Kawagishi and R. Noyori, Tetrahedron Lett., 22, 1809 (1981); D. J. Ager, I. Fleming and S. K. Patel, J. Chem. Soc., Perkin I, 2520 (1981).

(cyclohexenone)

1) $Me_3Si-Se\phi$
   ──────────
   TMSOTf (cat.)
2) $\phi CH(OMe)_2$
   TMSOTf (cat.)
3) H_2O_2

(cyclohexenone with CH(OMe)φ substituent)

75%

I.B.2.c-8 G. Wolff and G. Ourisson, Tetrahedron Lett., 22,
1441 (1981).

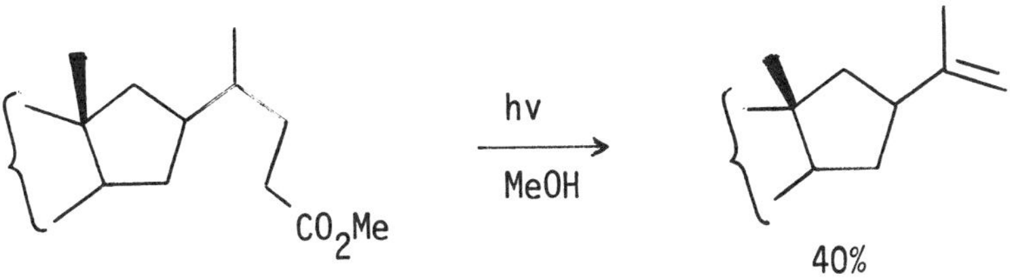

I.B.3. Other Carbon-Carbon Double Bond Forming Reactions

I.B.3-1 P. J. Kocienski, Chem. Ind., 548 (1981).

Review: "A New and Useful Olefin Synthesis Based on
Sulphones."

I.B.3-2 G. Melloni, G. Modena and U. Tonellato, Acct. Chem.
Res., 14, 227 (1981).

Review: "Relative Reactivities of Carbon-Carbon
Double and Triple Bonds toward Electrophiles."

I.B.3-3 R. M. Adlington and A. G. M. Barrett, Tetrahedron, 37,
3935 (1981); J. Chem. Soc., Perkin I, 2848 (1981); Chem.
Commun., 65 (1981); F. T. Bond and R. A. DiPietro, J. Org.
Chem., 46, 1315 (1981).

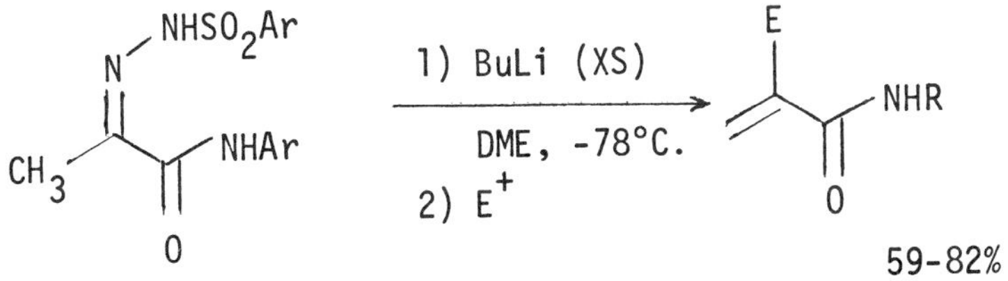

$E^+ = D_2O$, $(CH_2O)_n$, $RCHO$, $CH_2=CHCHO$, R_2CO.

I.B.3-4   J. E. Baldwin and J. C. Bottaro, Chem. Commun., 1121 (1981).

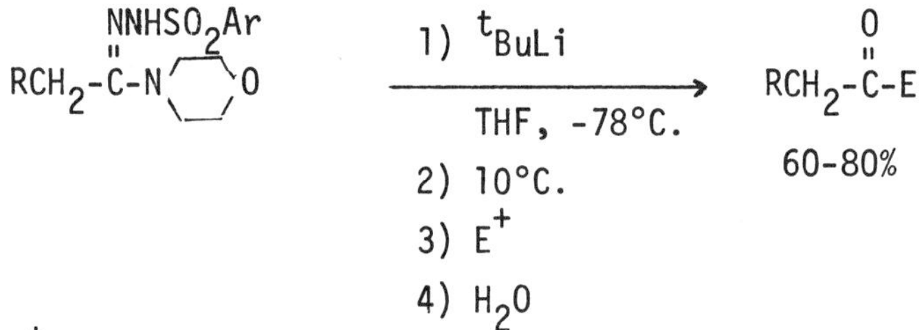

$$RCH_2\overset{NNHSO_2Ar}{\underset{\parallel}{\text{-C-N}}}\!\!\!\!\!\!\!\bigcirc\!\!O \quad \xrightarrow[\substack{\text{THF, -78°C.} \\ \text{2) 10°C.} \\ \text{3) } E^+ \\ \text{4) } H_2O}]{\text{1) } ^tBuLi} \quad RCH_2\overset{O}{\underset{\parallel}{\text{-C-E}}}$$

60-80%

$E^+ = RX$ or R_2CO.

I.B.3-5   Z. Rappoport, Acct. Chem. Res., 14, 7 (1981).

Review: "Nucleophilic Vinylic Substitution. A Single
          or a Multi-Step Process?"

I.B.3-6 R. Knorr and E. Lattke, Chem. Ber., 114, 2116 (1981);
B. Bogdanovic and B. Wermeckes, Angew. Chem., Int. Ed. Engl.,
20, 684 (1981); R. Sauvetre and J. F. Normant, Tetrahedron
Lett., 22, 957 (1981).

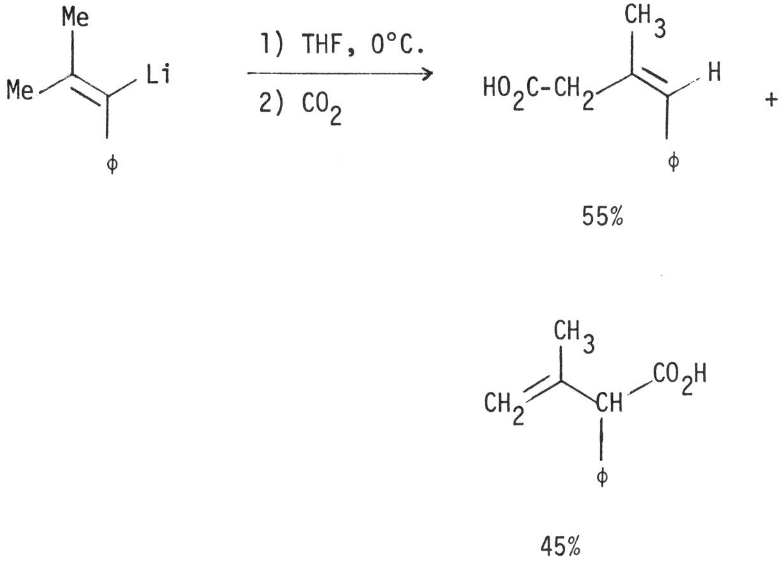

55%

45%

I.B.3-7 P. A. Grieco, P. A. Tuthill and H. L. Sham, J. Org.
Chem., 46, 5005 (1981); J. Coudane, Synthesis, 319 (1981).

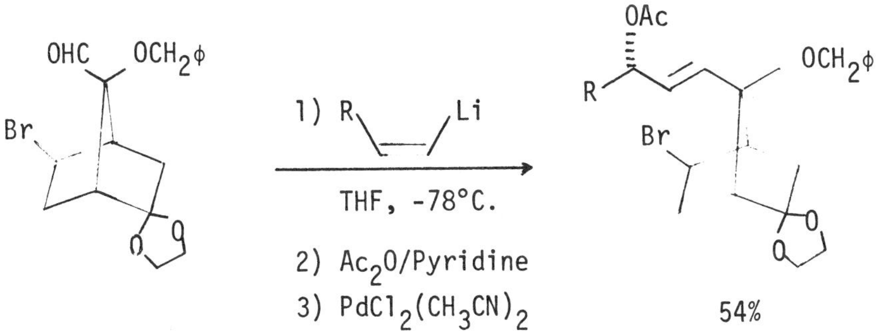

I.B.3-8 B. A. Feit et al, Tetrahedron, 37, 2143 (1981); B. A.
Feit, R. R. Schmidt et al, J. Chem. Soc., Perkin I, 1329
(1981); R. R. Schmidt and H. Speer, Tetrahedron Lett., 22,
4259 (1981).

$$\phi\diagdown\diagup CN \quad\quad \xrightarrow[\substack{Et_2O/Hexane \\ -75°C.\ to\ -115°C.}]{1)\ LDA} \quad\quad \phi\diagdown\diagup CN$$

1) LDA

Et$_2$O/Hexane

-75°C. to -115°C.

2) E$^+$ (ϕ_2CO)

43-61%

I.B.3-9 R. K. Boeckman, Jr. and K. J. Bruza, Tetrahedron, 37,
3997 (1981).

1) ^{t}BuLi

THF, -78°C.

2) 0°C.

3) E$^+$, -78°C.

30-86%

E$^+$ = RX, RCHO, R$_2$CO, CH$_2$=CHCHO (1,2-Addn.)

I.B.3-10 C. Shih and J. S. Swenton, Tetrahedron Lett., 22,
4217 (1981); J. C. Depezay and Y. Le Merrer, Bull. Soc.
Chim. Fr. II, 306 (1981); L. Duhamel and F. Tombret, J. Org.
Chem., 46, 3741 (1981).

1) 2 eq. nBuLi

THF

2) E$^+$

3) H$_3$O$^+$

66-86%

E$^+$ = RX, RCHO, R$_2$CO, CO$_2$, Me$_3$SiCl

I.B.3-11 M. Gill, H. P. Bainton and R. W. Rickards, Tetrahedron Lett., 22, 1437 (1981); Y. Torisawa, M. Shibasaki
and S. Ikegami, ibid, 2397.

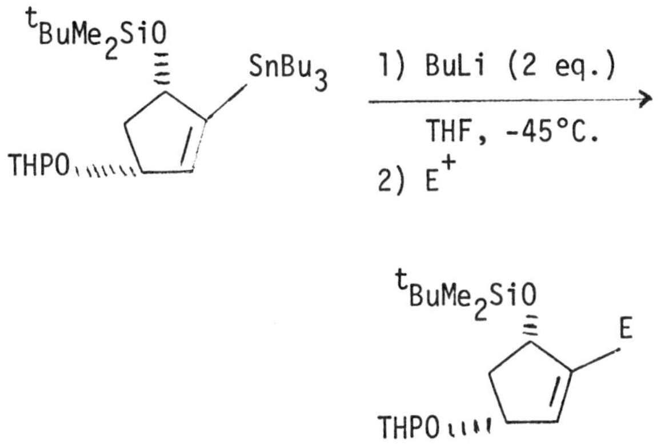

$$E^+ = RX, ClCO_2Et, HCO_2Et, \phi CHO \text{ and } Me_2CO.$$

I.B.3-12 R. M. Christie, M. Gill and R. W. Rickards, J. Chem.
Soc., Perkin I, 593 (1981); M. Gill and R. W. Rickards,
ibid, 599; C. P. Casey, C. R. Jones and H. Tukada, J. Org.
Chem., 46, 2089 (1981).

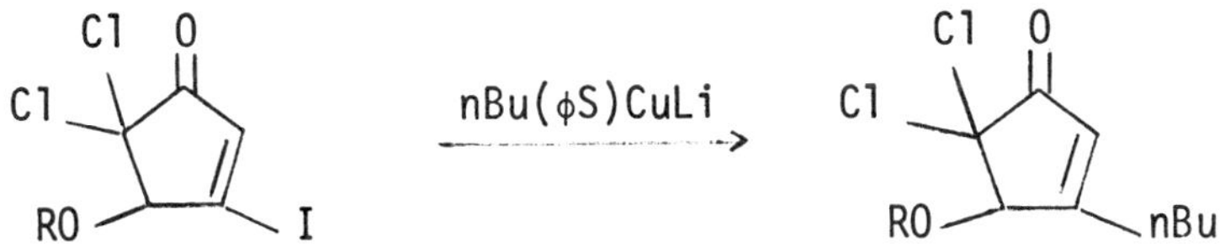

I.B.3-13 A. J. Dixon, R. J. K. Taylor and R. F. Newton, J.
Chem. Soc., Perkin I, 1407 (1981); P. Vermeer et al, Rec.
Trav. Chim., 100, 249 (1981).

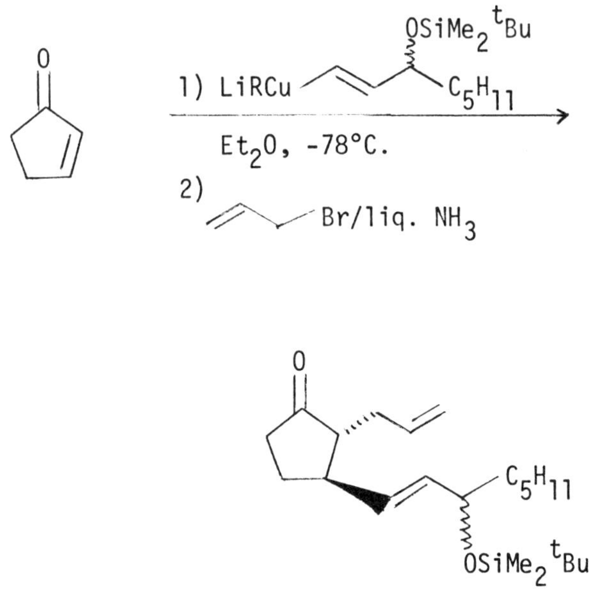

1) LiRCu $\overset{OSiMe_2{}^tBu}{\diagup\diagdown}$ C_5H_{11}

Et$_2$O, -78°C.

2) $\diagup\diagup$ Br/liq. NH$_3$

C_5H_{11}

OSiMe$_2{}^t$Bu

I.B.3-14 J. Schwartz et al, J. Amer. Chem. Soc., 103, 4466
(1981); Tetrahedron Lett., 22, 2629, 4655 (1981).

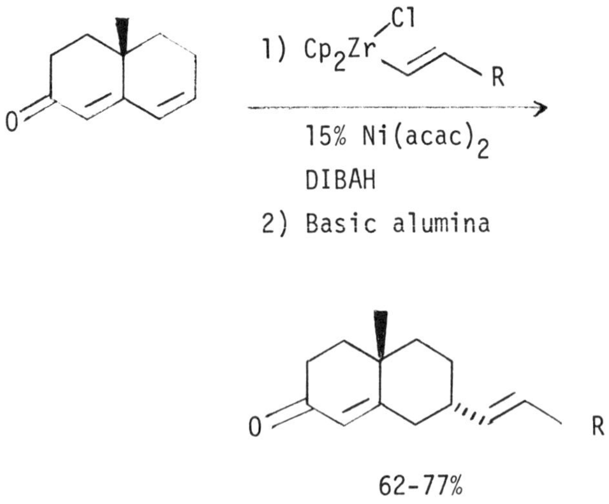

1) Cp$_2$Zr $\overset{Cl}{\diagdown}$ $\diagup\diagdown$ R

15% Ni(acac)$_2$
DIBAH

2) Basic alumina

R

62-77%

I.B.3-15   R. F. Heck et al, J. Org. Chem., <u>46</u>, 1061, 1067
(1981);   F. E. Ziegler, U. R. Chakraborty and R. B. Weisenfeld,
Tetrahedron, <u>37</u>, 4035 (1981);   G. P. Chiusoli et al, J. Organometal. Chem., <u>219</u>, C16 (1981).

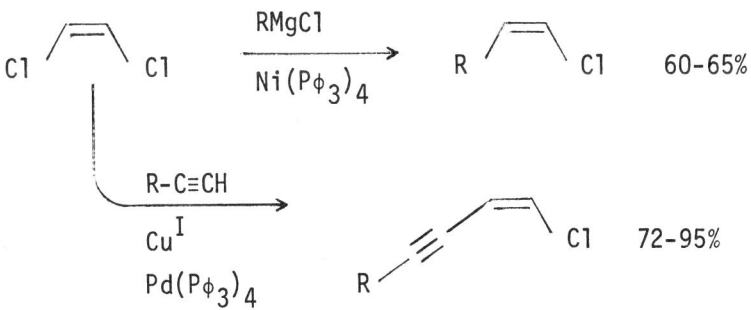

$$\text{Br} + \text{CH(OEt)}_2 \xrightarrow[\substack{\text{Pd(OAc)}_2 \\ \text{Ar}_3\text{P, 100°C.}}]{\text{1) Piperidine}}$$

2) $H_2O/5\%$ $(CO_2H)_2$

CHO

76%

Also, coupling with α,β-unsaturated esters.

I.B.3-16   V. Ratovelomanana and G. Linstrumelle, Tetrahedron
Lett., <u>22</u>, 315 (1981);   M. Kumada et al, ibid, 137;   H.
Brunner and M. Probster, J. Organometal. Chem., <u>209</u>, C1 (1981).

Cl   Cl $\xrightarrow[\text{Ni(P}\phi_3)_4]{\text{RMgCl}}$ R   Cl   60-65%

$\xrightarrow[\substack{\text{Cu}^{I} \\ \text{Pd(P}\phi_3)_4}]{\text{R-C}\equiv\text{CH}}$ R   Cl   72-95%

I.B.3-17 E. Negishi, H. Matsushita and N. Okukado, Tetrahedron
Lett., 22, 2715 (1981); J. F. Fauvarque and A. Jutand, J.
Organometal. Chem., 209, 109 (1981); A. Gaudemar et al, ibid,
393.

$$\phi CH_2 ZnBr \quad + \quad \overset{R^1}{\underset{I}{\diagup}} = \overset{R^2}{\underset{R^3}{\diagdown}} \quad \xrightarrow[\substack{(cat.) \\ iBu_2AlH}]{(\phi_3 P)_2 PdCl_2}$$

$$\phi CH_2 Br \quad + \quad \overset{R^1}{\underset{R_2Al}{\diagup}} = \overset{R^2}{\underset{R^3}{\diagdown}} \quad \xrightarrow[(cat.)]{(\phi_3 P)_4 Pd} \quad \overset{R^1}{\underset{\phi CH_2}{\diagup}} = \overset{R^2}{\underset{R^3}{\diagdown}}$$

I.B.3-18 N. Jabri, A. Alexakis and J. F. Normant, Tetrahedron
Lett., 22, 959 (1981); N. Miyaura, H. Suginome and A. Suzuki,
ibid, 127; B. Singaram, G. A. Molander and H. C. Brown,
Heterocycles, 15, 231 (1981).

$$\left(\overset{R^1}{\underset{R^2}{\diagup}} = \diagdown \right)_2 CuLi \quad + \quad \overset{I}{\diagup} = \overset{R^3}{\underset{R^4}{\diagdown}} \quad \xrightarrow[\substack{Pd^\circ (\phi_3 P)_4 \\ THF, -25°C. \\ 2) \text{ aq. } NH_4Cl}]{1) \text{ } ZnBr_2}$$

$$\overset{R^1}{\underset{R^2}{\diagup}} = \diagdown \diagup = \overset{R^3}{\underset{R^4}{\diagdown}}$$

80-96% (>98% Stereoselective)

I.B.3-19   A. Umani-Ronchi et al, Chem. Commun., 541 (1981).

$$R^1 \diagdown C = C \diagup H \diagdown R^2 \quad \xrightarrow[\text{Bu}_3\text{P}]{R^3 I, \text{ Pd-Graphite}} \quad R^1 \diagdown R^3 \diagup C = CHR^3 \quad 58\text{-}92\%$$

I.B.3-20   R. C. Larock and S. S. Hershberger, Tetrahedron Lett., 22, 2443 (1981).

$$\phi CH=CHHgCl \quad \xrightarrow[\substack{\text{HMPA/LiCl} \\ 70°C.}]{\text{MeRhI}_2(\text{P}\phi_3)_2} \quad \phi CH=CHCH_3 \quad 97\% \ (GC)$$

I.B.3-21   K. Oshima et al, Tetrahedron Lett., 22, 1609 (1981);
N. Fukamiya, M. Oki and T. Aratani, Chem. Ind., 606 (1981);
M. Kumada et al, Synthesis, 1001 (1981).

$$R^1 \diagdown H \diagup C = C \diagdown OP(O\phi)_2 \diagup S\phi \quad \xrightarrow[\phi H, \ 25°C.]{R^2_3 Al, \ (\phi_3P)_4Pd} \quad R^1 \diagdown H \diagup C = C \diagup R^2 \diagdown S\phi \quad 55\text{-}83\%$$

I.B.3-22   D. W. Cameron, G. I. Feutrill and J. M. Thiel, Austr. J. Chem., 34, 453 (1981).

$$\xrightarrow[\text{DMSO}]{CH_2=C(NMe\phi)_2}$$

52%

I.B.3-23   T. Hudlicky and T. Srnak, Tetrahedron Lett., 22, 3351 (1981);   M. E. Jung and R. W. Brown, ibid, 3355;   G. A. Olah and A. P. Fung, Synthesis, 473 (1981);   G. R. Knox and I.G. Thom, Chem. Commun., 373 (1981).

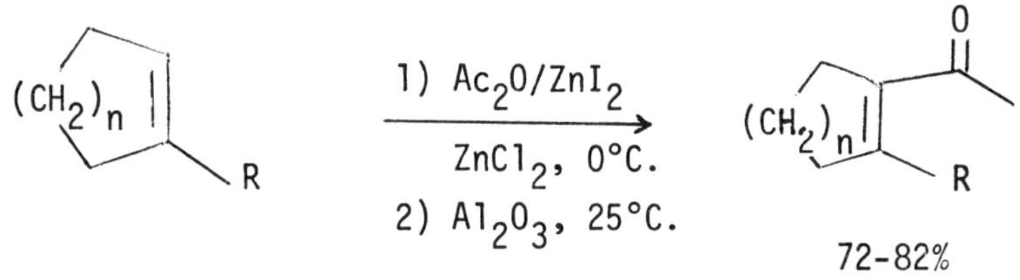

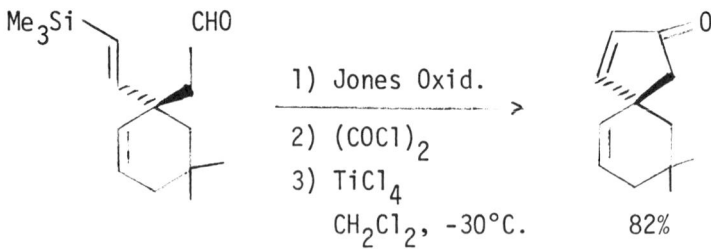

R = Me or H;   n = 1, 2 or 3.

I.B.3-24   T. Hirao et al, Tetrahedron Lett., 22, 3633 (1981); P. Miginiac et al, J. Chem. Res. (S), 343 (1981).

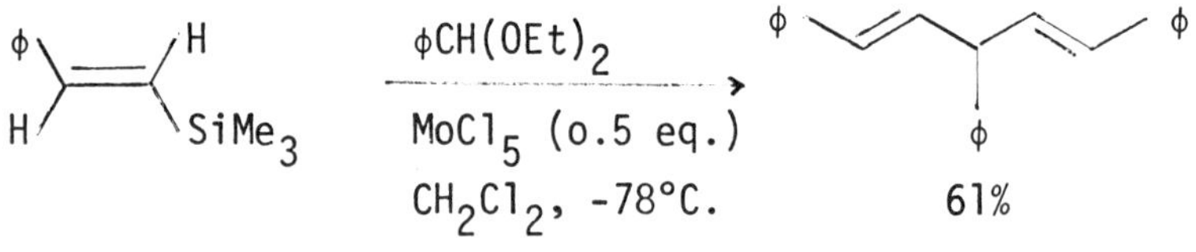

I.B.3-25   S. D. Burke et al, J. Org. Chem., 46, 2400 (1981).

I.B.3-26 J. F. Normant and A. Alexakis, Synthesis, 841 (1981).

Review: "Carbometallation (C-Metallation) of Alkynes: Stereospecific Synthesis of Alkenyl Derivatives."

I.B.3-27 I. Fleming, T. W. Newton and F. Roessler, J. Chem. Soc., Perkin I, 2527 (1981).

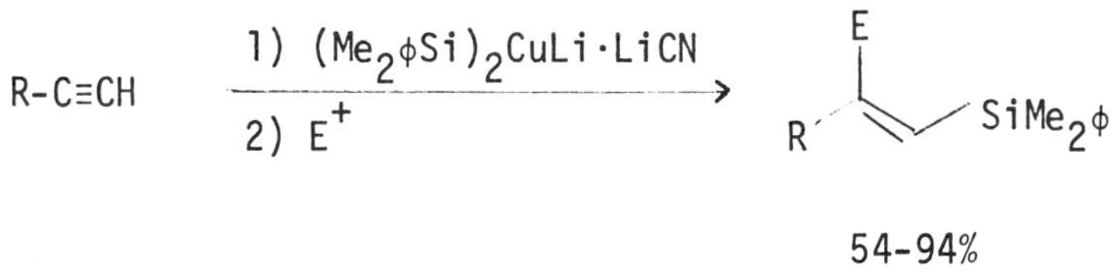

R-C≡CH
1) $(Me_2\phi Si)_2CuLi\cdot LiCN$
2) E^+
E
R
$SiMe_2\phi$
54-94%

E^+ = H_2O, D_2O, I_2, CO_2, RCOCl, RX, α,β-unsat. Ketones and Epoxides.

I.B.3-28 H. Nishiyama, M. Sasaki and K. Itoh, Chem. Lett., 905 (1981).

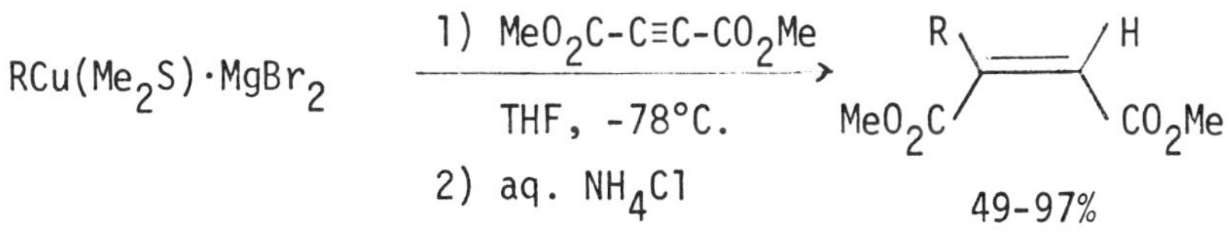

$RCu(Me_2S)\cdot MgBr_2$
1) $MeO_2C-C\equiv C-CO_2Me$
THF, -78°C.
2) aq. NH_4Cl
R
H
MeO_2C
CO_2Me
49-97%

I.B.3-29 J. P. Marino and R. J. Linderman, J. Org. Chem., 46, 3696 (1981).

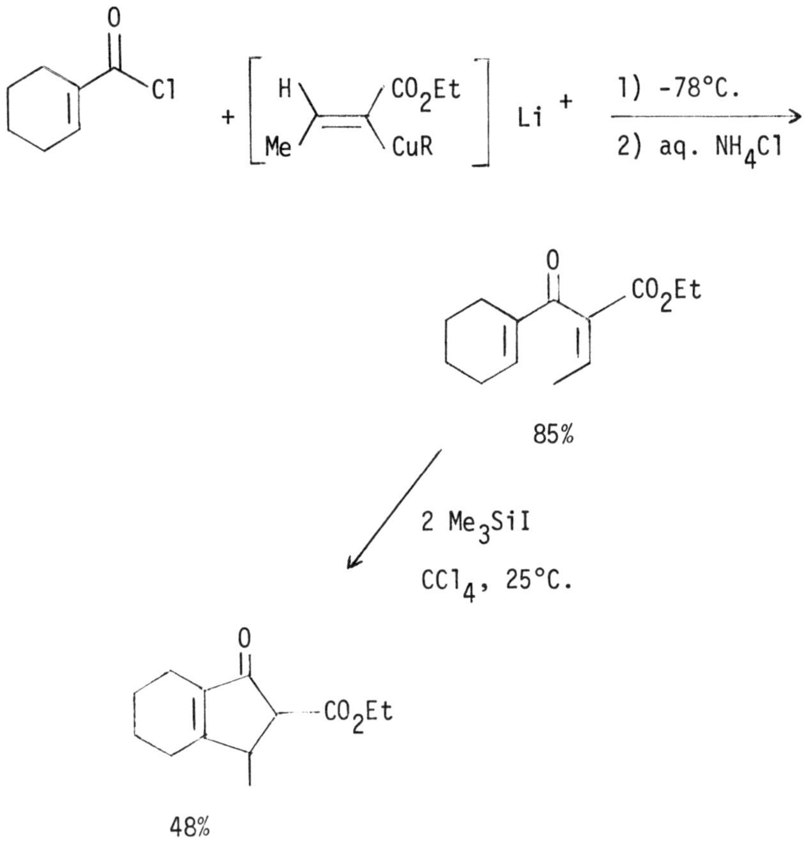
O
Cl
H
Me
CO_2Et
CuR
Li
+
1) -78°C.
2) aq. NH_4Cl
O
CO_2Et
85%
2 Me_3SiI
CCl_4, 25°C.
O
CO_2Et
48%

I.B.3-30 P. Vermeer et al, Rec. Trav. Chim., 100, 98 (1981); W. J. E. Parr, J. Chem. Res. (S), 354 (1981).

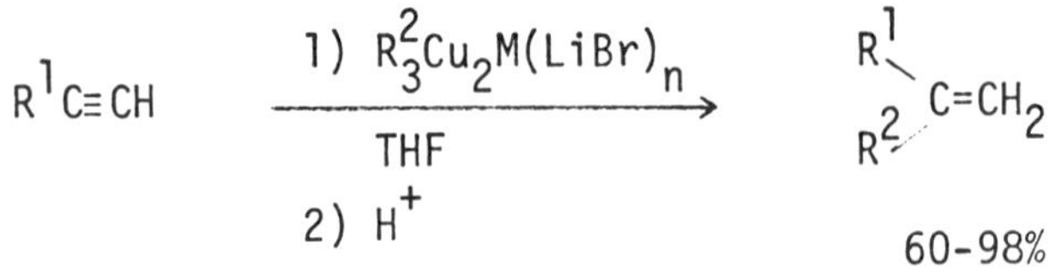
$R^1C\equiv CH$
1) $R^2_3Cu_2M(LiBr)_n$
THF
2) H^+
R^1
R^2
$C=CH_2$
60-98%

(M = Li, MgCl or MgBr)

I.B.3-31 V. H. M. Elferink, R. G. Visser and H. J. T. Bos,
Rec. Trav. Chim., 100, 414 (1981).

$$Et_2N-C{\equiv}C-Me \xrightarrow[CH_3CN/\Delta]{ArCS_2R} \underset{\underset{O}{\overset{\parallel}{Me_2NC}}}{\overset{Me}{\diagdown}}C{=}C\overset{SR}{\underset{Ar}{\diagup}}$$

50-76%

I.B.3-32 A. Pelter, et al, Chem. Commun., 164 (1981); A.
Pelter, K. Smith et al, J. Chem. Soc., Perkin I, 653 (1981).

$$R^1_3B-C{\equiv}C-R^2 \quad \begin{array}{l} 1)\ \text{[dithiole]}\ BF_4^- \\ \hline 2)\ iPrCO_2H,\ 25°C. \\ 3)\ HgO,\ BF_3 \end{array} \quad > \quad \overset{R^1}{\underset{H}{\diagdown}}{=}\overset{R^2}{\underset{CHO}{\diagup}}$$

71-87% (Before
Hydrol.)
(83-100% cis)

I.B.3-33 E. Negishi et al, J. Amer. Chem. Soc., 103, 2882,
4985 (1981); J. Org. Chem., 46, 4093 (1981).

$$\text{[enyne]} \quad \begin{array}{l} 1)\ Me_3Al \\ \hline Cl_2ZrCp_2 \\ 2)\ Cl{-}{=}{\diagup}_R \\ Pd(P\phi_3)_4\ (cat.) \\ THF \end{array} \quad \longrightarrow \quad \text{[diene-R]}$$

77%

I.B.3-34 M. D. Schiavelli, J. J. Plunkett and D. W. Thompson,
J. Org. Chem., 46, 807 (1981); A. M. Caporusso, G. Giacomelli
and L. Lardicci, J. Chem. Soc., Perkin I, 1900 (1981).

$$CH_3CH_2C\equiv CCH_2CH_2OH \quad \xrightarrow[\substack{CH_2Cl_2, \; 0°C. \\ 2) \; TiCl_4, \; -78°C. \\ 3) \; MeOH/H_2SO_4}]{1) \; Me_3Al}$$

81%

I.B.3-35 P. Vermeer et al, J. Organometal. Chem., 206, 257
(1981); Rec. Trav. Chim., 100, 337 (1981).

$$R^1-C\equiv C-CN \quad \xrightarrow[\substack{THF \\ 2) \; H_3O^+}]{1) \; R_2^2AgMgCl}$$

70-90%

I.B.3-36 F. Sato, et al, Tetrahedron Lett., 22, 85 (1981);
Chem. Commun., 718 (1981).

$$R^1-C\equiv C-R^2 \quad \xrightarrow[\substack{Cp_2TiCl_2 \\ 2) \; E^+}]{1) \; iBuMgBr}$$

80-90%

$E^+ = \phi CHO$ or MeI

I.B.3-37 J. M. Huggins and R. G. Bergman, J. Amer. Chem. Soc., 103, 3002 (1981).

1) $R^1C{\equiv}CR^2$
2) LAH or TsOH

$R^1CH{=}C\begin{smallmatrix}Me\\R^2\end{smallmatrix}$

47-100% (GC)

With unsymmetrical alkynes, R^1 is the largest group.

I.B.3-38 K. Itoh et al, Chem. Lett., 865 (1981).

$RCH_2CH{=}CH_2$ $\xrightarrow[\substack{Pd^* \\ \\ \phi H, \ 45°C.}]{MeO_2C\text{-}C{\equiv}C\text{-}CO_2Me}$

Pd* = Bis(maleic anhydride)(norbornene)palladium.

I.B.3-39 L. Castedo et al, J. Org. Chem., 46, 4292 (1981); W. H. Richardson, Synth. Commun., 11, 895 (1981); M. P. Fleming and J. E. McMurry, Org. Syn., 60, 113 (1981).

$$\underset{\text{ArCR}}{\overset{O}{\|}} \xrightarrow[\substack{\text{DME or THF}\\\text{Reflux}}]{\text{TiCl}_3/\text{Zn-Cu}} \underset{\text{Ar-C=C-Ar}}{\overset{R\ \ R}{\underset{|\ \ \ |}{}}}$$

60-97%

(R = H or Me)

I.B.3-40   S. K. Pradhan et al, J. Org. Chem., 46, 2622 (1981).

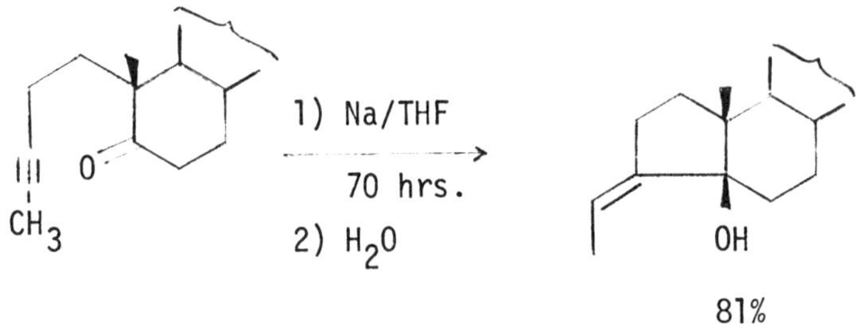

1) Na/THF
---------------->
70 hrs.

2) H₂O

81%

I.B.3-41   J. Barluenga et al, J. Org. Chem., 46, 2721 (1981).

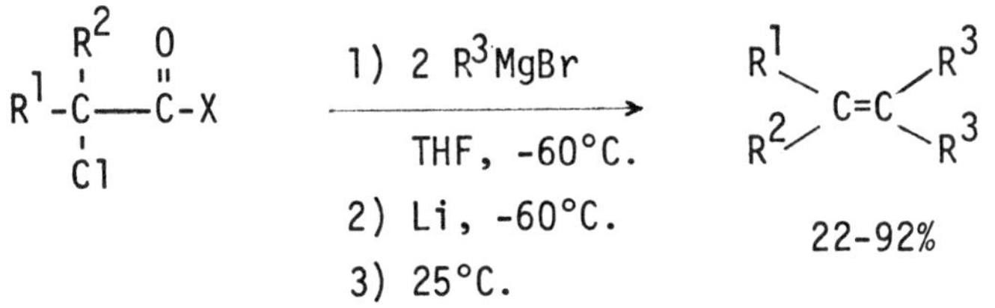

$R^1-\overset{R^2}{\underset{Cl}{\overset{|}{C}}}-\overset{O}{\overset{||}{C}}-X$

1) 2 R³MgBr
---------------->
THF, -60°C.

2) Li, -60°C.

3) 25°C.

$R^1\diagdown_{R^2}C=C\diagup^{R^3}_{R^3}$

22-92%

(X = Cl or OEt)

I.B.3-42   K. V. Scherer, Jr., T. F. Terranova and D. D. Lawson, J. Org. Chem., 46, 2379 (1981).

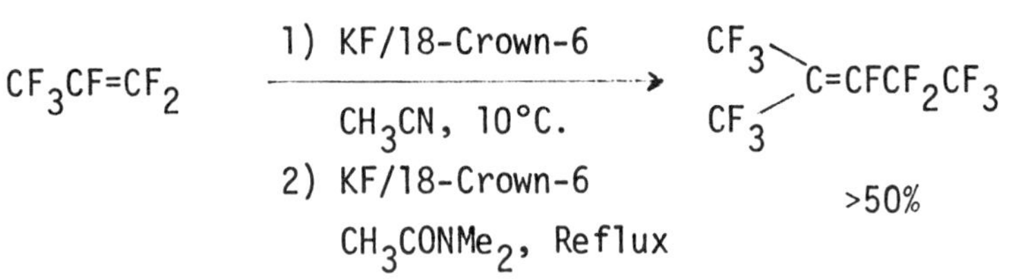

$CF_3CF=CF_2$

1) KF/18-Crown-6
---------------->
CH₃CN, 10°C.

2) KF/18-Crown-6

CH₃CONMe₂, Reflux

$\overset{CF_3}{\underset{CF_3}{\diagup}}C=CFCF_2CF_3$

>50%

I.B.3-43 M. P. Cooke, Jr., Tetrahedron Lett., 22, 381 (1981).

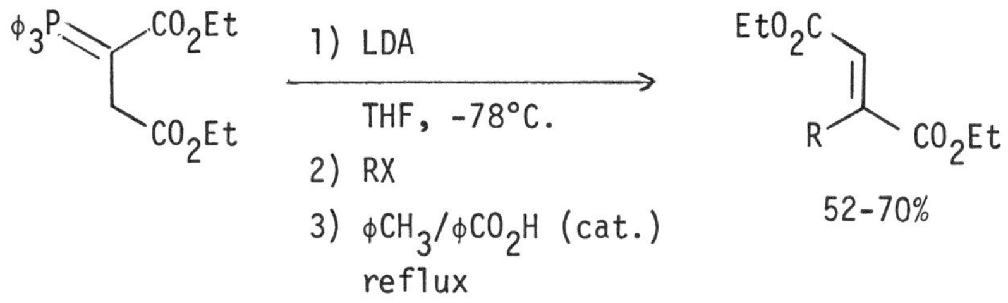

I.B.4. Allene Forming Reactions

I.B.4-1 J. C. Clinet and G.Linstrumelle, Synthesis, 875
(1981); P. Vermeer et al, J. Organometal. Chem., 217, 267
(1981).

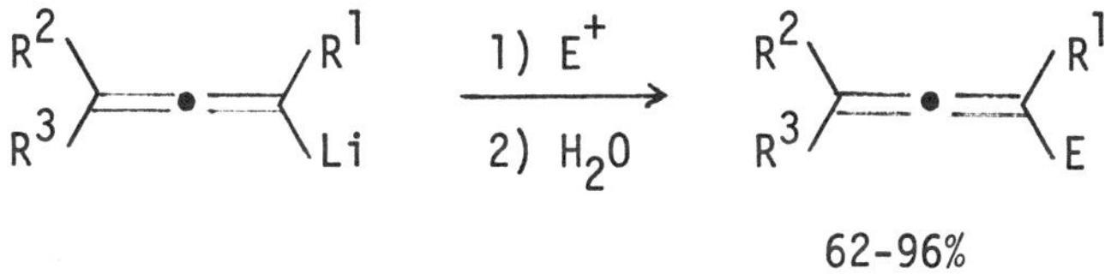

$$E^+ = R^2CO, CO_2, Me_2NCONMe_2, \text{Epoxides}, Me_3SiCl, MeSSMe.$$

I.B.4-2 L. Brandsma et al, Tetrahedron Lett., 22, 2827 (1981);
Rec. Trav. Chim., 100, 34 (1981).

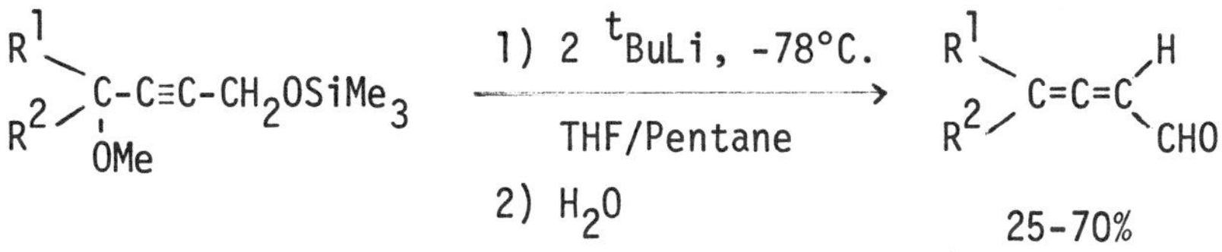

I.B.4-3 P. Vermeer et al, Tetrahedron Lett., 22, 1451, 2237 (1981).

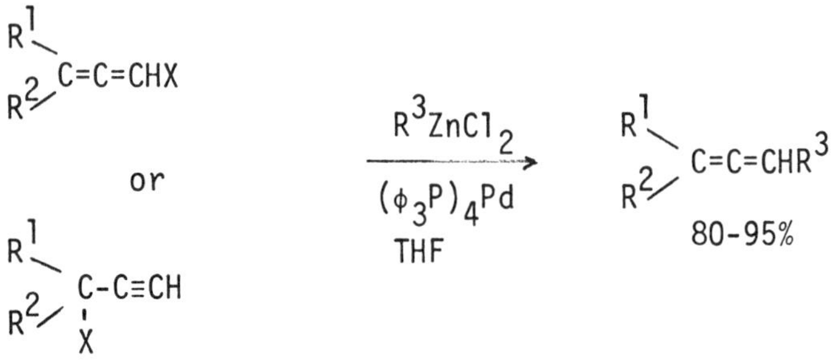

$$R^1 R^2 C=C=CHX$$

or

$$R^1 R^2 \underset{X}{C}-C\equiv CH$$

$$\xrightarrow[\substack{(\phi_3 P)_4 Pd \\ THF}]{R^3 ZnCl_2}$$

$$R^1 R^2 C=C=CHR^3$$

80-95%

X = Br, I or OAc.

I.B.4-4 J. Gore et al, J. Chem. Res. (S), 278 (1981); P. Place, C. Verniere and J. Gore, Tetrahedron, 37, 1359 (1981).

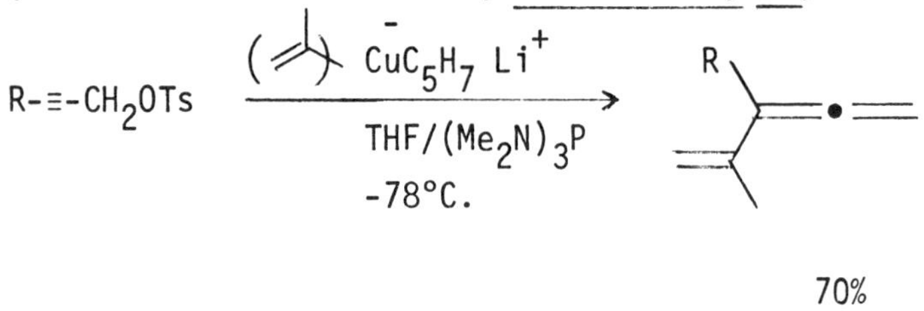

$$R-\equiv-CH_2 OTs \xrightarrow[\substack{THF/(Me_2 N)_3 P \\ -78°C.}]{CuC_5 H_7^- Li^+}$$

70%

I.B.4-5 L. Brandsma et al, Rec. Trav. Chim., 100, 244, 372 (1981).

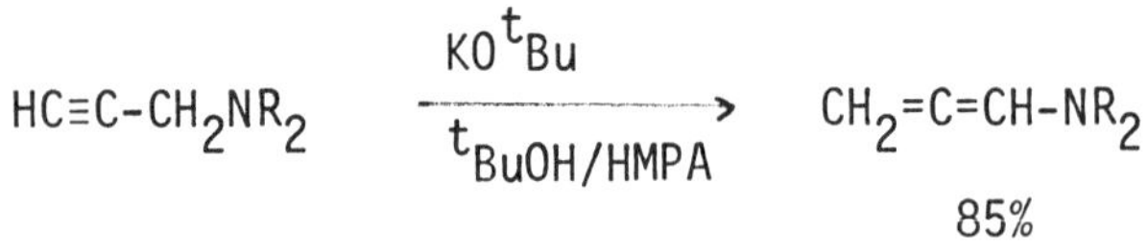

$$HC\equiv C-CH_2 NR_2 \xrightarrow[\substack{t_{BuOH}/HMPA}]{KO^t Bu} CH_2=C=CH-NR_2$$

85%

I.B.4-6   R. Weiss et al, J. Amer. Chem. Soc., 103, 6142 (1981).

$$R_2N \diagdown = \bullet = \diagup NR_2$$
$$\phi_3P_+ \qquad\qquad +P\phi_3$$

Synthesis

$$2I^-$$

I.B.4-7   D. Mesnard, J. P. Charpentier and L. Miginiac, J. Organometal. Chem., 214, 15, 23, 135 (1981).

$$Me_2NCH_2-C\equiv C-CH=CH-CH_2Y \quad \xrightarrow[\text{2) } H_2O]{\text{1) RM}}$$

(Y = OH, OMe, NMe_2)

(M = Li or MgBr)

$$Me_2NCH_2CH=C=CH-CH-CH_2Y$$
$$\overset{|}{R}$$

15-70%

I.B.4-8   J. Pornet et al, Tetrahedron Lett., 22, 3609, 1327 (1981);   J. P. Pillot et al, Tetrahedron Lett., 22, 3401 (1981).

$$Me_3SiCH_2-C\equiv C-CH_2SiMe_3 \quad \xrightarrow[\text{CH}_2Cl_2, \, -60°C.]{RCH(OR^1)_2/TiCl_4}$$

$$CH_2=C=C-CH_2SiMe_3$$
$$\underset{R\diagup \quad \diagdown OR^1}{\overset{|}{CH}}$$
25-77%

I.B.4-9   R. G. Daniels and L. A. Paquette, Tetrahedron Lett.,
22, 1579 (1981);   T. Yoshida and E. Negishi, J. Amer. Chem.
Soc., 103, 1276 (1981).

$$R^1\overset{\overset{\text{O}}{\|}}{C}R^2 \xrightarrow[\text{THF, 0°C.}]{Me_3Si-C\equiv C-CH_2ZnBr} \quad R^1\underset{R^2}{\overset{OH}{C}}\!\!-\!\!\equiv\!-SiMe_3 \qquad 61\text{-}76\%$$

$$\xrightarrow[\text{THF, 0°C.}]{Me_3Si-C\equiv C-CH_2Al_2/_3Br} \quad R^1\underset{R^2}{\overset{OH}{C}}\underset{Me_3Si}{\diagdown}\!\!=\!\!\bullet\!\!=\!\! \qquad 68\text{-}81\%$$

I.B.4-10   B. Wrackmeyer, J. Organometal. Chem., 205, 1 (1981).

$$Me_3Sn-C\equiv C-R^1 \quad \xrightarrow[\substack{2)\ Me_3SnC\equiv C-R^2 \\ 75\text{-}120°C.}]{1)\ Et_3B}$$

$$\underset{R^2}{\overset{Me_3Sn}{\diagdown}}\!\!=\!\!\bullet\!\!=\!\!\underset{R^1}{\overset{Et}{\diagup}}C\overset{BEt_2}{\underset{SnMe_3}{}}$$

I.B.4-11   D. J. Morgans, Jr., K. B. Sharpless and S. G.
Traynor, J. Amer. Chem. Soc., 103, 462 (1981);   P. Vermeer
et al, J. Organometal. Chem., 221, 117 (1981).

$$\xrightarrow[\text{CH}_2\text{Cl}_2,\ 25°C.]{Ti(OiPr)_4}$$

I.B.4-12 A. M. Moiseenkov et al, <u>Tetrahedron Lett.</u>, <u>22</u>, 151 (1981).

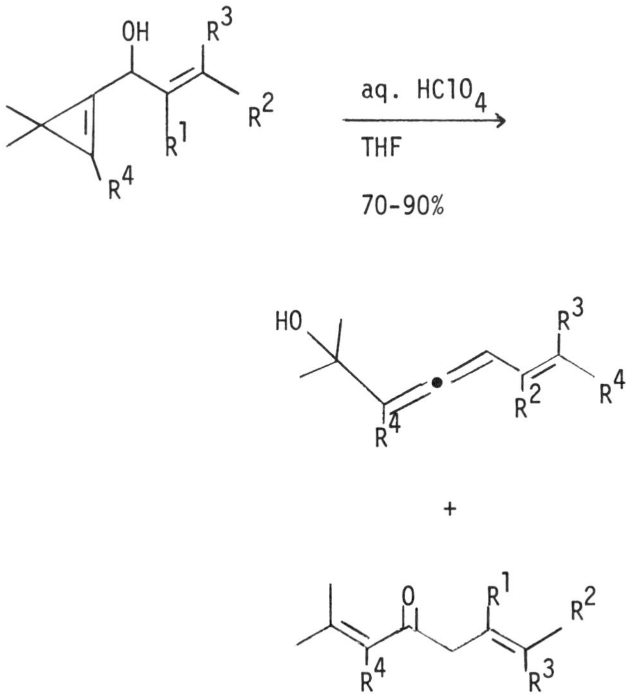

I.B.4-13 G. Buono, <u>Synthesis</u>, 872 (1981).

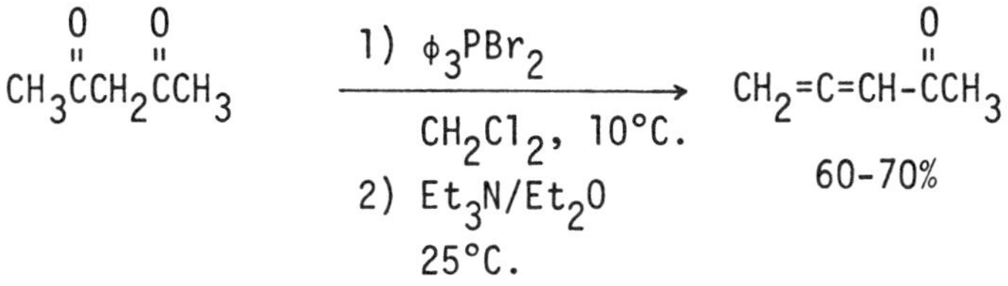

I.B.4-14 D. M. Hilvert, M. D. Jacobs and T. H. Morton, Org.
Prep. Proc. Int., 13, 197 (1981).

$$
\begin{array}{c}
\text{Me} \diagdown \quad \overset{\text{O}}{\underset{\text{CH-C-CH}}{\|}} \quad \diagup \text{Me} \\
\text{Me} \diagup \qquad\qquad \diagdown \text{Me}
\end{array}
\qquad
\begin{array}{l}
\text{1) } NH_2NH_2 \\
\text{2) } I_2, Et_3N \\
\text{3) NaOEt/DMSO}
\end{array}
\qquad
\begin{array}{c}
\text{Me} \diagdown \qquad \diagup \text{Me} \\
\diagup \!\!=\!\! \bullet \!\!=\!\! \diagdown \\
\text{Me} \diagup \qquad \diagdown \text{Me} \\
 31\%
\end{array}
$$

I.C. Carbon-Carbon Triple Bonds
      (see also: VI.A.16).

I.C-1 B. A. Trofimov, Russian Chem. Rev., 50, 138 (1981).

Review: "Reactions of Acetylene in Superbasic Media."

I.C-2 D. Mesnard, F. Bernadou and L. Miginiac, J. Chem. Res.
(S), 270 (1981).

$$ ArC\equiv CH \qquad\qquad\qquad ArC\equiv CR $$

"Best" methods for preparing pure alkynes containing
one aryl group.

I.C-3 L. Brandsma et al, Synthesis, 459 (1981); E. Nagashima,
K. Suzuki and M. Sekiya, Chem. Pharm. Bull., 29, 1274 (1981);
T. Prange et al, Chem. Commun., 363 (1981).

$$
R\text{-}C\equiv C\text{-}Li
\quad
\xrightarrow{\quad\text{1) Hexane/THF}\quad}
$$

2) LiBr (0.5 eq.)

3) Cyclohexanone

4) MeI/Hexane

   THF/DMSO

$$
\begin{array}{c}
\text{C}\equiv\text{C-R} \\
\text{OMe}
\end{array}
$$

86-92%

I.C-4 G. Himbert and L. Henn, Tetrahedron Lett., 22, 2637
(1981); R. A. Earl and L. B. Townsend, Org. Syn., 60, 81
(1981); G. Himbert and W. Schwickerath, Justus Liebigs Ann.
Chem., 1844 (1981); G. Himbert, M. Feustel and M. Jung,
ibid, 1907.

$$\text{MeO-C} \equiv \text{C-Sn}\phi_3 \xrightarrow{\text{RCOX}} \overset{\overset{\displaystyle O}{\displaystyle \|}}{\text{MeO-C} \equiv \text{C-C-R}}$$

27-60%

I.C-5 R. Locher and D. Seebach, Angew. Chem., Int. Ed. Engl.,
20, 569 (1981).

1) LiC≡C-R

THF, -45°C.

2) 25°C.

3) H₂O

32-89%

I.C-6 M. M. Midland, C. A. Brown et al, Tetrahedron Lett.,
22, 4171 (1981).

KAPA (3 eq.)

25°C.

KAPA = Potassium 3-aminopropylamide.

I.C-7 E. V. Dehmlow and M. Lissel, Tetrahedron, 37, 1653 (1981); A. M. Zvonok, N. M. Kuzemenok and L. S. Statnishevskii, J. Org. Chem. (USSR), 17, 1053 (1981).

$$CH_2-CH-CH(OEt)_2 \atop \;\; Br \;\; Br$$

$$\xrightarrow[\substack{Pet.\ Ether \\ 20\text{-}90°C.}]{\substack{KOH\ (solid) \\ (nC_8H_{17})_4N^+\ Br^-\ (cat.)}}$$

$$HC{\equiv}C\text{-}CH(OEt)_2$$

79-98%

I.C-8 M. Ladika and P. J. Stang, Chem. Commun., 459 (1981); Synthesis, 29 (1981).

$$^tBu\text{-}CH{=}C\text{-}C{\equiv}C\text{-}C{\equiv}C\text{-}SiMe_3 \atop \qquad OSO_2CF_3$$

$$\xrightarrow[\substack{Glyme,\ 50°C.}]{KOAr}$$

$$^tBu{+}C{\equiv}C{)}_3H$$

85-90%

I.C-9 A. S. Kende, P. Fludzinski and J. H. Hill, J. Amer. Chem. Soc., 103, 2904 (1981); A. Roedig, C. Ibis and G. Zaby, Chem. Ber., 114, 684 (1981); U. Stampfli and M. Neuenschwander, Chimia, 35, 336 (1981).

I.C-10 D. H. Wadsworth and B. A. Donatelli, Synthesis, 285 (1981).

$$Ar^1 \overset{Ar^2}{\underset{O}{\bigtriangleup}} \xrightarrow[\substack{\text{o-dichlorobenzene} \\ \Delta}]{Al_2O_3} Ar^1\text{-}C\equiv C\text{-}Ar^2$$

79-95% (Crude)

I.C-11 M. Shimizu, R. Ando and I. Kuwajima, J. Org. Chem., 46, 5246 (1981); T. Frejd, J. O. Karlsson and S. Gronowitz, ibid, 3132.

$$\xrightarrow[\text{THF}]{\text{NaH}}$$

40-84%

I.C-12 Y. Thebtaranonth et al, Chem. Lett., 1241 (1981).

1) R^2MgX
2) pTsOH/φH, Δ
3) 500°C./0.05 mm Hg

$R^1\text{-}C\equiv C\text{-}R^2$

60-100%

I.C-13 P. G. Karmarkar, A. A. Thakar and M. S. Wadia, Tetrahedron Lett., 22, 2301 (1981).

$$\xrightarrow[\text{Et}_3\text{N}]{\text{H}_2\text{O}_2}$$

Ar-≡-Ar

30%

I.C-14 F. M. Simmross and P. Weyerstahl, Synthesis, 72 (1981).

$$\underset{\text{O}}{\overset{\text{O}}{CH_3\overset{\|}{C}CH_2CO_2Et}} \quad \xrightarrow[\begin{array}{c}\text{EtOH}\\\text{2) } Br_2\\\text{3) } OH^-\\\text{4) } H_3O^+\end{array}]{\text{1) } NH_2NH_2} \quad CH_3-C\equiv C-CO_2H$$

63%

I.C-15 M. J. Robins and P. J. Barr, Tetrahedron Lett., 421
(1981); T. Mitsudo et al, Chem. Commun., 496 (1981).

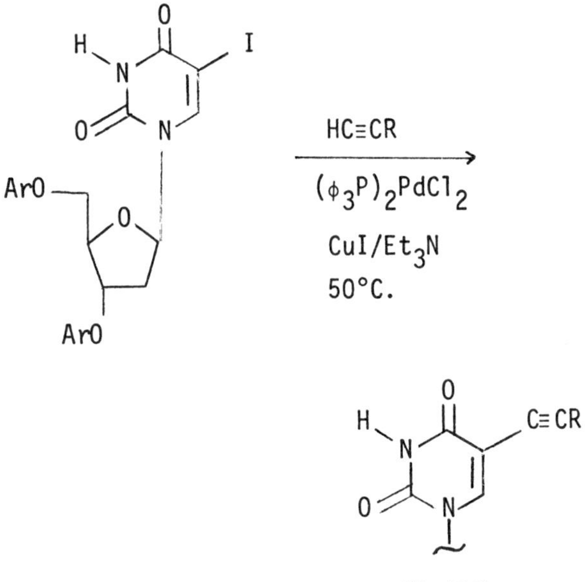

72-91%

I.C-16   G. Zweifel and N. R. Pearson, <u>J. Org. Chem.</u>, <u>46</u>, 829
(1981); H. C. Brown and G. A. Molander, <u>ibid</u>, 645.

1) 2 ClCH$_2$C≡CLi

THF, -78°C.

2) R^{3}CHO, -78°C.

3) H$_2$O$_2$/NaOH

61-80%

1,2,4-Trienols also prepared.

I.D. Cyclopropanations

I.D.1.   Carbene or Carbenoic Additions to Multiple Bonds

(see also: VI.A.7).

I.D.1-1   T. Fujita et al, <u>Synthesis</u>, 1004 (1981); L. Anke, D.
Reinhard and P. Weyerstahl, <u>Justus Liebigs Ann. Chem.</u>, 591
(1981); P. Fischer and G. Schaefer, <u>Angew. Chem., Int. Ed.
Engl.</u>, <u>20</u>, 863 (1981); R. Barlet, R. LeGoaller and C. Gey,
<u>Can. J. Chem.</u>, <u>59</u>, 621 (1981); J. M. Birchall, R. N.
Haszeldine et al, <u>J. Chem. Soc., Perkin I</u>, 2080 (1981).

CHCl$_3$

Q$^+$ Cl$^-$

50% aq. NaOH

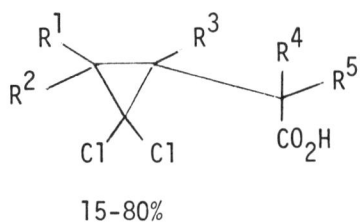

15-80%

I.D.1-2 H. Matsuda and H. Kanai, Chem. Lett., 395 (1981);
N. Kawabata et al, Bull. Chem. Soc. Jpn., 54, 2539 (1981).

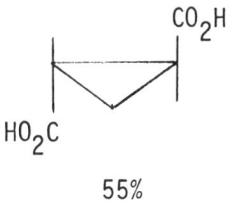

55%

(47% Optical Yield)

R* = (+) -Bornyl.

I.D.1-3   J. H. Babler and B. J. Invergo, Tetrahedron Lett., 22, 2743 (1981).

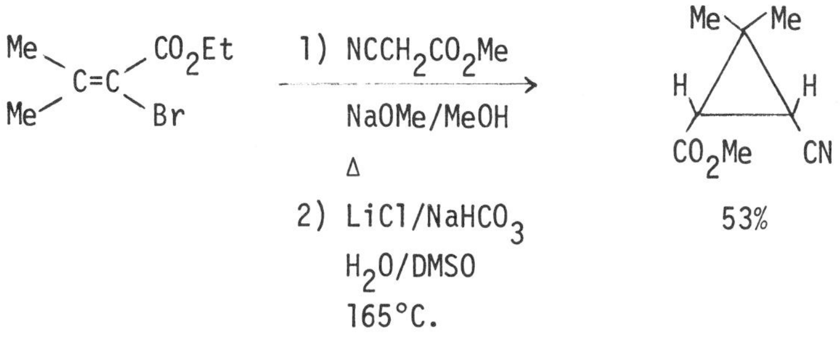

I.D.1-4   T. Agawa et al, J. Chem. Soc., Perkin I, 751 (1981); R. S. Tewari and A. K. Awasthi, Ind. J. Chem., 20B, 168 (1981).

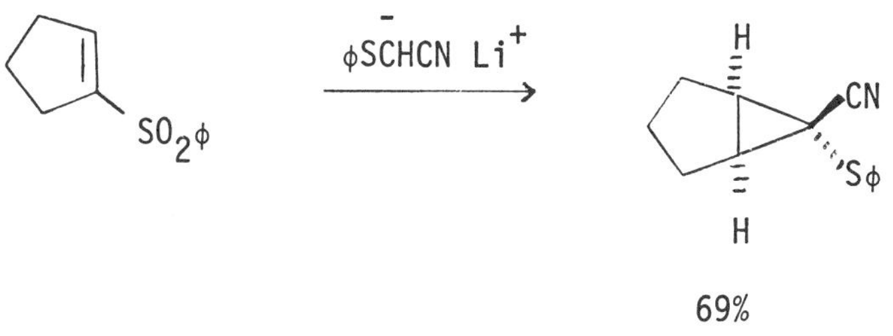

I.D.1-5   R. Carrie et al, Tetrahedron Lett., 22, 1961 (1981).

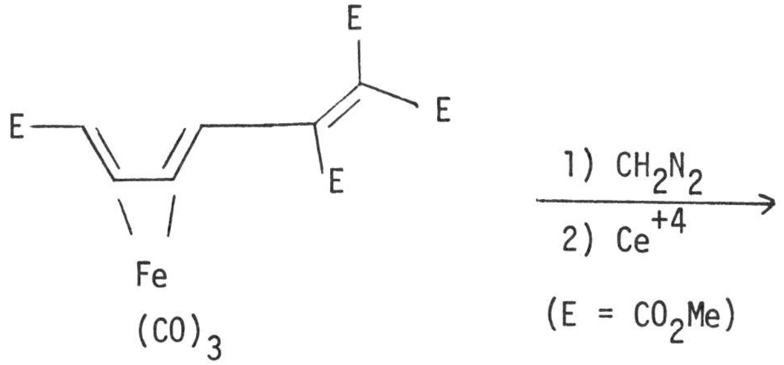

E
E
E
H

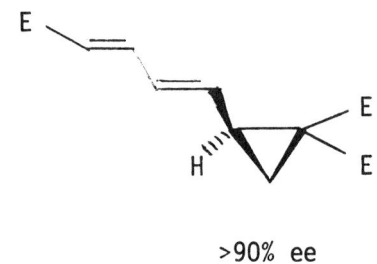

>90% ee

I.D.1-6 K. A. M. Kremer, P. Helquist and R. C. Kerber, J. Amer. Chem. Soc., 103, 1862 (1981).

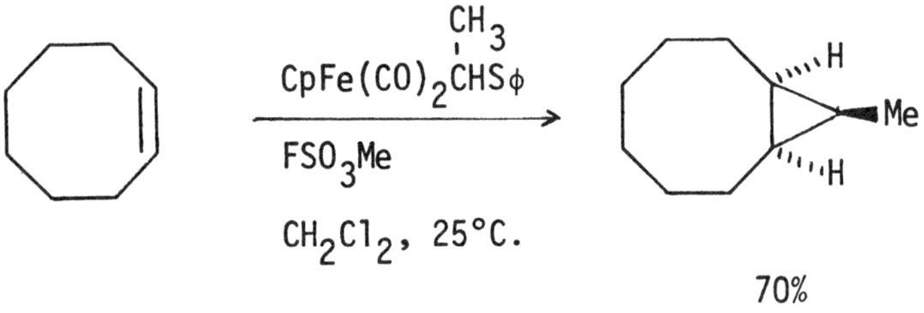

CH3
CpFe(CO)2CHSϕ
FSO3Me
CH2Cl2, 25°C.
H
Me
H
70%

I.D.1-7 M. Suda, Synthesis, 714 (1981).

R1
R2
C=CH2
CH2N2
Pd(OAc)2 (cat.)
Et2O, 0°C.
R1
R2
C
CH2
CH2
63-97%

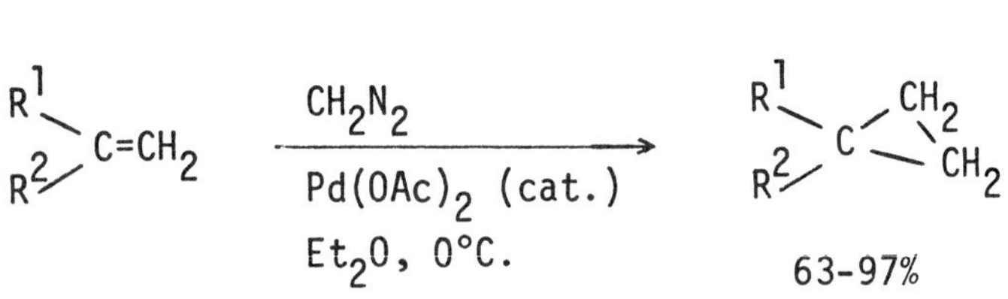

I.D.1-8 M. P. Doyle et al, <u>Tetrahedron Lett.</u>, <u>22</u>, 1783 (1981);
<u>Synthesis</u>, 787 (1981); J. D. White et al, <u>J. Amer. Chem. Soc.</u>,
<u>103</u>, 1808, 1813 (1981).

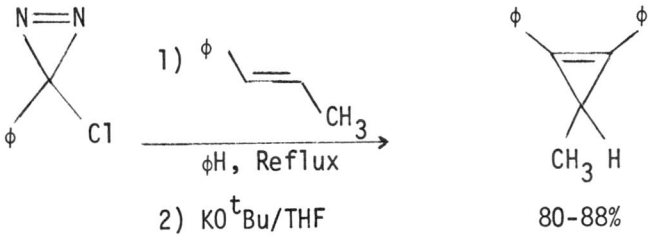

59-88%

I.D.1-9 A. Padwa, M. J. Pulwer and T. J. Blacklock, <u>Org. Syn.</u>,
<u>60</u>, 53 (1981); R. A. Moss et al, <u>J. Amer. Chem. Soc.</u>, <u>103</u>,
6164 (1981).

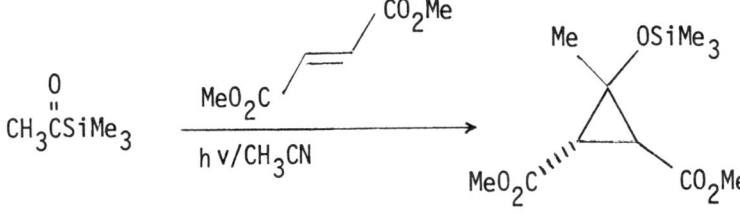

I.D.1-10 J. C. Dalton and R. A. Bourque, <u>J. Amer. Chem. Soc.</u>,
<u>103</u>, 699 (1981); P. J. Kropp et al, <u>Tetrahedron</u>, <u>37</u>, 3229
(1981).

I.D.2. Other Cyclopropanations

I.D.2-1 D. Arlt, M. Jautelat and R. Lantzsch, Angew. Chem.,
Int. Ed. Engl., 20, 703 (1981).

Review: "Synthesis of Pyrethroid Acids."

I.D.2-2 Y. Gaoni, Tetrahedron Lett., 22, 4339 (1981); G.
Procter et al, ibid, 1751; T. Durst et al, Can. J. Chem., 59,
1415 (1981).

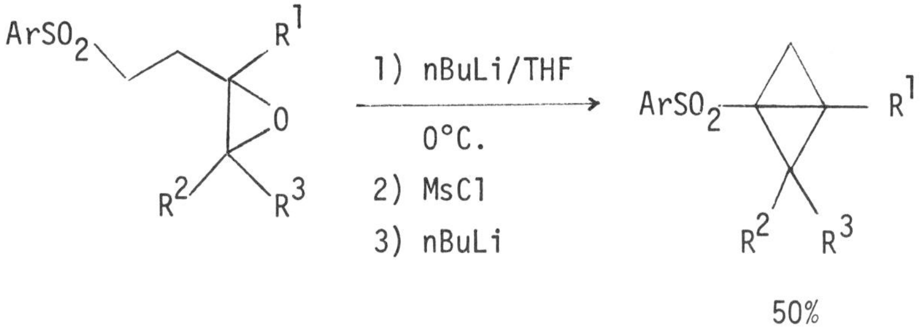

I.D.2-3 M. Madesclaire et al, Synthesis, 828 (1981); R. Verhe
et al, Org. Prep. Proc. Int., 13, 13 (1981); T. Sakai, T.
Katayama and A. Takeda, J. Org. Chem., 46, 2924 (1981).

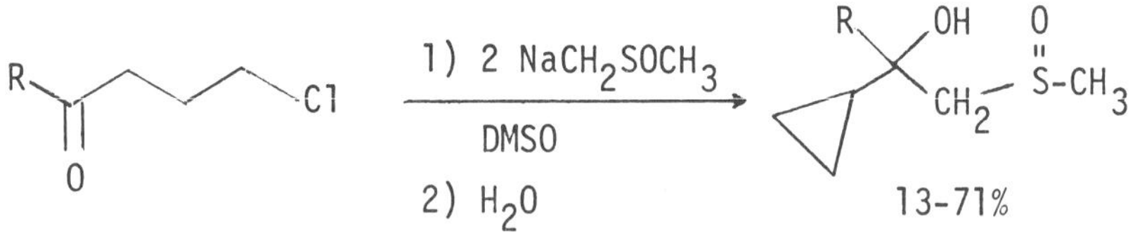

I.D.2-4   D. Bellus et al, Helv. Chim. Acta, 64, 2812 (1981);
J. P. Genet and F. Piau, J. Org. Chem., 46 2414 (1981);   R. A.
J. Smith et al, Austr. J. Chem., 34, 181 (1981).

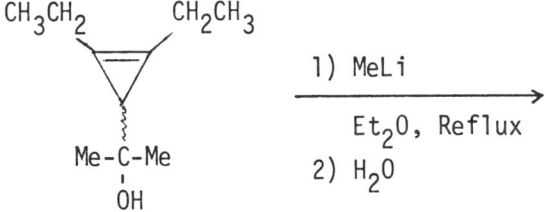

I.D.2-5   M. Vincens, C. Dumont and M. Vidal, Tetrahedron, 37,
2683 (1981);   N. S. Zefirov et al, J. Org. Chem. (USSR), 17,
1291 (1981).

70% (80% E)

I.D.2-6   E. Elkik, M. Imbeaux-Oudotte and H. Normant, Compt.
Rend. (II), 292, 1023 (1981).

$$(R^1O)_2 \overset{\overset{O}{\|}}{P} \overset{-}{\underset{F}{C}} -CO_2R^2 \ Na^+ \quad \xrightarrow{\quad \overset{R^3}{\underset{R^4}{}}\triangle O \quad} \quad \overset{R^3}{\underset{R^4}{}} \triangle \overset{CO_2R^1}{\underset{F}{}}$$

9-40%

I.D.2-7   P. Callant, H. DeWilde and M. Vandewalle, Tetrahedron,
37, 2079 (1981);   P. Callant, R. Ongena and M. Vandewalle,
ibid, 2085.

1) Et$_3$N/CH$_3$CN

   pTosSO$_2$N$_3$
   0°C.

2) Cu(II)acetylacetonate
   ϕCH$_3$, Reflux

65%

I.D.2-8   J. P. Barnier, J. Champion and J. M. Conia, Org. Syn.,
60, 25 (1981).

1) LAH

2) BF$_3$·Et$_2$O

$\triangleright$—CHO

52-76%

I.D.2-9 F. G. Klarner, W. Rungeler and W. Maifeld, Angew.
Chem., Int. Ed. Engl., 20, 595 (1981).

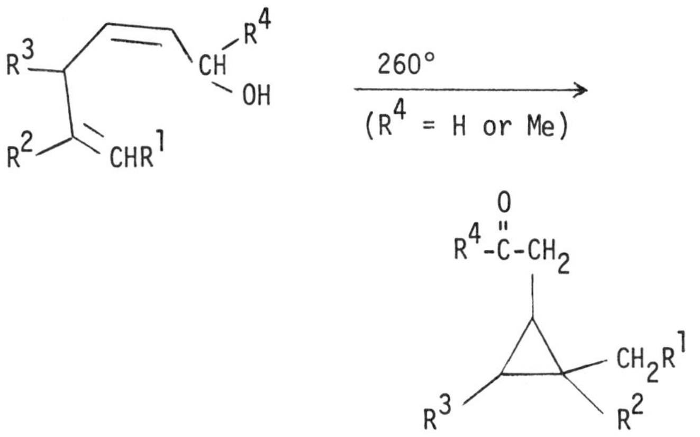

75-90%

I.D.2-10 K. Dietliker and H. J. Hansen, Chimia, 35, 52 (1981);
T. A. Lyle and B. Frei, Helv. Chim. Acta, 64, 2598 (1981).

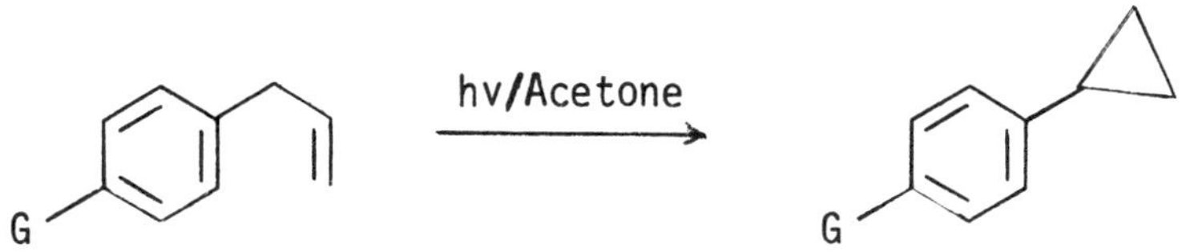

G = CN, CO_2Me, CHO   71-94%

I.D.2-11 R. Noyori et al, J. Org. Chem., 46, 2846, 2854 (1981).

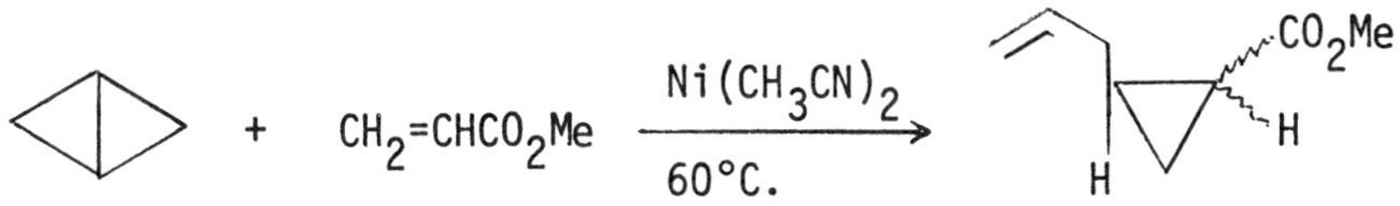

95%

(65% cis)

I.D.2-12 M. E. Jason, P. R. Kurzweil and C. C. Cahn, Synth. Commun., 11, 865 (1981).

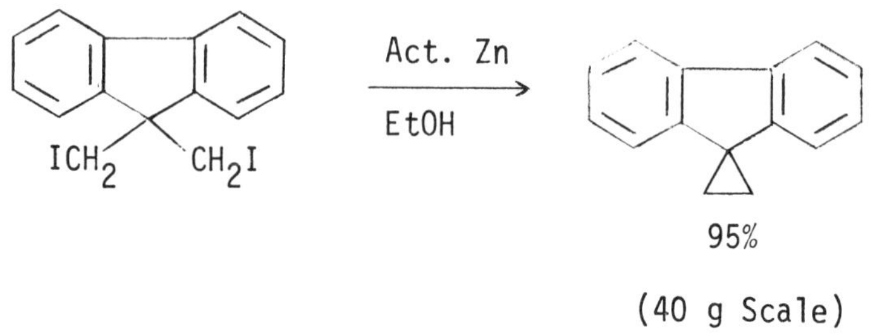
Act. Zn
EtOH
ICH_2
CH_2I
95%
(40 g Scale)

I.D.2-13 E. Vilsmaier et al, Synthesis, 721, 724, 726, 206, 207 (1981).

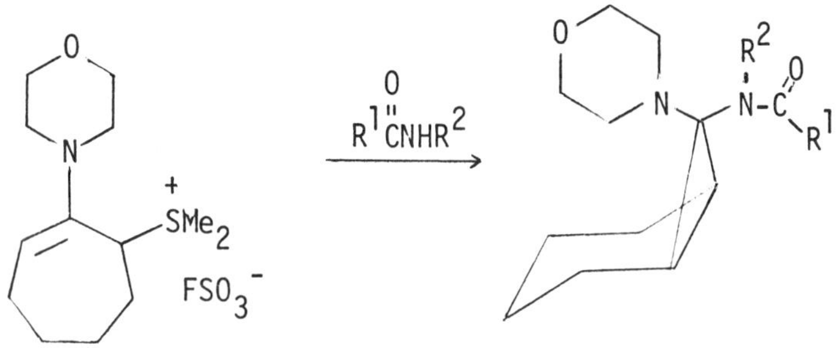
O
N
$\overset{+}{S}Me_2$
FSO_3^-
$R^1\overset{O}{\overset{\|}{C}}NHR^2$
O
N
R^2
N-C
O
R^1

I.D.2-14 O. G. Kulinkovich, I. G. Tishchenko and S. V. Danilovich, J. Org. Chem. (USSR), 17, 1057 (1981); E. P. Kundig and C. Perret, Helv. Chim. Acta, 64, 2606 (1981).

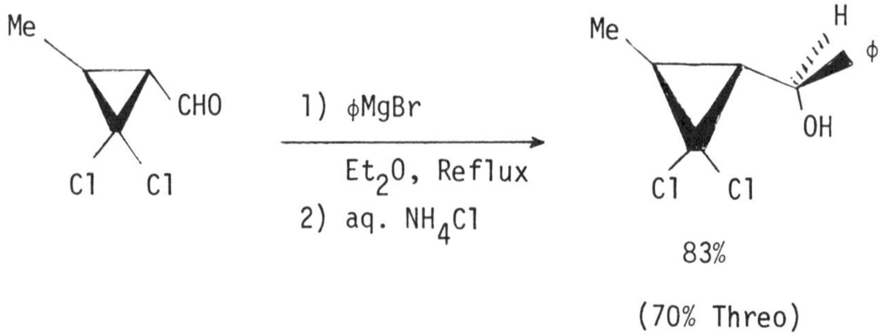
Me
CHO
Cl Cl
1) ϕMgBr
Et_2O, Reflux
2) aq. NH_4Cl
Me
H
ϕ
OH
Cl Cl
83%
(70% Threo)

I.D.2-15 F. Huet, A. Lechevallier and J. M. Conia, Tetrahedron
Lett., 22, 3585 (1981).

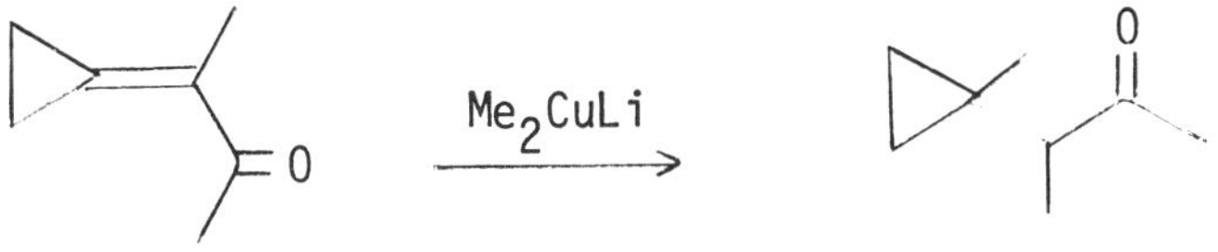

67%

I.D.2-16 A. Krief et al, Tetrahedron Lett., 22, 4341, 4737
(1981).

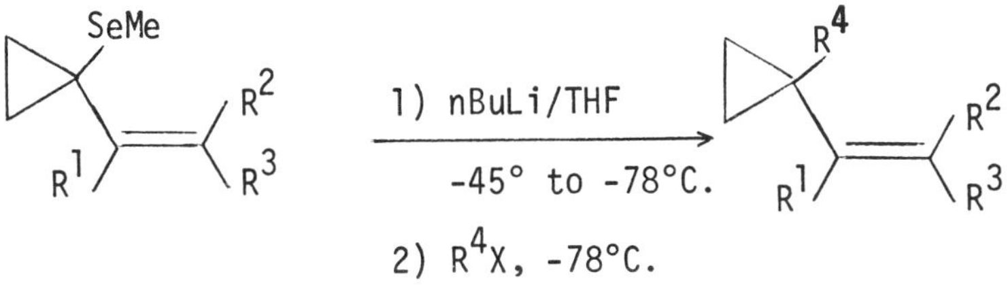

1) nBuLi/THF
   -45° to -78°C.
2) R^4X, -78°C.

65-94%

Cyclobutanones obtained from starting material plus
pTsOH.

I.D.2-17 S. Halazy and A. Krief, Tetrahedron Lett., 22, 1829,
1833, 2135 (1981).

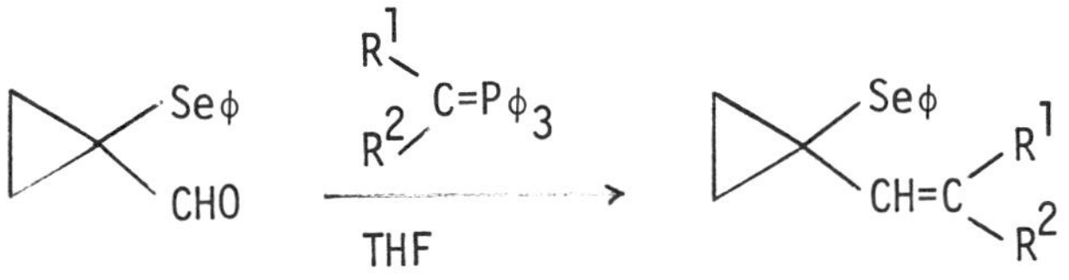

THF

85-91%

I.D.2-18   T. Hiyama et al, Bull. Chem. Soc. Jpn., 54, 2151 (1981).

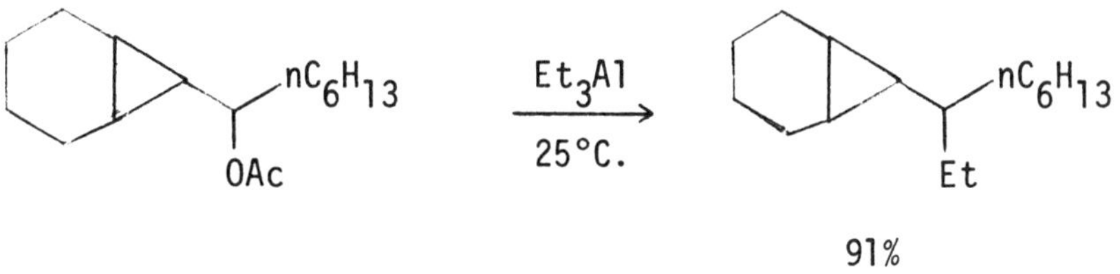

91%

Phenyl substituted cyclopropylmethyl acetates undergo alkylation with ring-opening.

I.D.2-19   J. Arct, B. Migaj and A. Leonczynski, Tetrahedron, 37, 3689 (1981); W. E. Billups et al, ibid, 3215.

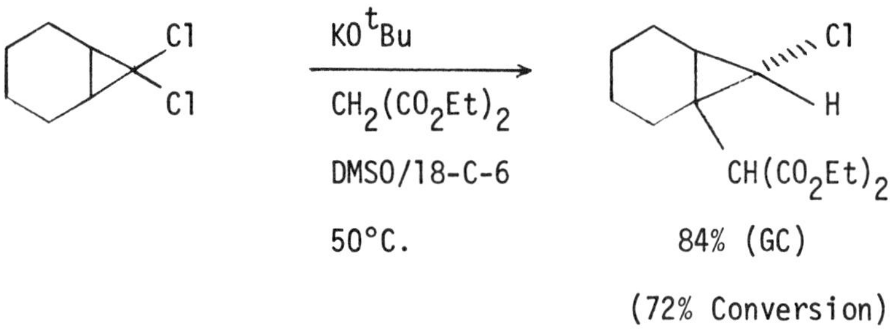

84% (GC)

(72% Conversion)

I.E.   Thermal Reactions

I.E.1.   Cycloadditions

I.E.1-1   M. Petrzilka and J. I. Grayson, Synthesis, 753 (1981).

Review: "Preparation and Diels-Alder Reactions of Heterosubstituted 1,3-Dienes."

I.E.1-2 S. Danishefsky, Acct. Chem. Res., 14, 400 (1981).

Review: "Siloxy Dienes in Total Synthesis."

I.E.1-3 K. N. Houk et al, J. Org. Chem., 46, 2338 (1981);
Tetrahedron Lett., 22, 2043, 2047 (1981).

Theory of Diels-Alder Cycloadditions to Benzoquinones.

I.E.1-4 H. D. Scharf et al, Justus Liebigs Ann. Chem., 306
(1981).

"On the Difficulty of Classifying Diels-Alder Reactions
into 'Normal' and 'Inverse'."

I.E.1-5 E. J. Corey and H. Estreicher, Tetrahedron Lett., 22,
603 (1981); M. R. Bell, J. L. Herrmann and V. Akullian,
Synthesis, 357 (1981).

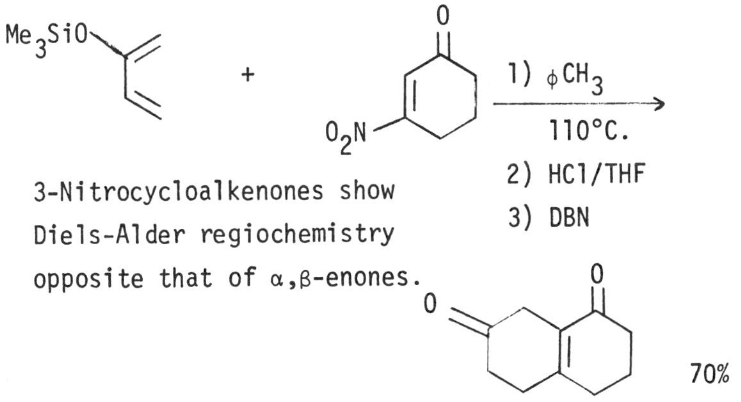

3-Nitrocycloalkenones show
Diels-Alder regiochemistry
opposite that of α,β-enones.

70%

I.E.1-6 D. Liotta, M. Saindane and C. Barnum, J. Amer. Chem.
Soc., 103, 3224 (1981); J. A. Miller et al, J. Chem. Soc.,
Perkin I, 1096 (1981).

98-100%

X = H or Seφ.

I.E.1-7 D. J. Bellville, D. D. Wirth and N. L. Bauld, J. Amer.
Chem. Soc., 103, 718 (1981); M. Christl, U. Lanzendorfer and
S. Freund, Angew. Chem., Int. Ed. Engl., 20, 674 (1981).

40%

Cation-Radical Catalyzed Diels-Alder

I.E.1-8 W. Oppolzer et al, Tetrahedron Lett., 22, 2545 (1981);
Helv. Chim. Acta, 64, 2802 (1981);J. Bachner, U. Huber and G.
Buchbauer, Monat. Chem., 112, 679 (1981); D. Horton and T.
Machinami, Chem. Commun., 88 (1981); Z. M. Ismail and H. M. R.
Hoffmann, J. Org. Chem., 46, 3549 (1981).

Lewis Acid

Study of Chiral Induction

I.E.1-9 D. I. Davies and R. F. Newton, J. Chem. Soc., Perkin I, 146 (1981); G. Mehta and A. V. Reddy, Chem. Commun., 756 (1981); H. Parlar and R. Baumann, Angew. Chem., Int. Ed. Engl., 20, 1014 (1981); A. Weber and M Neuenschwander, ibid, 774.

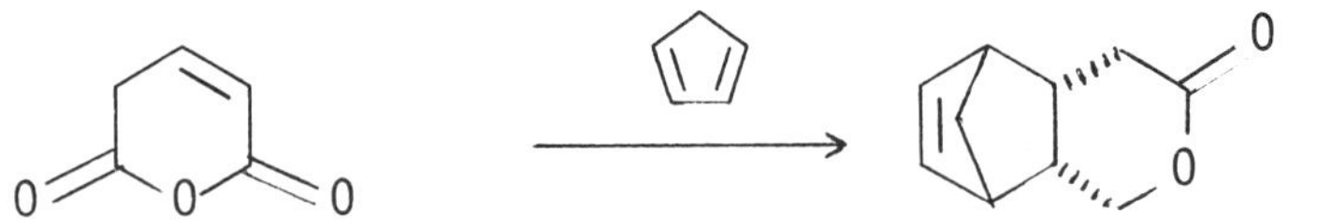

I.E.1-10 P. Yates and I. Gupta, Chem. Commun., 449 (1981); W. Kirmse and K. Loosen, Chem. Ber., 114, 400 (1981).

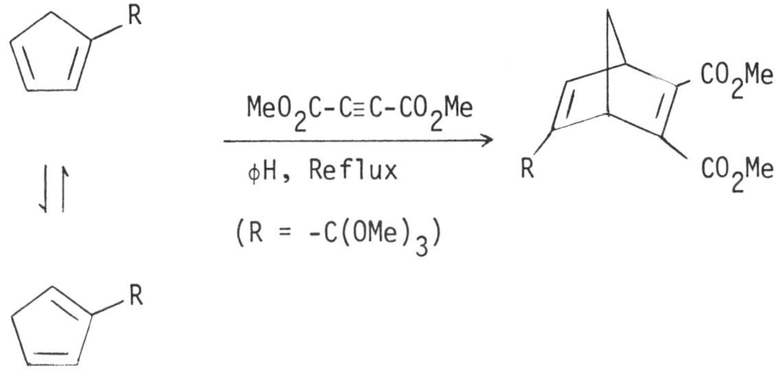

I.E.1-11 A. P. Kozikowski, K. Sugiyama and J. P. Springer, J. Org. Chem., 46, 2426 (1981); D. A. Jackson and R. J. Stoodley, Chem. Commun., 478 (1981); K. Krohn, Justus Liebigs Ann. Chem., 2285 (1981).

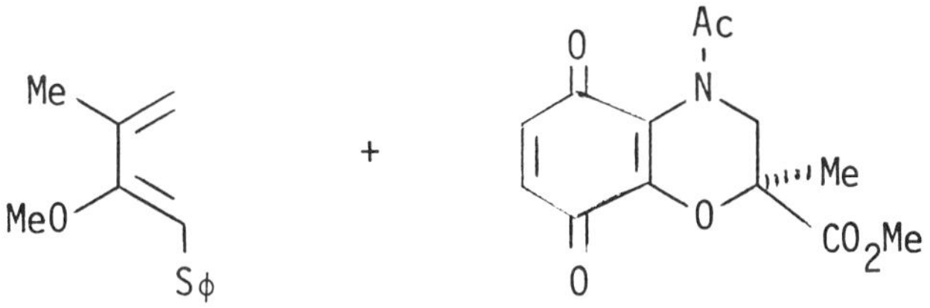

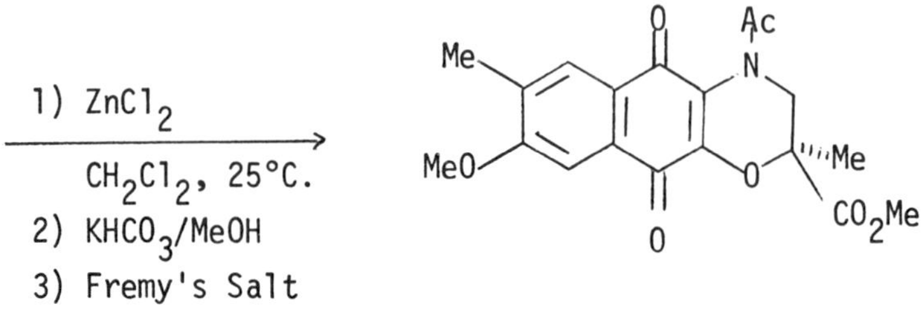
1) ZnCl2
CH2Cl2, 25°C.
2) KHCO3/MeOH
3) Fremy's Salt
Me
MeO
O
O
Ac
N
O
Me
CO2Me

Excellent yield

I.E.1-12 R. Gompper and M. Sramek, <u>Synthesis</u>, 649 (1981); Y. Tamura et al, <u>Tetrahedron Lett.</u>, <u>22</u>, 4283 (1981); W. B. Manning, <u>ibid</u>, 1571.

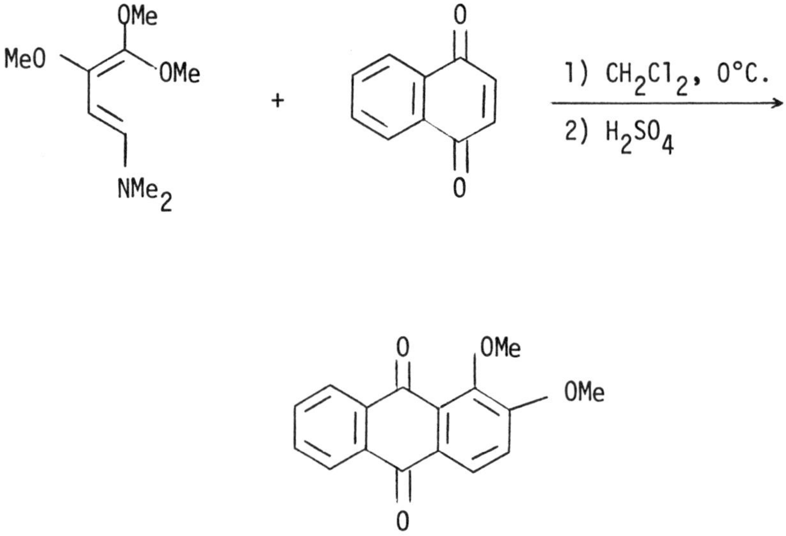
OMe
MeO
OMe
NMe2
+
O
O
1) CH2Cl2, 0°C.
2) H2SO4
O
OMe
OMe
O

70%

I.E.1-13 M. P. Cava et al, J. Amer. Chem. Soc., 103, 1992 (1981).

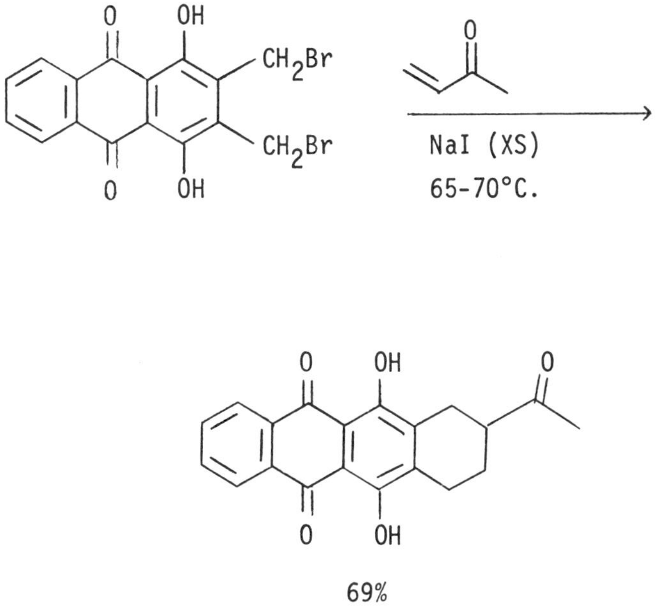

69%

I.E.1-14 Y. Kishi et al, J. Amer. Chem. Soc., 103, 4248 (1981); F. Farina et al, J. Chem. Res. (S), 316 (1981).

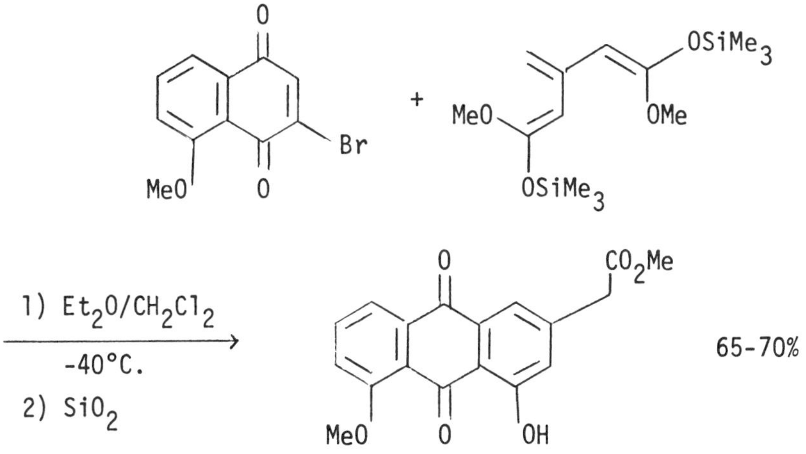

1) Et$_2$O/CH$_2$Cl$_2$

   -40°C.

2) SiO$_2$

65-70%

I.E.1-15 J. A. Kloek, J. Org. Chem., 46, 1951 (1981); J. L.
Pyle, A. A. Shaffer and J. S. Cantrell, ibid, 115; H. Hart
and S. Shamouilian, ibid, 4874.

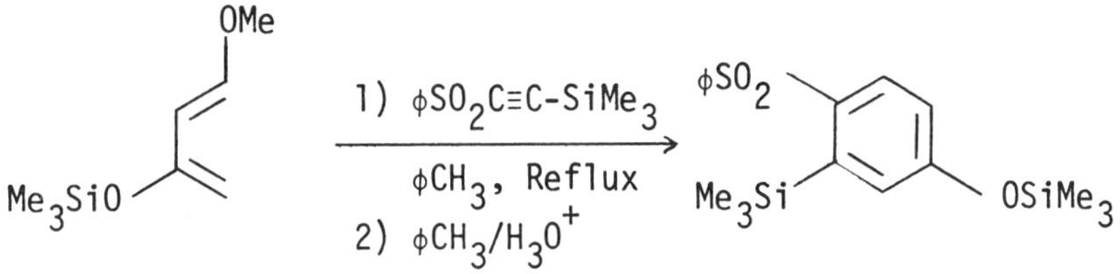

63%

I.E.1-16 H. Hopf et al, Org. Syn., 60, 41 (1981); Justus
Liebigs Ann. Chem., 165 (1981); T. Saegusa et al, J. Org.
Chem., 46, 1043 (1981); S. Misumi et al, Chem. Lett., 627
(1981).

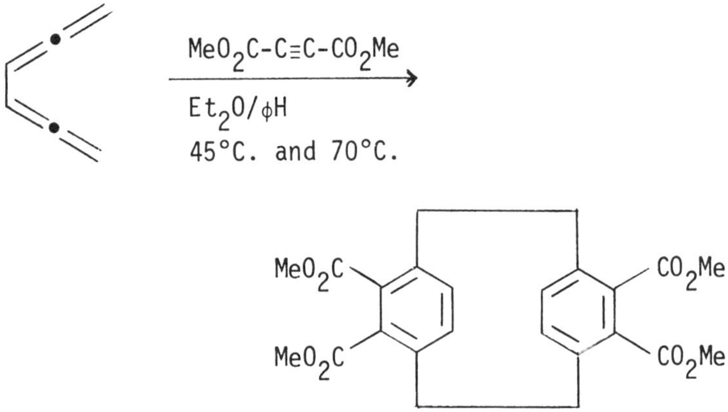

40-50%

I.E.1-17 M. E. Jung et al, J. Amer. Chem. Soc., 103, 6677
(1981); D. W. Cameron, G. I. Feutrill and P. Perlmutter,
Tetrahedron Lett., 22, 3273 (1981); G. Roberge and P.
Brassard, J. Org. Chem., 46, 4161 (1981); Synthesis, 381
(1981).

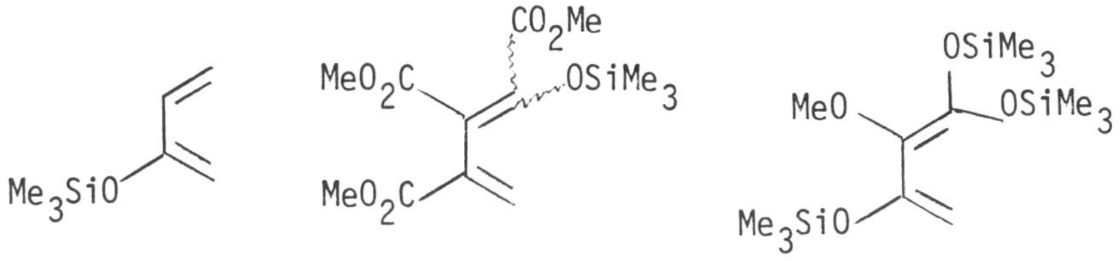

Diels-Alder Dienes.

I.E.1-18 A. P. Kozikowski, K. Sugiyama and E. Huie, Tetrahedron Lett., 22, 3381 (1981); W. Abele and R. R. Schmidt,
ibid, 4807; D. H. R. Barton et al, J. Chem. Soc., Perkin I,
1582 (1981).

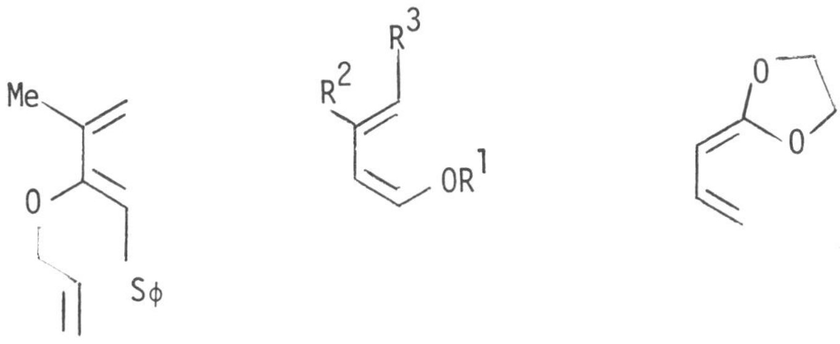

Diels-Alder Dienes

I.E.1-19 S. Danishefsky and T. A. Craig, Tetrahedron, 37, 4081
(1981); C. Brisson and P. Brassard, J. Org. Chem., 46, 1810
(1981); K. Krageloh and G. Simchen, Synthesis, 30 (1981).

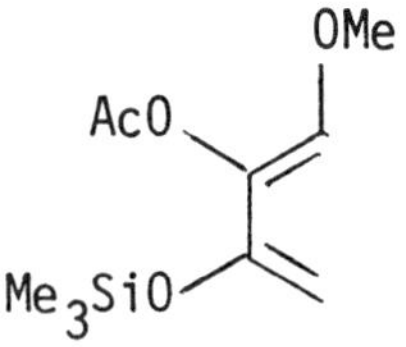

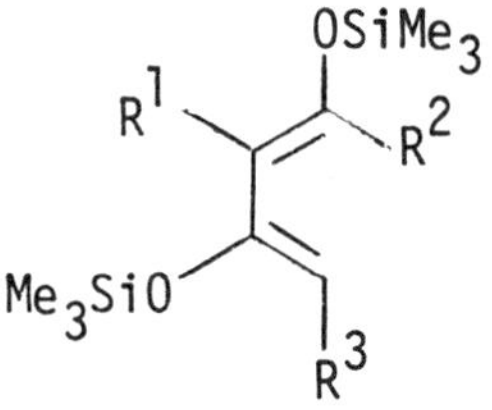

Diels-Alder Dienes.

I.E.1-20 L. E. Overman et al, J. Amer. Chem. Soc., 103, 2807,
2816 (1981); R. R. Schmidt and A. Wagner, Synthesis, 273
(1981); Y. Nomura et al, Bull. Chem. Soc. Jpn., 54, 2779
(1981).

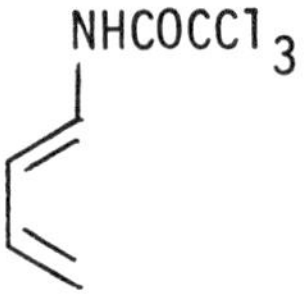

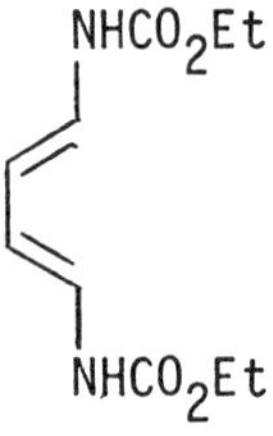

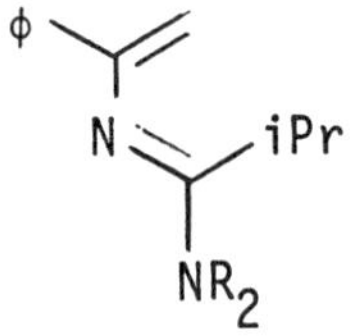

Diels-Alder Dienes.

I.E.1-21 M. J. Carter, I. Fleming and A. Percival, J. Chem.
Soc., Perkin I, 2415 (1981); M. E. Garst and P. Arrhenius,
Synth. Commun., 11, 481 (1981).

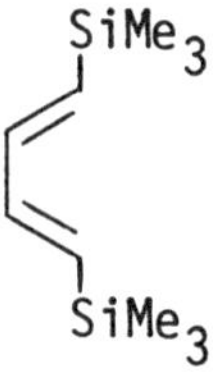

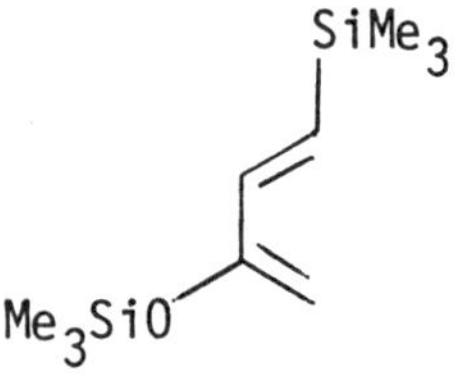

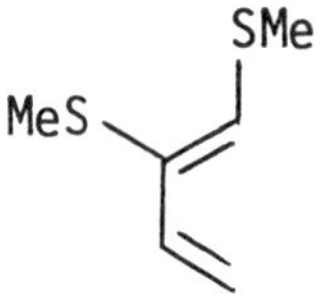

Diels-Alder Dienes.

I.E.1-22 M. Franck-Neumann, D. Martina and F. Brion, Angew.
Chem., Int. Ed. Engl., 20, 864 (1981); D. P. G. Hamon and P.
R. Spurr, Synthesis, 873 (1981); J. P. Gesson, J. C. Jacquesy
and M. Mondon, Tetrahedron Lett., 22, 1337 (1981).

Diels-Alder Dienes.

I.E.1-23 L. K. Bee and P. J. Garratt, J. Chem. Res. (S), 368
(1981); M. C. Lasne, J. L. Ripoll and J. M. Denis, Tetrahedron, 37, 503 (1981); O. Pilet and P. Vogel, Helv. Chim. Acta,
64, 2563 (1981).

Diels-Alder Dienes.

I.E.1-24 A. Murai, S. Sato and T. Masamune, Chem. Lett., 429
(1981); Chem. Commun., 904 (1981); R. K. Geïvandov and E. I.
Kovshev, J. Org. Chem. (USSR), 17, 461 (1981).

Diels-Alder Dienes.

I.E.1-25 H. Auksi and P. Yates, Can. J. Chem., 59, 2510
(1981); S. Yamamura and M. Niwa, Chem. Lett., 625 (1981); C.
W. Rees et al, J. Chem. Soc., Perkin I, 3214, 3221, 3225, 3234
(1981).

Diels-Alder Dienes.

I.E.1-26 D. B. MacLean et al, Can. J. Chem., 59, 1247 (1981);
P. Vogel et al, Helv. Chim. Acta, 64, 1818 (1981); G. Grundke
and H. M. R. Hoffmann, J. Org. Chem., 46, 5428 (1981).

Diels-Alder Dienes.

I.E.1-27 M. Mori, A. Hayamizu and K. Kanematsu, J. Chem. Soc.,
Perkin I, 1259 (1981); T. K. Bandyopadhyay and A. J.
Bhattacharya, Ind. J. Chem., 20B, 91, 95 (1981); R. N.
Warrener et al, Austr. J. Chem., 34, 131, 397, 1223 (1981).

Diels-Alder Dienes.

I.E.1-28 A. Oku et al, J. Org. Chem., 46, 4152 (1981); S. Knapp, R. Lis and P. Michna, ibid, 624.

$$R^1 \diagdown \!\! \underset{R^2}{\diagup} C=C \underset{OCR^3}{\overset{CN}{\diagup}}$$

Diels-Alder Dienophiles.

I.E.1-29 R. J. Ardecky, F. A. J. Kerdesky and M. P. Cava, J. Org. Chem., 46, 1483 (1981); J. Tamariz and P. Vogel, Helv. Chim. Acta, 64, 188 (1981).

$$Me_3SiO \diagdown \overset{O}{\underset{\|}{C}} \diagup CH_3$$

$$RO \diagdown \overset{O}{\underset{\|}{C}}$$

Diels-Alder Dienophiles.

I.E.1-30 M. Shen and A. G. Schultz, Tetrahedron Lett., 22, 3347 (1981); L. A. Paquette and R. V. Williams, Tetrahedron Lett., 22, 4643 (1981).

$$\phi SO_2 - \!\!\equiv\!\! - CO_2Et \qquad\qquad \phi SO_2 - \!\!\equiv\!\! - SiMe_3$$

Diels-Alder Dienophiles.

I.E.1-31 T. Sasaki et al, J. Amer. Chem. Soc., 103, 565 (1981).

Diels-Alder Dienophile.

I.E.1-32   F. Farina et al, <u>J. Chem. Res. (S)</u>, 370 (1981); M.
Oda and Y. Kanao, <u>Chem. Lett.</u>, 1547 (1981).

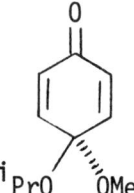

Diels-Alder Dienophiles.   (Second entry reacts at
internal double bond.)

I.E.1-33   D. N. Gupta, P. Hodge and N. Khan, <u>J. Chem. Soc.</u>,
<u>Perkin I</u>, 689 (1981); K. Kanematsu et al, <u>J. Amer. Chem. Soc.</u>,
<u>103</u>, 5211 (1981).

Diels-Alder Dienophiles.

I.E.1-34   G. Helmchen and R. Schmierer, <u>Angew. Chem., Int. Ed.
Engl.</u>, <u>20</u>, 205 (1981); S. Danishefsky and M. Kahn, <u>Tetrahedron
Lett.</u>, <u>22</u>, 489 (1981).

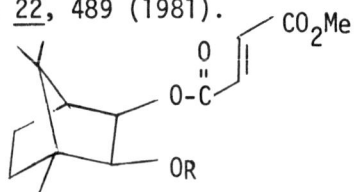
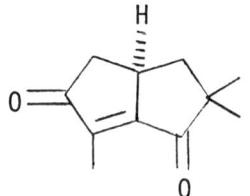

Diels-Alder Dienophiles.

I.E.1-35   T. Kametani and H. Nemoto, <u>Tetrahedron</u>, <u>37</u>, 3 (1981).

Review: "Recent Advances in the Total Synthesis of
          Steroids via Intramolecular Cycloaddition
          Reactions."

I.E.1-36   W. R. Roush et al, <u>J. Amer. Chem. Soc.</u>, <u>103</u>, 6696,
5200 (1981);   D. F. Taber et al, <u>Tetrahedron Lett.</u>, <u>22</u>, 5141
(1981).

MeO$_2$C

THPO   Me

1) 210°C.
2) H$_3$O$^+$

MeO$_2$C   Me   OH

64%

I.E.1-37   M. P. Edwards, S. V. Ley and S. G. Lister, <u>Tetrahe-</u>
<u>dron Lett.</u>, <u>22</u>, 361 (1981);   K. C. Nicolaou and R. L. <u>Magolda</u>,
<u>J. Org. Chem.</u>, <u>46</u>, 1506 (1981).

EtO$_2$C

MEMO

Et

φCH$_3$
Reflux

CO$_2$Et
H   H

MEMO

H

60%

I.E.1-38 T. Mukaiyama and N. Iwasawa, <u>Chem. Lett.</u>, 29 (1981).

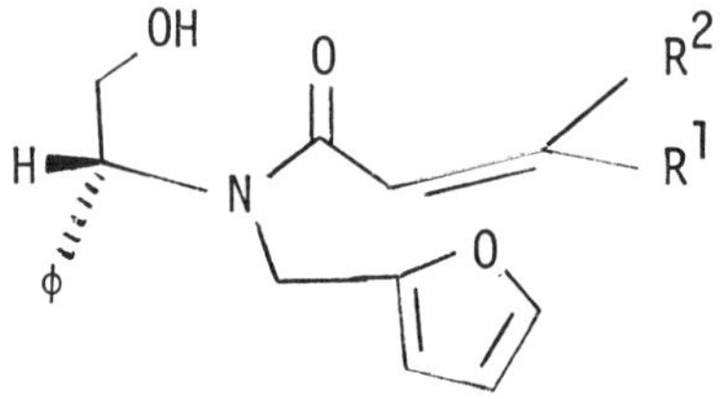

Asymmetric Diels-Alder

I.E.1-39 M. E. Jung and K. M. Halweg, <u>Tetrahedron Lett.</u>, <u>22</u>, 3929 (1981); S. A. Bal and P. Helquist, <u>ibid</u>, 3933.

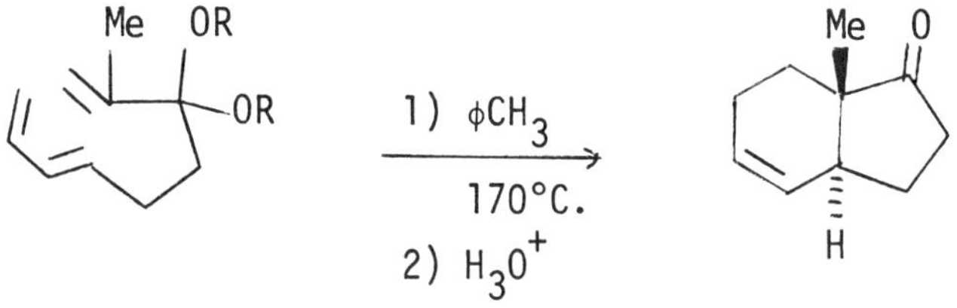
Me OR
OR
1) ϕCH3
170°C.
2) H3O+
Me O
H

88% (72% trans)

I.E.1-40 W. M. Best and D. Wege, <u>Tetrahedron Lett.</u>, <u>22</u>, 4877 (1981).

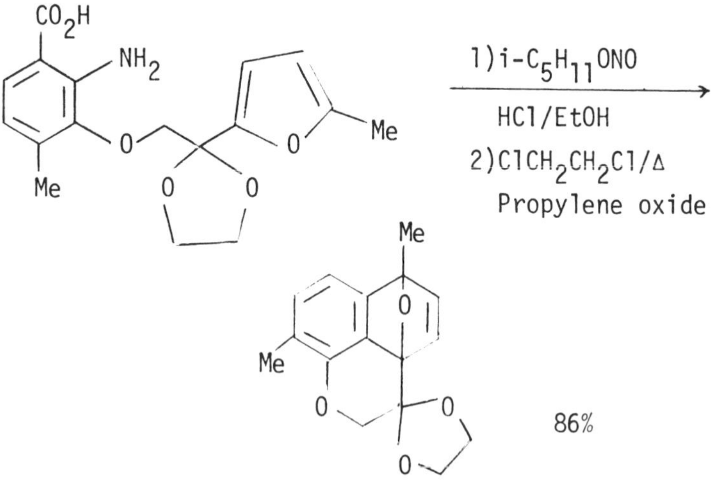
CO2H
NH2
Me
O
O
O
O
Me
1)i-C5H11ONO
HCl/EtOH
2)ClCH2CH2Cl/Δ
Propylene oxide
Me
O
Me
O
O
O

86%

I.E.1-41 J. D. White and B. G. Sheldon, <u>J. Org. Chem.</u>, <u>46</u>, 2273 (1981); S. F. Martin and C. Y. Tu, <u>ibid</u>, 3763.

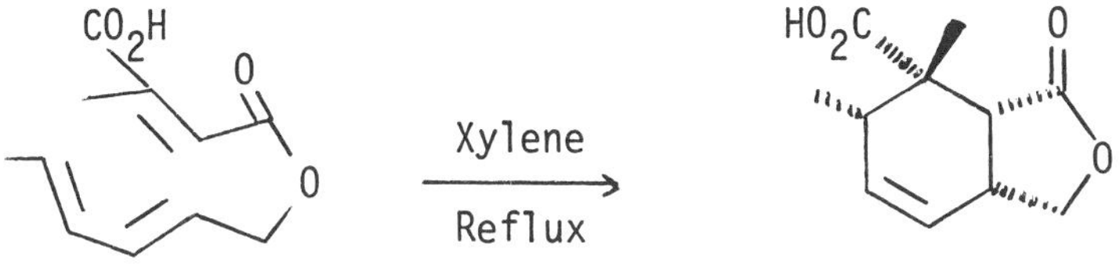

Xylene
Reflux

32%

I.E.1-42 J. Tsuji et al, <u>Tetrahedron Lett.</u>, <u>22</u>, 1357 (1981); T. Kametani et al, <u>J. Chem. Soc., Perkin I</u>, 1383, 1386 (1981); <u>Tetrahedron</u>, <u>37</u>, 2547, 2555, 3813 (1981).

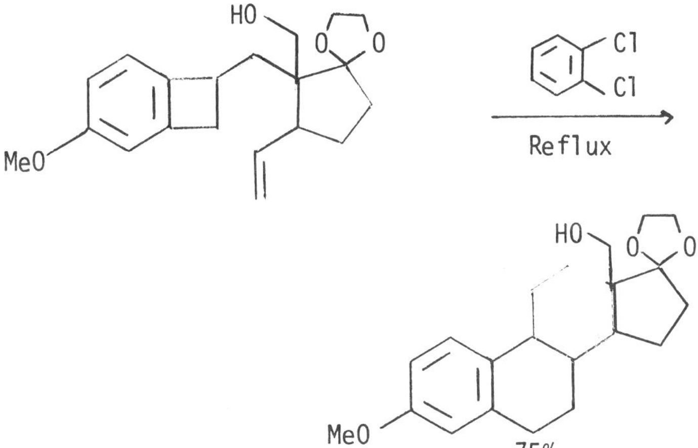

Reflux

75%

I.E.1-43 Y. Ito, M. Nakatsuka, and T. Saegusa, J. Amer. Chem. Soc., 103, 476 (1981).

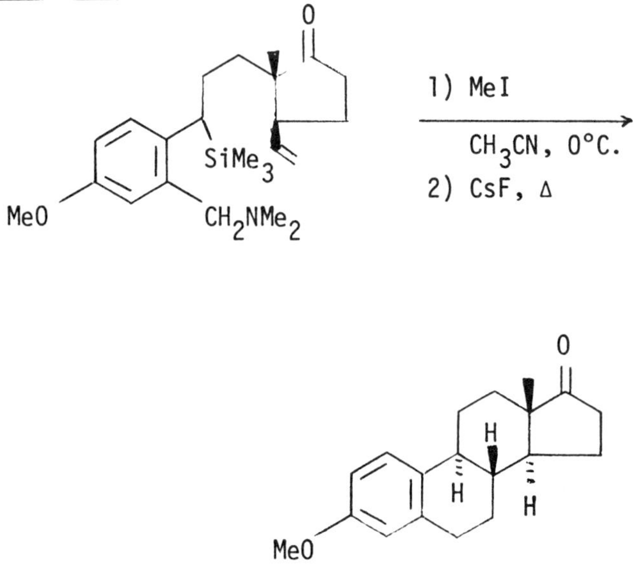

86%

I.E.1-44 G. Stork, G. Clark and C. S. Shiner, J. Amer. Chem. Soc., 103, 4948 (1981).

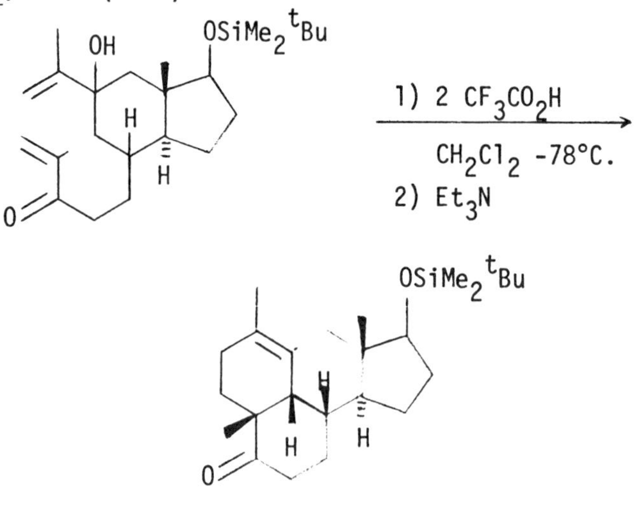

~ 70%

I.E.1-45   R. L. Snowden, <u>Tetrahedron Lett.</u>, <u>22</u>, 97, 101 (1981);
G. Ohloff et al, <u>Helv. Chim. Acta</u>, <u>64</u>, 1387 (1981).

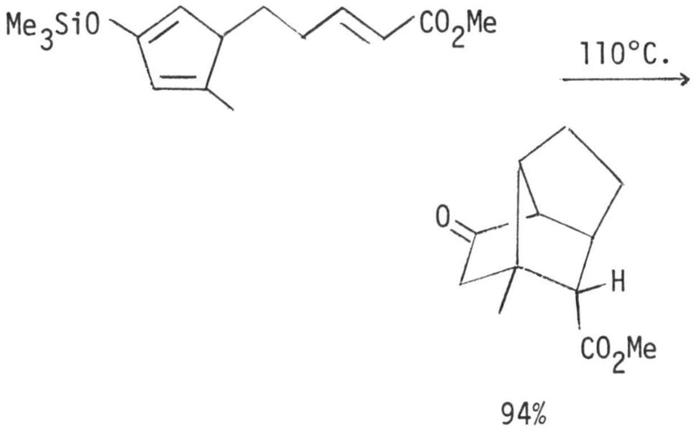

$$Me_3SiO \quad \xrightarrow{\quad CO_2Me \quad} \quad \xrightarrow{110°C.}$$

94%

I.E.1-46   W. R. Roush and A. G. Myers, <u>J. Org. Chem.</u>, <u>46</u>, 1509
(1981).

Lewis Acid Catalyzed Intramolecular Diels-Alder
Reactions.

I.E.1-47   B. Zwanenburg et al, <u>Tetrahedron Lett.</u>, <u>22</u>, 4553,
4557 (1981);   M. Adamczyk and M. Mokrosz, <u>Synthesis</u>, 802
(1981).

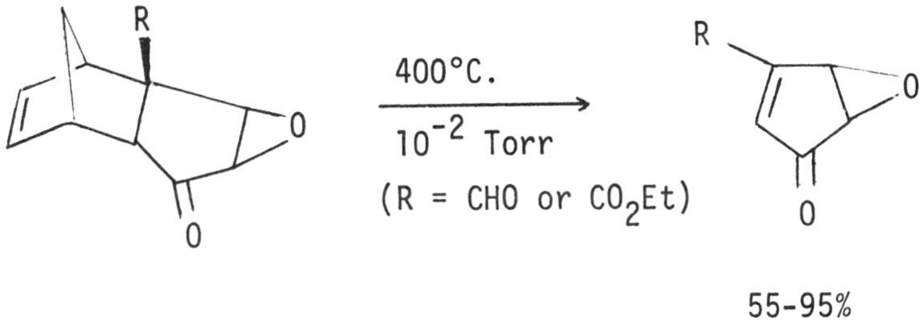

$$\xrightarrow[\substack{10^{-2}\ Torr \\ (R = CHO\ or\ CO_2Et)}]{400°C.}$$

55-95%

I.E.1-48   J. S. H. Kueh, M. Mellor and G. Pattenden, J. Chem.
Soc., Perkin I, 1052 (1981), G. Pattenden and D. Whybrow,
ibid, 1046;   G. Kaupp and M. Stark, Chem. Ber., 114, 2217
(1981).

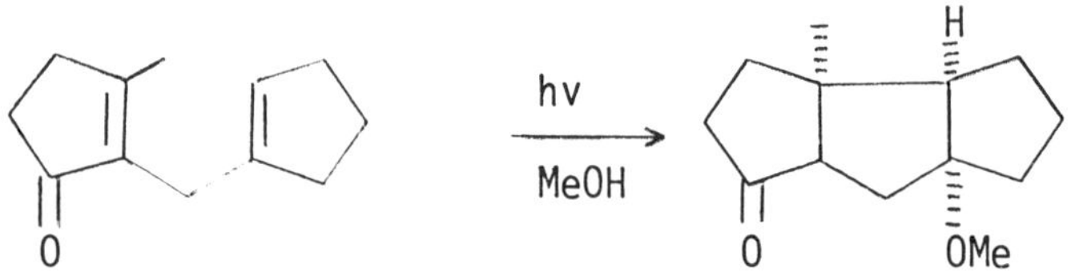

90%

I.E.1-49   K. Mizuno, H. Ueda and Y. Otsuji, Chem. Lett., 1237
(1981);   L. Eisenhuth, H. Siegel and H. Hopf, Chem. Ber., 114,
3772 (1981).

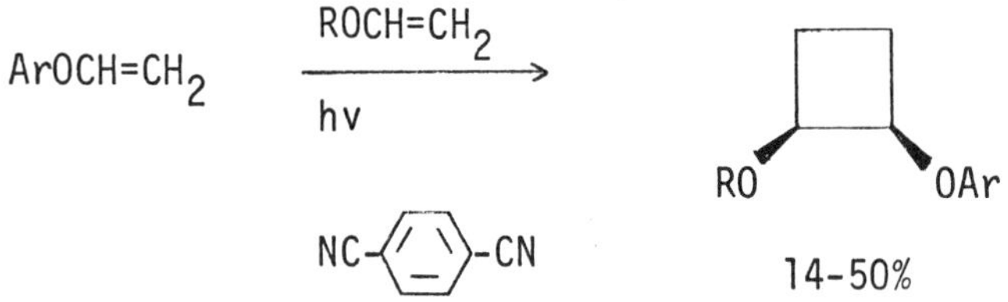

14-50%
(73-98% cis)

I.E.1-50   I. A. Akhtar and J. J. McCullough, J. Org. Chem., 46,
1447 (1981);   A. W. H. Jans and J. Cornelisse, Rec. Trav.
Chim., 100, 213 (1981).

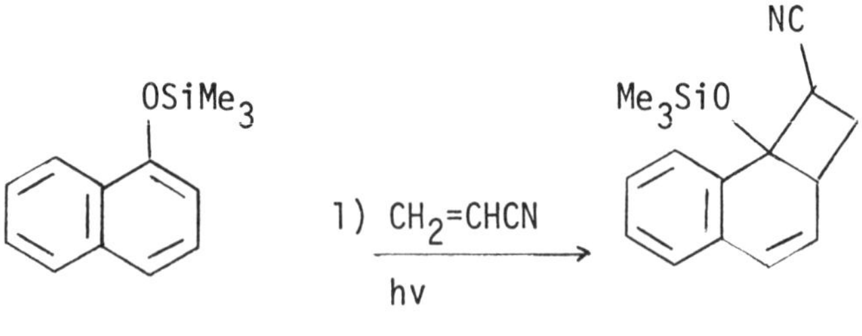

24%

I.E.1-51 M. C. Pirrung, J. Amer. Chem. Soc., 103, 82 (1981); A. H. White et al, J. Chem. Soc., Perkin II, 1473 (1981); M. Fetizon et al, Chem. Commun., 953 (1981); W. Oppolzer, L. Gorrichon and T. G. C. Bird, Helv. Chim. Acta, 64, 186 (1981).

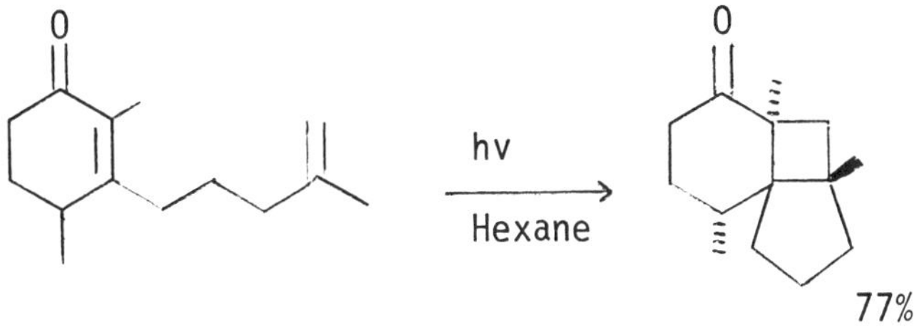

I.E.1-52 J. R. Williams and C. Lin, Chem. Commun., 752 (1981).

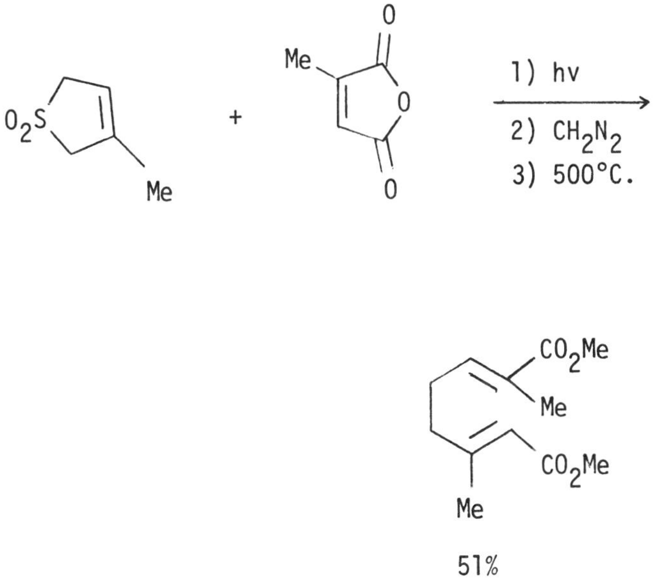

I.E.1-53 M. Van Audenhove, D. De Keukeleire and M. Vandewalle,
Bull. Soc. Chim. Belg., 90, 255 (1981).

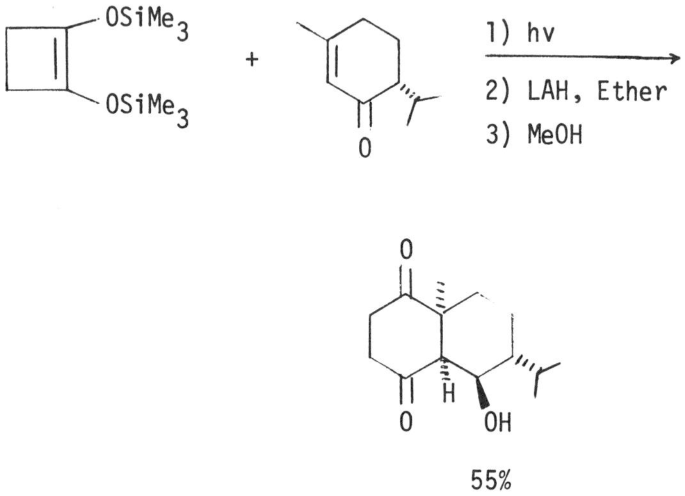

55%

I.E.1-54 W. T. Brady, Tetrahedron, 37, 2949 (1981).

Review: "Synthetic Applications Involving
         Halogenated Ketenes."

I.E.1-55 H. W. Moore and M. D. Gheorghiu, Chem. Soc. Rev., 10,
289 (1981).

Review: "Cyanoketenes: Synthesis and Cycloadditions."

I.E.1-56 L. Ghosez et al, J. Amer. Chem. Soc., 103, 4616 (1981).

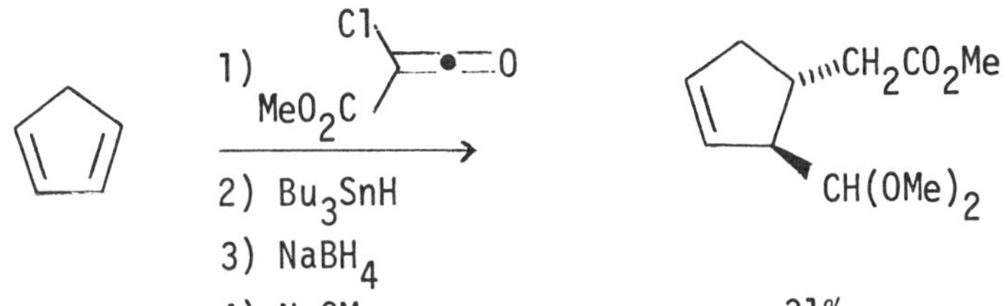

31%

I.E.1-57 I. Fleming and R. V. Williams, J. Chem. Soc., Perkin I, 684 (1981); W. T. Brady and R. M. Lloyd, J. Org. Chem., 46, 1322 (1981); P. Martin, H. Greuter and D. Bellus, Helv. Chim. Acta, 64, 64 (1981).

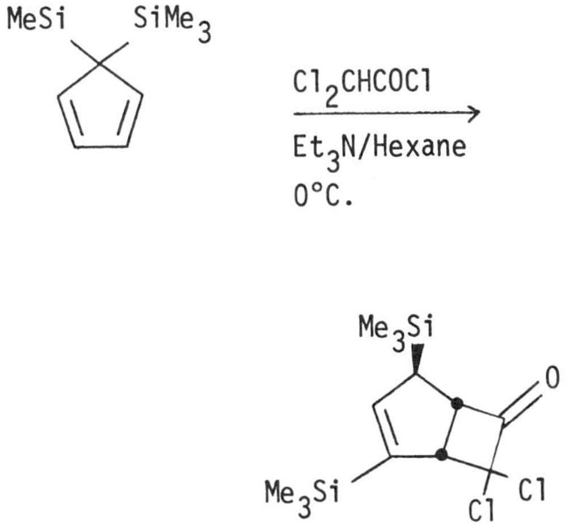

72%

I.E.1-58   J. S. Swenton et al, <u>Chem. Commun.</u>, 179 (1981).

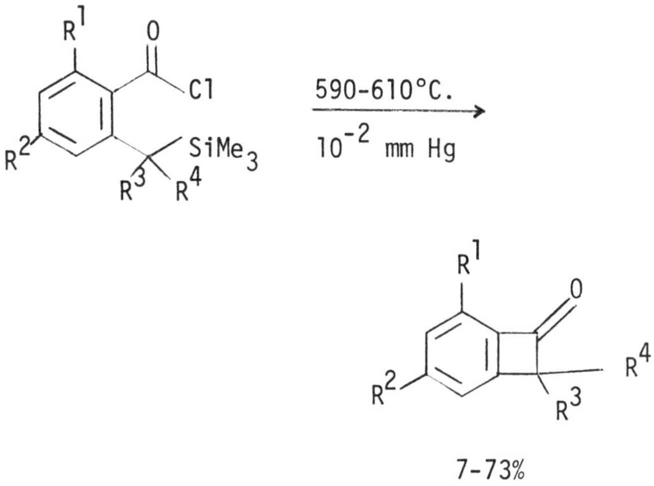

$$\xrightarrow[10^{-2} \text{ mm Hg}]{590\text{-}610^\circ C.}$$

7-73%

I.E.1-59   H. M. R. Hoffmann, Z. M. Ismail and A. Weber,
<u>Tetrahedron Lett.</u>, <u>22</u>, 1953 (1981);   M. Ochiai, M. Arimoto and
E. Fujita, <u>Chem. Commun.</u>, 460 (1981).

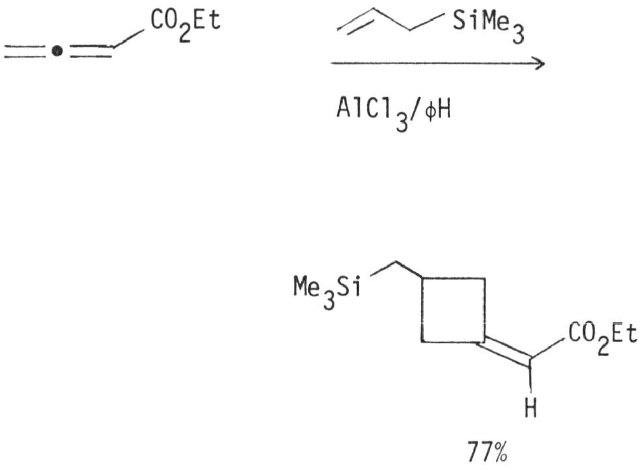

$$\xrightarrow[AlCl_3/\phi H]{}$$

77%

I.E.1-60 M. E. Jung and K. M. Halweg, <u>Tetrahedron Lett.</u>, <u>22</u>, 2735 (1981).

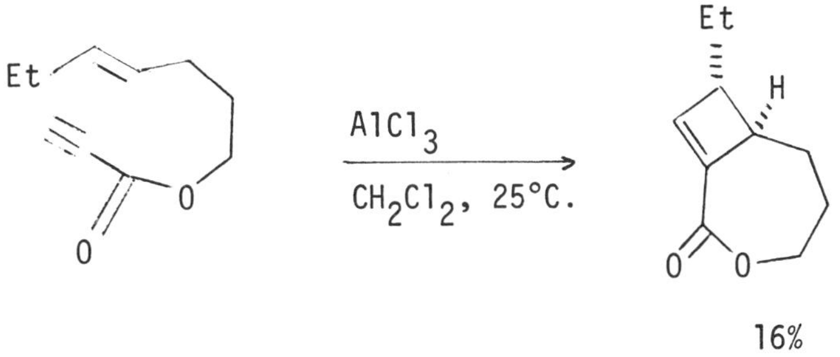

16%

I.E.1-61 J. Ficini, A. Krief et al, <u>Tetrahedron Lett.</u>, <u>22</u>, 725 (1981); L. Henn and G. Himbert, <u>Chem. Ber.</u>, <u>114</u>, 1015 (1981); K. H. Dotz, B. Trenkle and U. Schubert, <u>Angew. Chem.</u>, Int. Ed. Engl., <u>20</u>, 287 (1981).

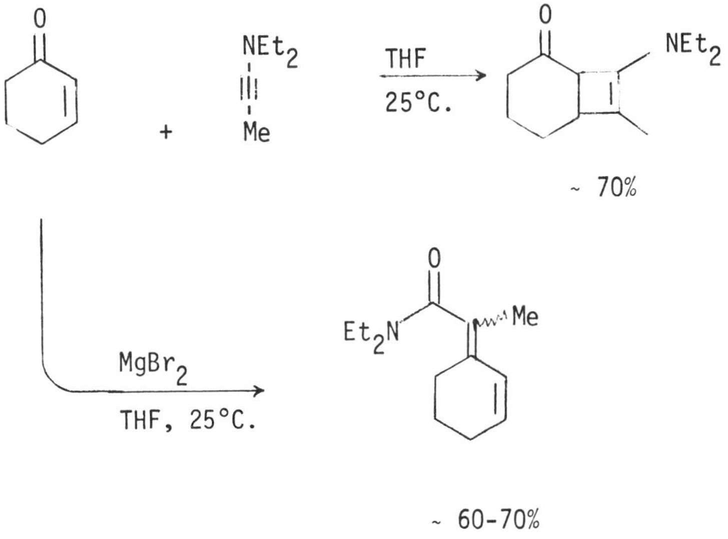

~ 70%

~ 60-70%

I.E.1-62 H. G. Heine and W. Hartmann, <u>Angew. Chem., Int. Ed.</u>
<u>Engl.</u>, <u>20</u>, 782 (1981); <u>Synthesis,</u> 706 (1981); L. Ghosez
et al, <u>Angew. Chem., Int. Ed. Engl.</u>, <u>20,</u> 879 (1981).

$$Me_2C=C=NMe_2 \quad ZnCl_3^-$$

1) $\overset{CO_2Me}{\diagup}$

CH$_2$Cl$_2$, Reflux

2) H$_2$O

57%

I.E.1-63 H. Mayr et al, <u>J. Org. Chem.</u>, <u>46</u>, 1041 (1981).

$$\phi-C\equiv C-\underset{X}{\overset{R^1}{\underset{|}{\overset{|}{C}}}}-R^1$$

1) ZnX$_2$/Et$_2$O

CH$_2$Cl$_2$, -78°C.

R$_2$C=CMe$_2$

2) 0°C.

26-62%

I.E.1-64 T. Kametani et al, <u>J. Amer. Chem. Soc.</u>, <u>103</u>, 1256
(1981).

CH$_2$SO$_2$Ar

nBuLi

THF, -30°C.

CH$_2$SO$_2$Ar

90%

I.E.1-65 P. Binger et al, <u>Chem. Ber.</u>, <u>114</u>, 3313, 3325 (1981);
J. Organometal. Chem., <u>221</u>, C33 (1981).

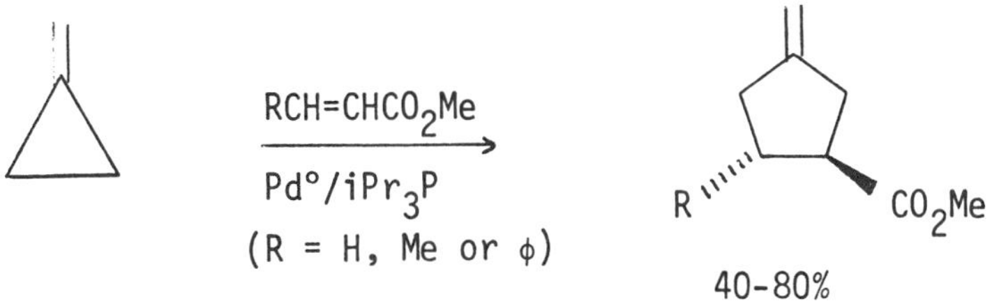

$$\text{RCH=CHCO}_2\text{Me} \xrightarrow[\text{(R = H, Me or } \phi)]{\text{Pd}^\circ/\text{iPr}_3\text{P}}$$

R''''' CO_2Me

40-80%

I.E.1-66 R. L.Danheiser, D. J. Carini and A. Basak, <u>J. Amer.</u>
<u>Chem. Soc.</u>, <u>103</u>, 1604 (1981); B. M. Trost and D. M. T. Chan,
<u>ibid</u>, 5972.

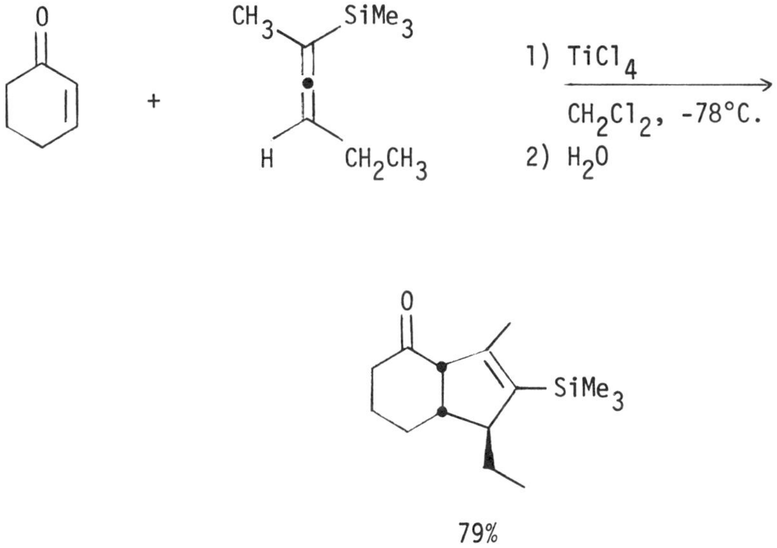

$$\text{CH}_3 \quad \text{SiMe}_3$$

$$\text{H} \quad \text{CH}_2\text{CH}_3$$

1) TiCl_4

$\xrightarrow{\text{CH}_2\text{Cl}_2, \ -78^\circ\text{C}.}$

2) H_2O

SiMe_3

79%

I.E.1-67 P. G. Wiering and H. Steinberg, J. Org. Chem., 46, 1663 (1981); P. G. Wiering, J. W. Verhoeven and H. Steinberg, J. Amer. Chem. Soc., 103, 7675 (1981).

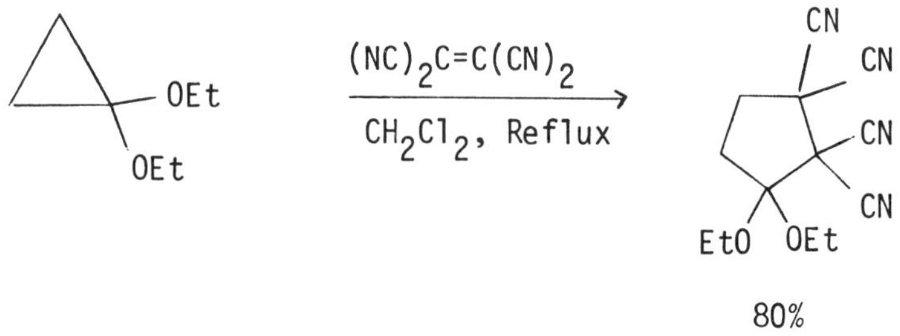
OEt
OEt
$(NC)_2C=C(CN)_2$
CH_2Cl_2, Reflux
CN
CN
CN
CN
EtO
OEt
80%

I.E.1-68 M. Yasunami et al, Chem. Lett., 555 (1981).

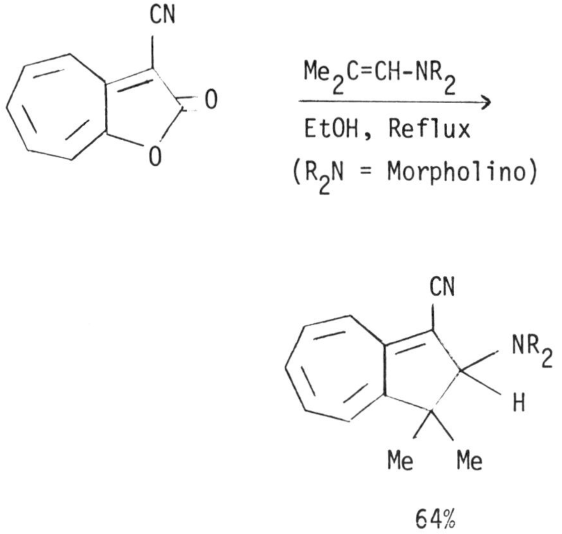
CN
O
O
$Me_2C=CH-NR_2$
EtOH, Reflux
(R_2N = Morpholino)
CN
NR_2
H
Me
Me
64%

I.E.1-69 H. Mayr, et al, Chem. Commun., 683 (1981); Tetrahedron Lett., 22, 925 (1981); Angew. Chem., Int. Ed. Engl., 20,
1027 (1981).

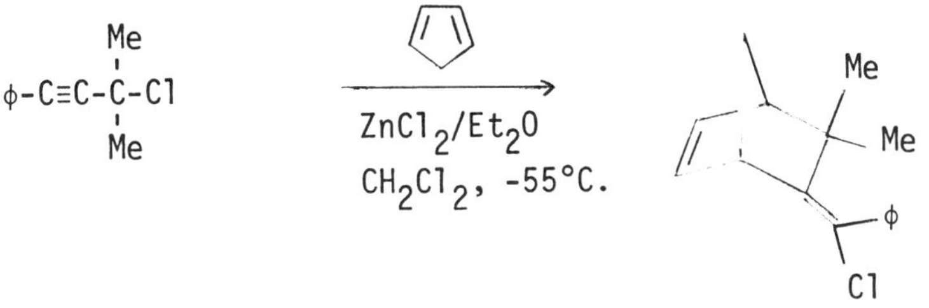

65%

I.E.1-70 H. M. R. Hoffmann and H. Vathke-Ernst. Chem. Ber.,
114, 1182, 1548 (1981).

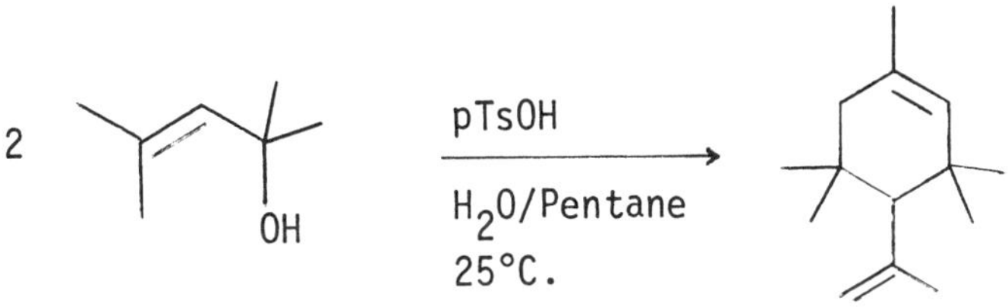

~ 70%

I.E.2. Other Thermal Reactions.

I.E.2-1 A. Viola, J. J. Collins and N. Filipp, Tetrahedron,
37, 3765 (1981).

Review: "Intramolecular Pericyclic Reactions of
Acetylenic Compounds."

I.E.2-2 W. R. Dolbier, Jr., Acct. Chem. Res., 14, 195 (1981).

    Review: "Thermal Rearrangements of gem-Difluorocyclo-
              propanes."

I.E.2-3 W. Oppolzer, Pure Appl. Chem., 53, 1181 (1981).

    Review: "Regio- and Stereo-Selective Synthesis of
              Cyclic Natural Products by Intramolecular
              Cycloaddition- and Ene-Reactions."

I.E.2-4 A. D. Batcho, D. E. Berger and M. R. Uskokovic, J.
Amer. Chem. Soc., 103, 1293 (1981); W. G. Dauben and T.
Brookhart, ibid, 237; M. R. Uskokovic et al, Helv. Chim. Acta,
64, 1682 (1981); S. N. Pardo, S. Ghosh and R. G. Salomon,
Tetrahedron Lett., 22, 1885 (1981).

$$\text{HC} \equiv \text{C-CO}_2\text{Me} \xrightarrow[\text{CH}_2\text{Cl}_2]{\text{EtAlCl}_2}$$

89%

I.E.2-5 Y. Tamura et al, Tetrahedron Lett., 22, 1343 (1981).

$$\overset{\displaystyle O}{\underset{\displaystyle}{\text{Me}\overset{\|}{\text{S}}\text{CH}_2\text{CO}_2\text{Et}}} \xrightarrow[\substack{\text{CF}_3\text{CO}_2\text{H, 0°C.} \\ \text{RCH}_2\text{CH}=\text{CH}_2}]{(\text{CF}_3\text{CO})_2\text{O}}$$

72-79%

I.E.2-6 O. Achmatowicz, Jr. and M. Pietraszkiewicz, <u>Tetra-</u>
<u>hedron Lett.</u>, <u>22</u>, 4323 (1981); <u>J. Chem. Soc., Perkin I</u>, 2680
(1981).

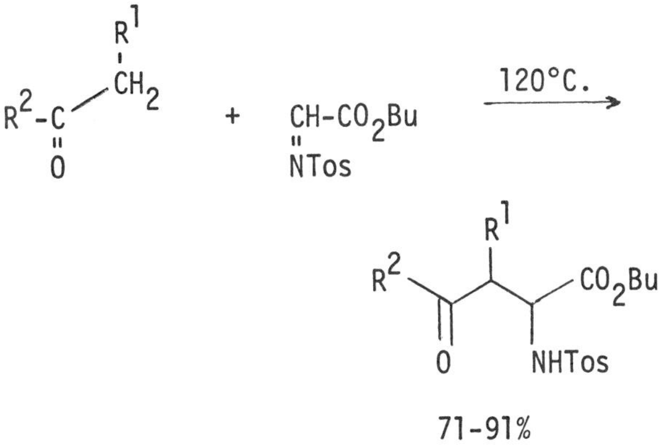

71-91%

I.E.2-7 W. Oppolzer et al, <u>Helv. Chim. Acta</u>, <u>64</u>, 1575, 2489
(1981); M. Karpf and A. S. Dreiding, <u>ibid</u>, 1123.

73%

I.E.2-8 A. Padwa and W. F. Rieker, <u>J. Amer. Chem. Soc.</u>, <u>103</u>,
1859 (1981).

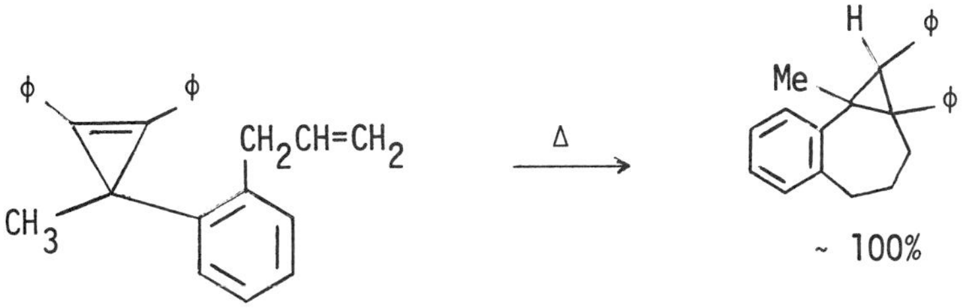

~ 100%

I.E.2-9 M. Bertrand, M. L. Roumestant and P. Sylvestre-
Panthet, Tetrahedron Lett., 22, 3589 (1981).

70%

I.E.2-10 S. Sarel, A. Schlossman and M. Langbeheim, Tetrahedron Lett., 22, 691 (1981); G. Pattenden and D. Whybrow, J.
Chem. Soc., Perkin I, 3147 (1981).

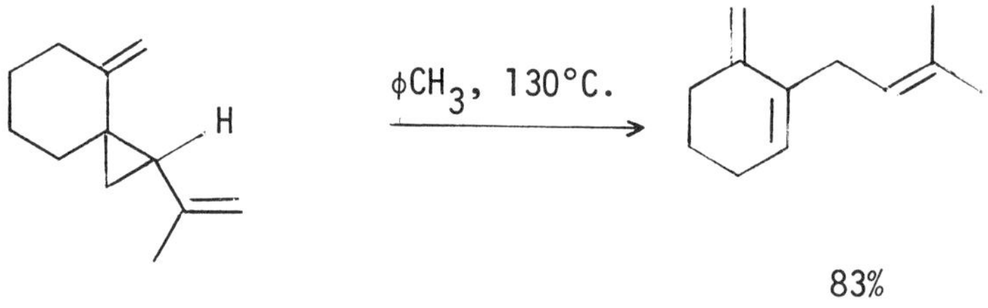

83%

I.E.2-11 R. L. Snowden and K. H. Schulte-Elte, Helv. Chim.
Acta, 64, 2193 (1981).

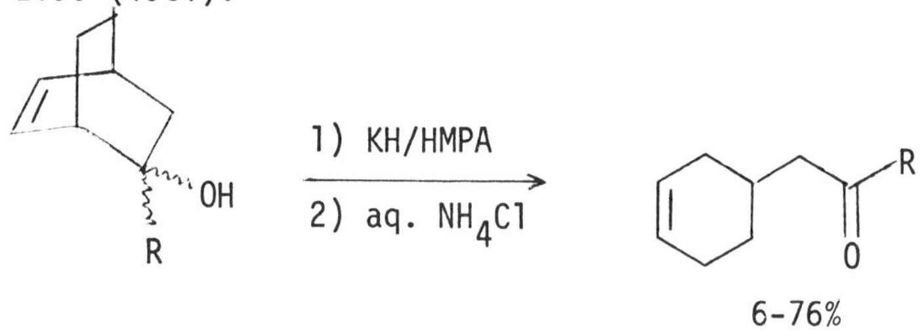

6-76%

I.E.2-12 W. G. Dauben and A. Chollet, <u>Tetrahedron Lett.</u>, <u>22,</u>
1583 (1981).

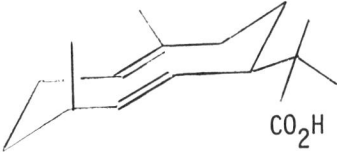

$$\begin{array}{c}
\text{CF}_3\text{CO}_2\text{H (1 eq.)} \\
\hline
\text{CH}_2\text{Cl}_2, \ 25°\text{C.} \\
\text{15 min.}
\end{array}$$

74%

I.E.2-13 S. Raucher et al, <u>J. Amer. Chem. Soc.</u>, <u>103</u>, 1853
(1981); F. E. Ziegler and T. F. Wang, <u>Tetrahedron Lett.</u>, <u>22,</u>
1179 (1981).

1) 210°C.

1,2,4-Trichlorobenzene

$\text{CH}_3\text{C(OSiMe}_3)\text{NSiMe}_3$

2) KF·2H$_2$O/HMPA

3) aq. KOH

CO$_2$H

50%

I.E.2-14 P. A. Wender et al, <u>Tetrahedron</u>, <u>37</u>, 3967 (1981).

$$\begin{array}{c}
\text{KH} \\
\hline
\text{THF, } 25°\text{C.}
\end{array}$$

90%

I.E.2-15   S. L. Schreiber and C. Santini, Tetrahedron Lett.,
22, 4651 (1981);   S. G. Levine and R. L. McDaniel, Jr., J. Org.
Chem., 46, 2199 (1981).

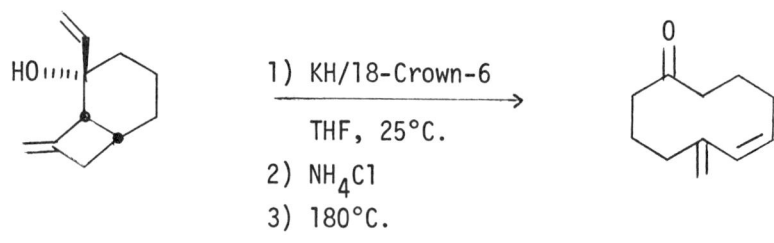

1) KH/18-Crown-6

THF, 25°C.

2) NH$_4$Cl

3) 180°C.

56%

I.E.2-16   G. D. Crouse and L. A. Paquette, J. Org. Chem., 46,
4272 (1981); J. Amer. Chem. Soc., 103, 6235 (1981); Tetrahedron Lett.. 22, 3167 (1981).

1) KH

Et$_2$O, -78°C.

2) Me$_3$SiCl/Et$_3$N

3) EtMgBr

(ϕ_3P)$_2$NiCl$_2$

46%

I.E.2-17   C. M. Tice and C. H. Heathcock, J. Org. Chem., 46, 9
(1981).

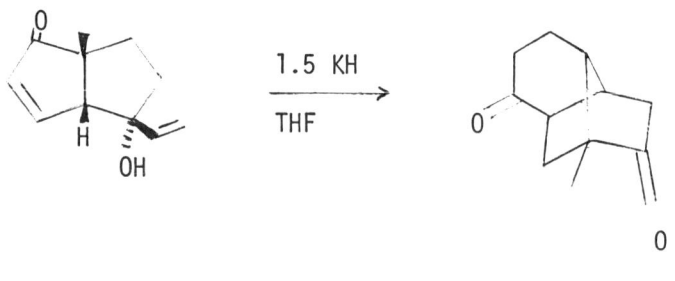

1.5 KH

THF

42%

Use of KOH/MeOH gives bicyclic enedione product.

I.E.2-18 C. J. Burrows and B. K. Carpenter, J. Amer. Chem.
Soc., 103, 6983, 6984 (1981); P. A. Bartlett and C. F. Pizzo,
J. Org. Chem., 46, 3896 (1981); R. C. Cambie et al, Austr. J.
Chem., 34, 819, 1079 (1981).

   Various Studies of the Claissen Rearrangement.

I.E.2-19 J. M. Bruce and Y. Roshan-Ali,J. Chem. Soc., Perkin
I, 2677 (1981); A. G. Schultz and J. Napier, Chem. Commun.,
224 (1981).

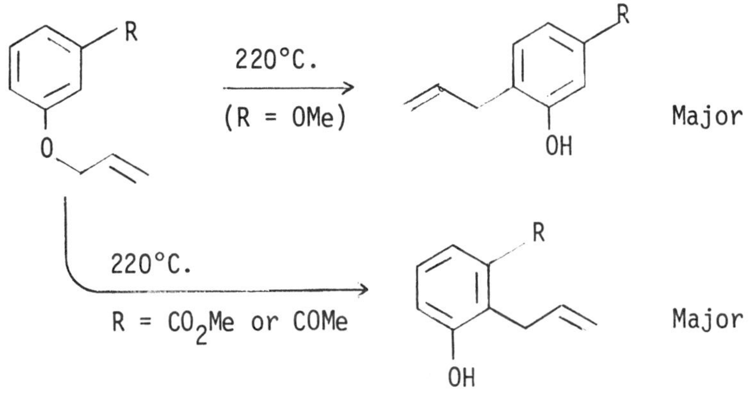

I.E.2-20 K. A. Parker and J. J. Petraitis, Tetrahedron Lett.,
22, 397 (1981).

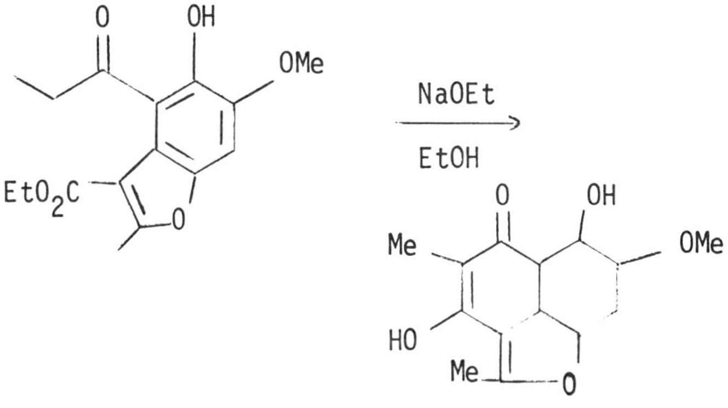

I.E.2-21 K. Oshima et al, Tetrahedron Lett., 22, 3985 (1981); T. Morimoto and M. Sekiya, Synthesis, 308 (1981).

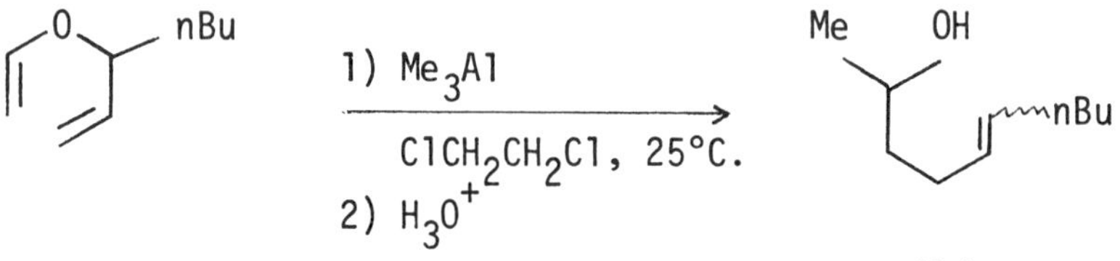

O
nBu
1) Me_3Al
$ClCH_2CH_2Cl$, 25°C.
2) H_3O^+
Me
OH
nBu

91%

(E:Z = 47:53)

I.E.2-22 T. Nakai, et al, Tetrahedron Lett., 22, 4097 (1981); Chem. Lett., 1289 (1981).

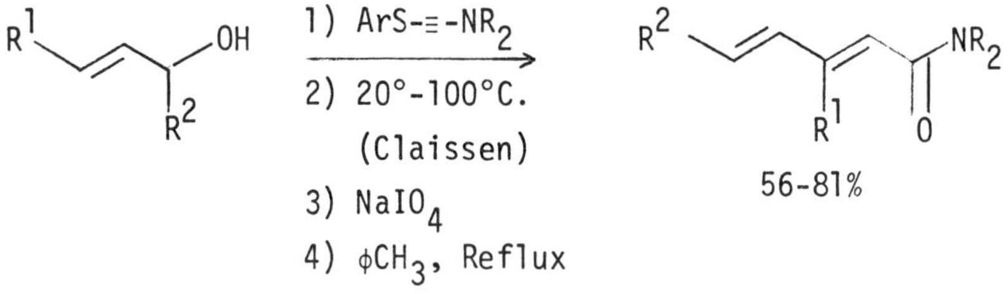

R^1
OH
R^2
1) ArS-≡-NR_2
2) 20°-100°C.
(Claissen)
3) $NaIO_4$
4) ϕCH_3, Reflux
R^2
NR_2
R^1
O
56-81%

I.E.2-23 J. E. McMurry et al, Tetrahedron, 37 (Suppl. 1), 319 (1981); J. Tsuji et al, J. Amer. Chem. Soc., 103, 5259 (1981).

O
NaO^{t}Pentyl (cat)
ϕCH_3, 220°C.
CHO

60%

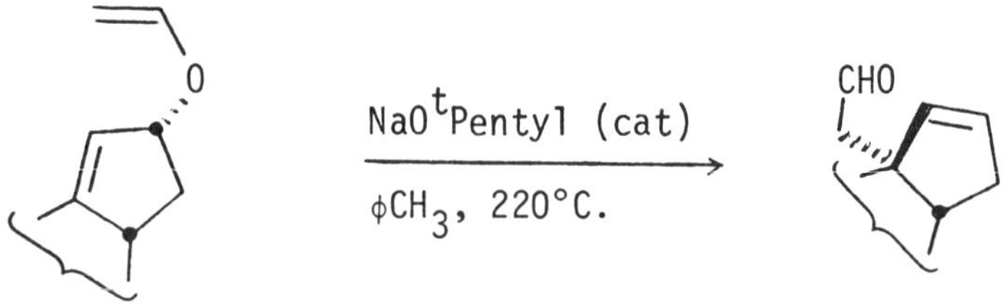

I.E.2-24 Y. Tamaru, M. Kagotani and Z. Yoshida, Tetrahedron Lett., 22, 4245 (1981); E. Nagashima, K. Suzuki and M. Sekiya, ibid, 2587.

$$CH_3-C=N-\phi \quad \xrightarrow[\text{1 hr.}]{200°C.} \quad CH_2=CHCH_2CH_2\overset{\overset{\displaystyle S}{\|}}{C}NH\phi$$
$$\underset{SCH_2CH=CH_2}{}$$

76%

I.E.2-25 D. Cooper and S. Trippett, J. Chem. Soc., Perkin I, 2127 (1981); J. C. Depezay and Y. Le Merrer, Bull. Soc. Chim. Fr. II, 435 (1981).

$$(MeO)_2\overset{\overset{\displaystyle O}{\|}}{P}\underset{\overset{|}{OH}}{C}HCH=CH\phi \quad \xrightarrow[\substack{CH_3CH_2CO_2H \text{ (cat)} \\ 130-175°C.}]{CH_3C(OEt)_3} \quad$$

$$(MeO)_2\overset{\overset{\displaystyle O}{\|}}{P} \diagup\!\!\diagdown\phi$$
$$EtO_2C$$

~ 50%

I.E.2-26 G. W. Daub et al, J. Org. Chem., 46, 1485 (1981); S. Raucher, J. E. Macdonald and R. F. Lawrence, J. Amer. Chem. Soc., 103, 2419 (1981).

$$R^1 \diagdown\!\!\diagup \overset{\diagup OH}{\underset{\diagdown R^2}{}} \quad + \quad \underset{CH_2OCH_3}{\overset{C(OMe)_3}{|}} \quad \xrightarrow[100°C.]{H^+}$$

$$R^1 \diagup\!\!\diagdown \underset{\overset{|}{R^2}}{} \overset{CO_2Me}{\diagdown} \!\!\sim\!\!\sim OMe$$

20-55%

I.E.2-27 J. P. Marino and M. P. Ferro, J. Org. Chem., 46, 1912 (1981).

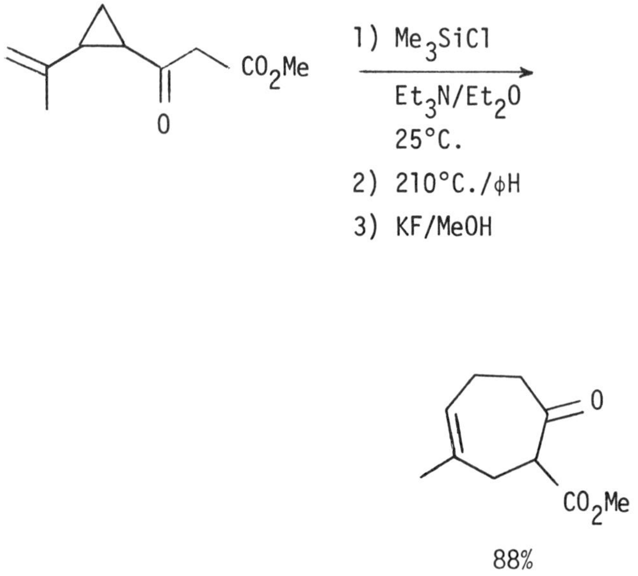
CO2Me
O
1) Me3SiCl
Et3N/Et2O
25°C.
2) 210°C./ϕH
3) KF/MeOH
O
CO2Me
88%

I.E.2-28 T. Nakai et al, J. Org. Chem., 46, 5447 (1981); Chem. Lett., 1721 (1981).

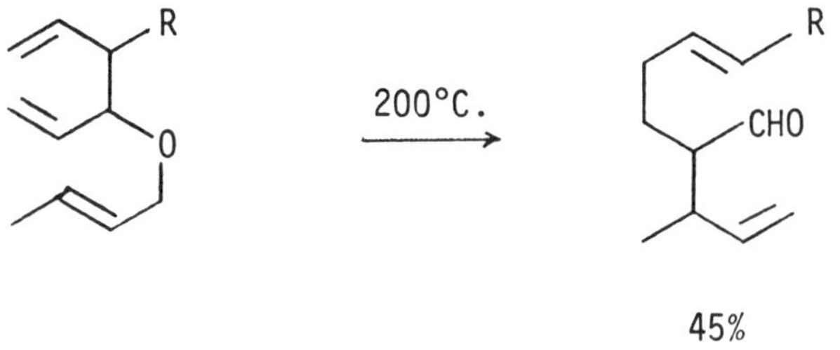
R
O
200°C.
R
CHO
45%

Tandem Oxy-Cope-Claissen Rearrangement.

I.E.2-29   T. Nakai et al, <u>J. Amer. Chem. Soc.</u>, <u>103</u>, 6492
(1981); <u>Tetrahedron Lett.</u>, <u>22</u>, 69 (1981).

1) nBuLi

THF, -85°C.

2) 0°C.

3) H_3O^+

OH

79%

I.E.2-30   W. D. Ollis et al, <u>J. Chem. Soc., Perkin I</u>, 1969,
2930 (1981);   R. Faragher, T. L. Gilchrist and I. W. Southon,
<u>ibid</u>, 2352.

+ NMe_2 Br^-

$\phi-\equiv$—

NaOMe

MeOH,

Reflux

ϕ—C—NMe_2

30%

I.E.2-31   G. S. Bates and S. Ramaswamy, <u>Can. J. Chem.</u>, <u>59</u>,
3120 (1981);   M. T. Zoeckler and B. K. Carpenter, <u>J. Amer.
Chem. Soc.</u>, <u>103</u>, 7661 (1981);   T. Toda et al, <u>Chem. Lett.</u>,
1535 (1981).

EtS   SEt

CHO

130-170°C.

EtS   SEt

CHO

Quantitative

I.E.2-32   A. Arcoleo et al, <u>Chem. Ind.</u>, 471 (1981).

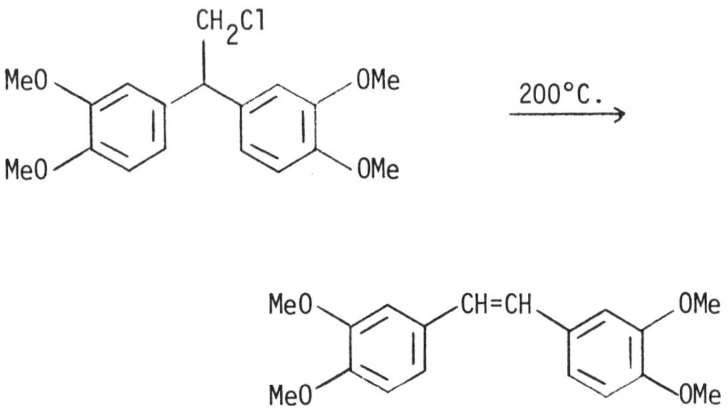

$$CH_2Cl$$

MeO —⟨ ⟩— CH —⟨ ⟩— OMe $\xrightarrow{200°C.}$

MeO   OMe

MeO —⟨ ⟩— CH=CH —⟨ ⟩— OMe

MeO   OMe

33%

I.F.   Aromatic Substitutions Forming a New Carbon-Carbon Bond

I.F.1.   Friedel-Crafts Type Aromatic Substitution Reactions

I.F.1-1   G. Casiraghi et al, <u>Synthesis,</u> 143 (1981);   A.
Burmester and H. B. Stegmann, <u>ibid,</u> 125.

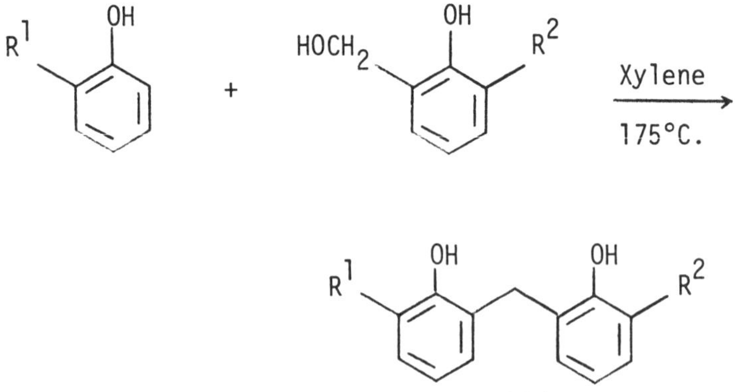

R^1 —⟨ ⟩— OH   + $HOCH_2$ —⟨ ⟩— R^2 $\xrightarrow[175°C.]{Xylene}$
                              OH

R^1 —⟨ ⟩— CH_2 —⟨ ⟩— R^2
       OH   OH

63-97%

I.F.1-2   T. Toyoda, K. Sasakura and T. Sugasawa, J. Org. Chem.,
46, 189 (1981);   V. R. Ranade, K. D. Deodhar and R. A. Kulkarni,
Ind. J. Chem., 20B, 500 (1981).

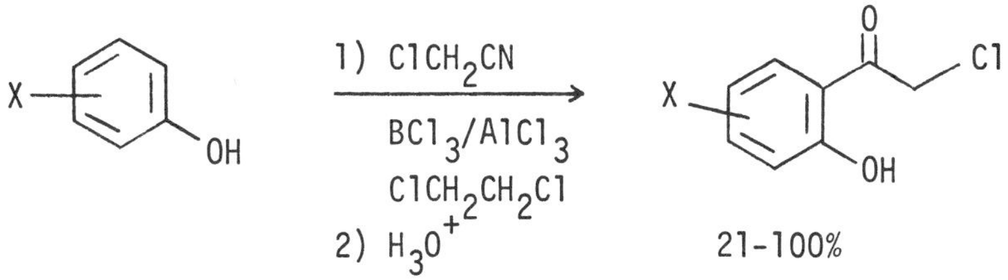

X = H, Cl, Me, OMe.

21-100%

I.F.1-3   A. S. Talmaki, G. P. Dhareshwar and B. D. Hosangadi,
Ind. J. Chem., 20B, 63 (1981);   G. G. Yakobson et al, J. Org.
Chem. (USSR), 17, 1340 (1981).

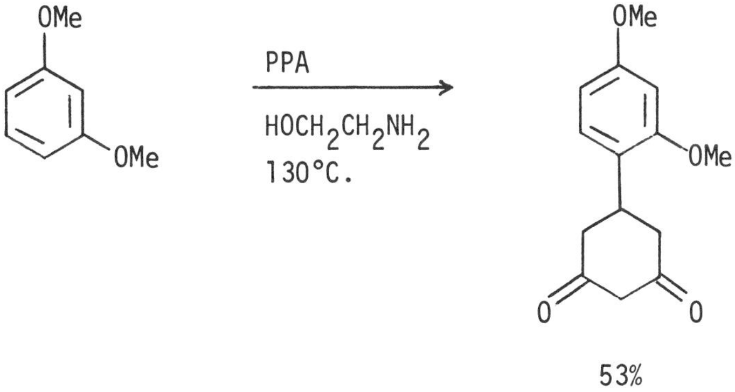

53%

I.F.1-4   F. Effenberger, G. Konig and H. Klenk, Chem. Ber., 926
(1981);   M. Tashiro et al, Org. Prep. Proc. Int., 13, 93 (1981).

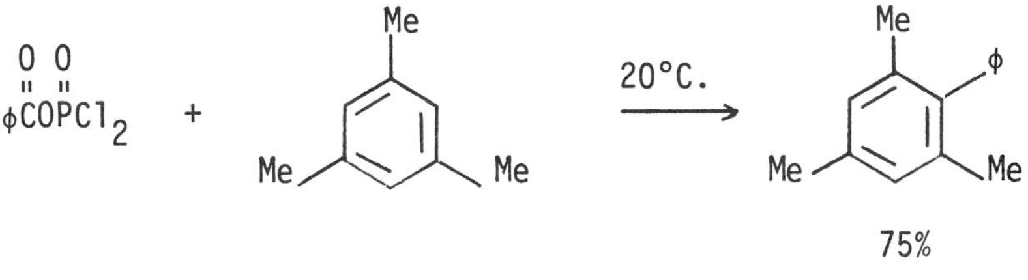

75%

I.F.1-5 Y. V. Pozdnyakovich, J. Org. Chem. (USSR), 17, 1386 (1981).

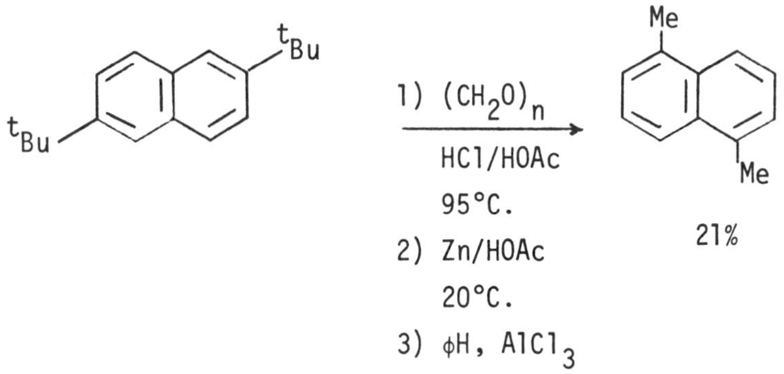

1) (CH₂O)ₙ
HCl/HOAc
95°C.
2) Zn/HOAc
20°C.
3) φH, AlCl₃

21%

I.F.1-6 R. Martin, Monat. Chem., 112, 1155 (1981).

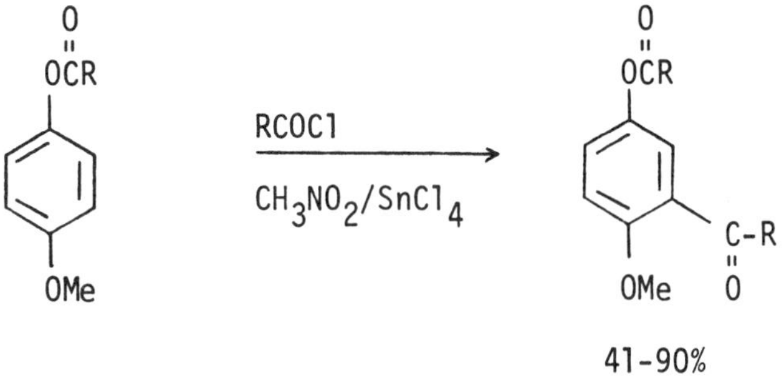

RCOCl
CH₃NO₂/SnCl₄

41-90%

I.F.1-7 G. A. Olah, M. R. Bruce and F. L. Clouet, J. Org. Chem., 46, 438 (1981).

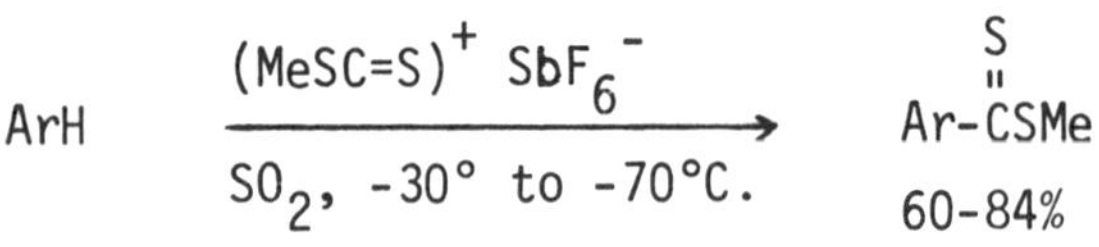

$$ArH \xrightarrow[SO_2, -30° \text{ to } -70°C.]{(MeSC=S)^+ SbF_6^-} Ar-\overset{S}{\overset{\|}{C}}SMe$$

60-84%

I.F.1-8 Y. Tamura et al, Tetrahedron Lett., 22, 81 (1981).

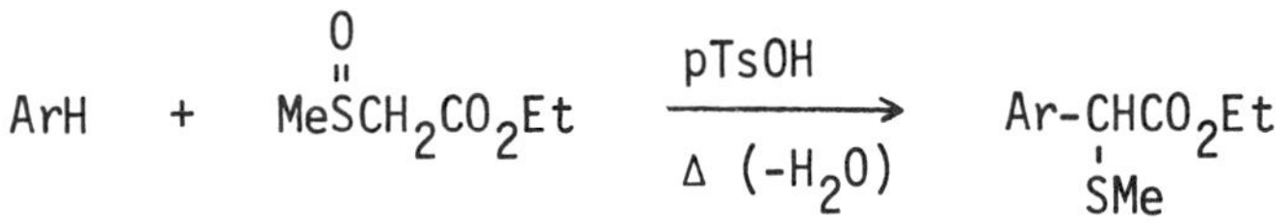

$$ArH \quad + \quad \overset{\overset{\text{O}}{\|}}{MeSCH_2CO_2Et} \quad \xrightarrow[\Delta \ (-H_2O)]{pTsOH} \quad Ar\text{-}\underset{\underset{SMe}{|}}{CH}CO_2Et$$

55-89%

I.F.1-9 S. Tanimoto et al, Bull. Chem. Soc. Jpn., 54, 2120 (1981).

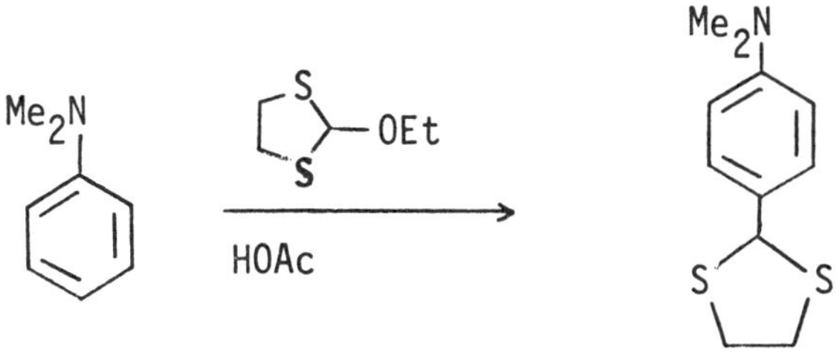

81%

I.F.1-10 D. P. Chakraborty, A. K. Mandal and S. K. Roy, Synthesis, 977 (1981).

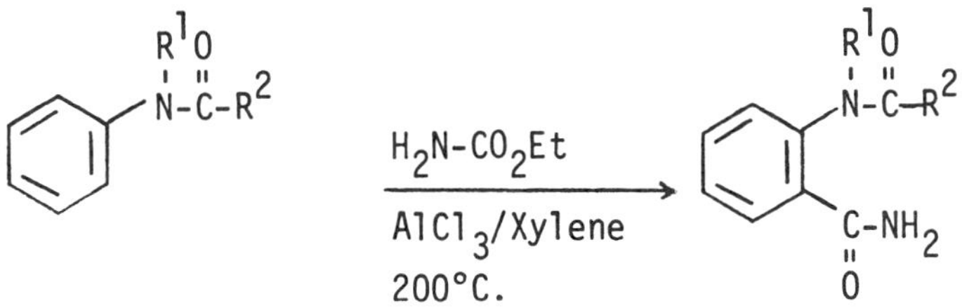

29-40%

I.F.1-11 M. A. El-Hashash et al, Synthesis, 798 (1981).

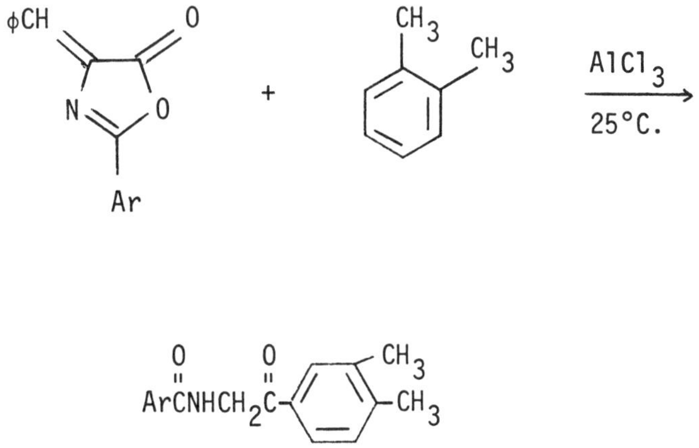

62%

I.F.1-12 G. Casiraghi et al, Synthesis, 310 (1981).

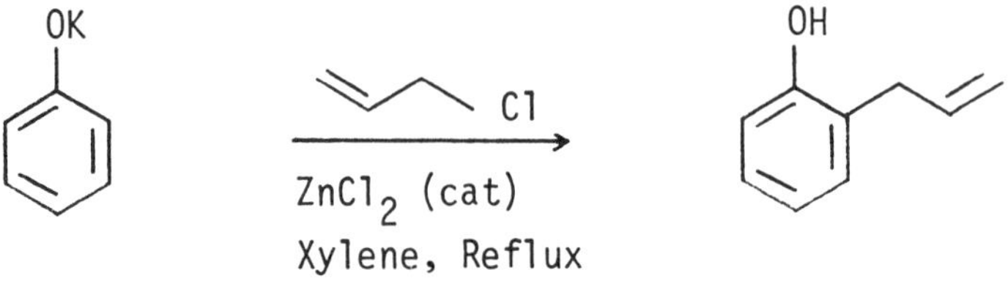

54%

I.F.1-13 P. G. Karmarkar, V. R. Chinchore and M. S. Wadia, Synthesis, 228 (1981); A. S. Dinge and S. K. Paknikar, Ind. J. Chem., 20B, 161 (1981).

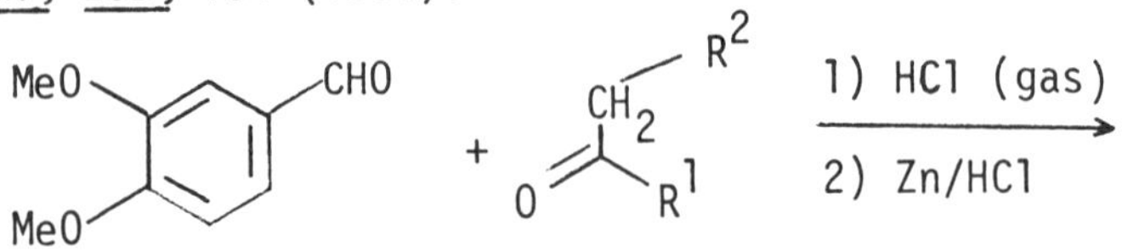

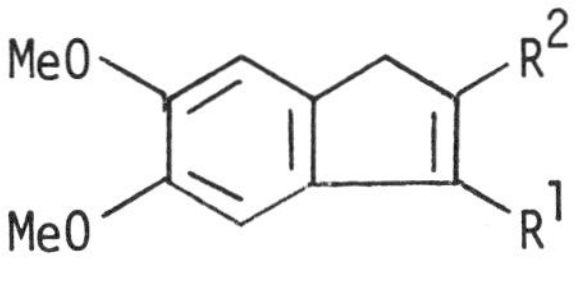

70-77%

I.F.1-14 H. W. Pinnick et al, J. Org. Chem., 46, 3758 (1981);
J. R. Merchant and R. B. Upasani, Ind. J. Chem., 20B, 241
(1981).

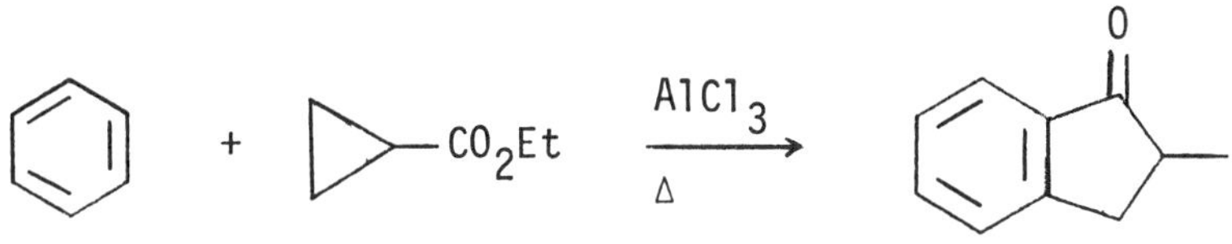

93%

I.F.1-15 A. Padwa, T. J. Blacklock and R. Loza, J. Amer. Chem.
Soc., 103, 2404 (1981); V. Premasagar, V. A. Palaniswamy and
E. J. Eisenbraun, J. Org. Chem., 46, 2974 (1981); S.
Bhattacharyya and D. Mukherjee, Synth. Commun., 11, 993 (1981);
M. O. Fatope and J. I. Okogun, Bull. Soc. Chim. Belg., 90, 847
(1981).

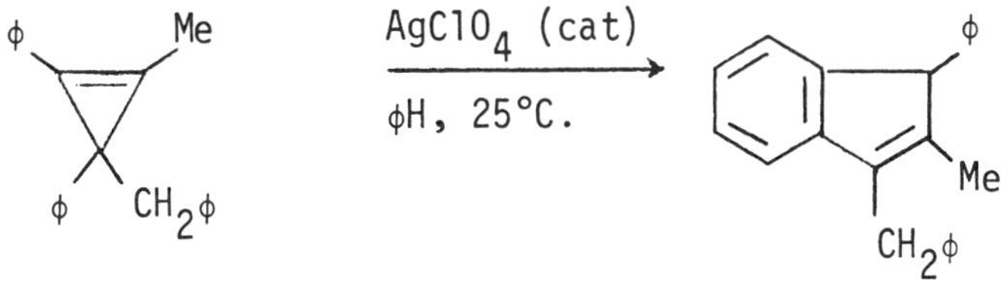

I.F.1-16 D. E. McClure et al, J. Org. Chem., 46, 2431 (1981).

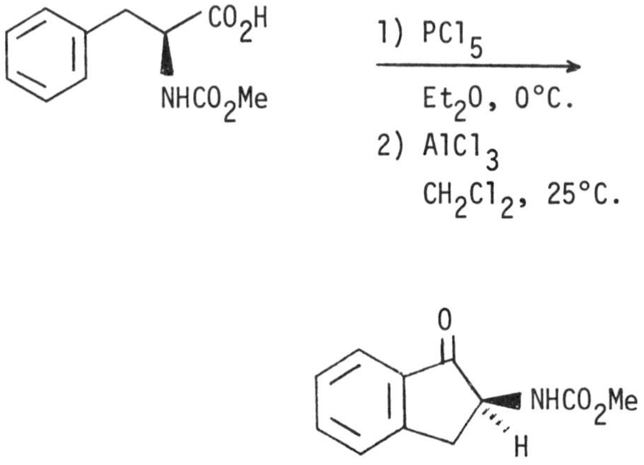
CO_2H
$NHCO_2Me$
1) PCl_5
Et_2O, 0°C.
2) $AlCl_3$
CH_2Cl_2, 25°C.
O
$NHCO_2Me$
H

55-75%

(≥98% ee)

I.F.1-17 R. Fusco and F. Sannicolo, J. Org. Chem., 46, 83, 90 (1981).

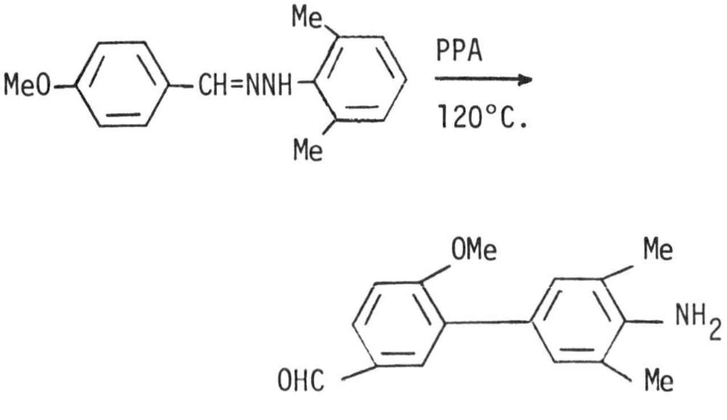
Me
MeO
CH=NNH
Me
PPA
120°C.
OMe
Me
NH_2
OHC
Me

I.F.1-18 D. A. Evans, S. P. Tanis and D. J. Hart, <u>J. Amer.</u>
<u>Chem. Soc.</u>, <u>103</u>, 5813 (1981); W. S. Murphy and S. Wattanasin,
<u>J. Chem. Soc., Perkin I</u>, 2920 (1981).

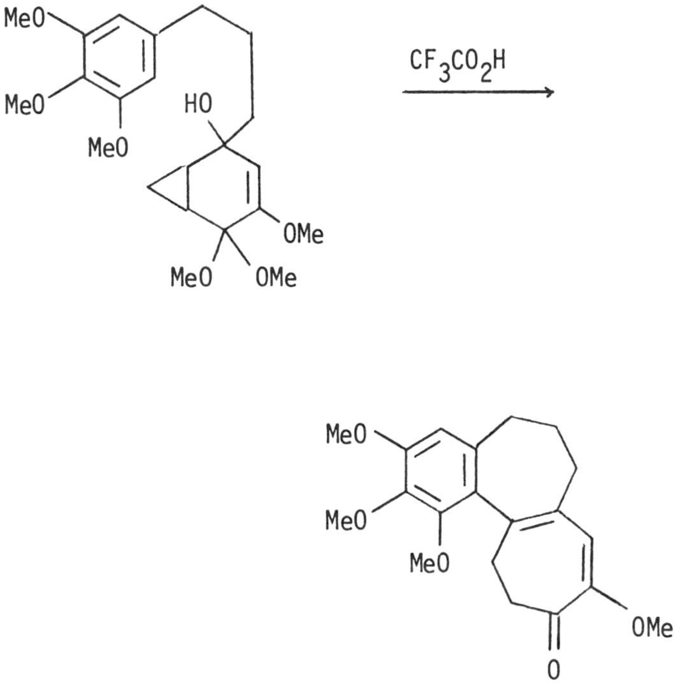

68%

I.F.1-19 P. N. Confalone and G. Pizzolato, <u>J. Amer. Chem. Soc.</u>
<u>103</u>, 4251 (1981); K. L. Platt and F. Oesch, <u>J. Org. Chem.</u>,
<u>46</u>, 2601 (1981).

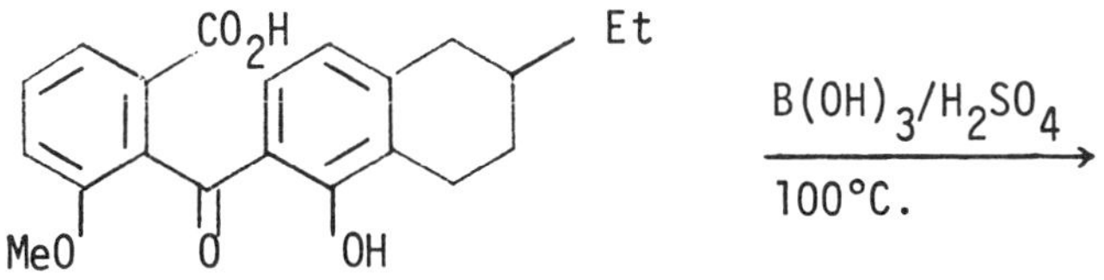

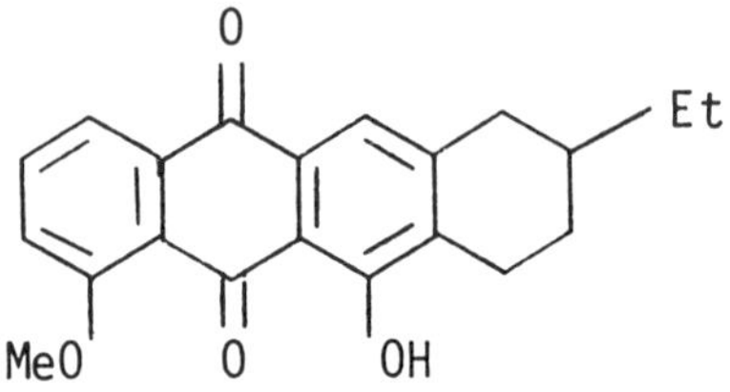

57%

I.F.1-20 B. W. Axon, B. R. Davis and P. D. Woodgate, J. Chem. Soc., Perkin I, 2956 (1981); K. D. Krautwurst and W. Tochtermann, Chem. Ber., 114, 214 (1981).

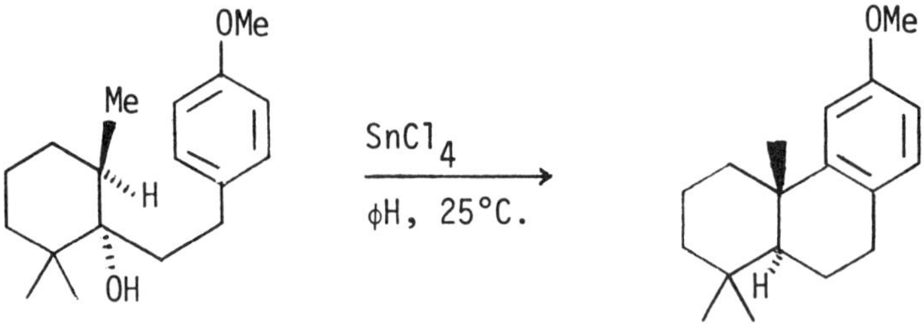

83%

I.F.1-21 E. E. van Tamelen et al, J. Amer. Chem. Soc., 103, 4615 (1981); S. K. Taylor et al, J. Org. Chem., 46, 2709 (1981).

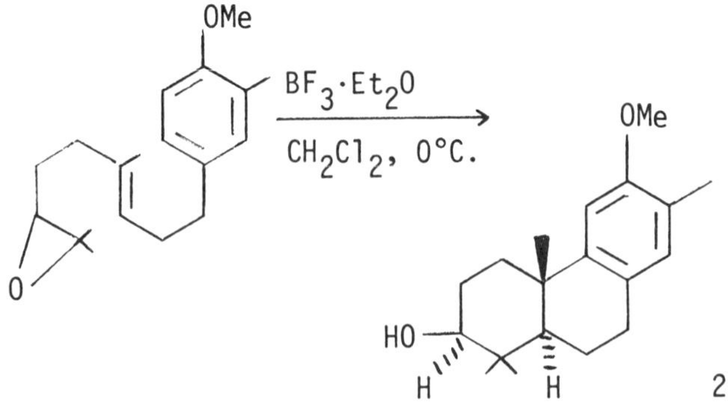

25-50%

I.F.1-22 E. Lee-Ruff, A. C. Hopkinson and L. H. Dao, <u>Can. J.</u>
<u>Chem.</u>, <u>59</u>, 1675 (1981); A. I. A. Broess, N. P. Van Vliet and
F. J. Zeelen, <u>J. Chem. Res. (S)</u>, 20 (1981).

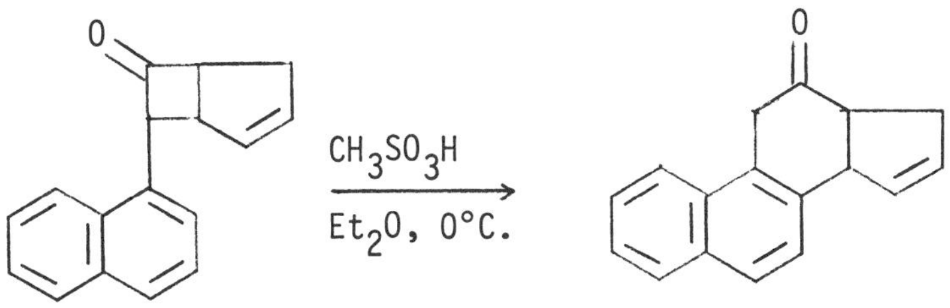

85%

I.F.2. Coupling Reactions to Form an Aromatic Carbon-Carbon
Bond

I.F.2-1 T. L. Ho, <u>Synth. Commun.</u>, <u>11</u>, 579 (1981); C.
Georgoulis et al, <u>Tetrahedron Lett.</u>, <u>22</u>, 2479 (1981).

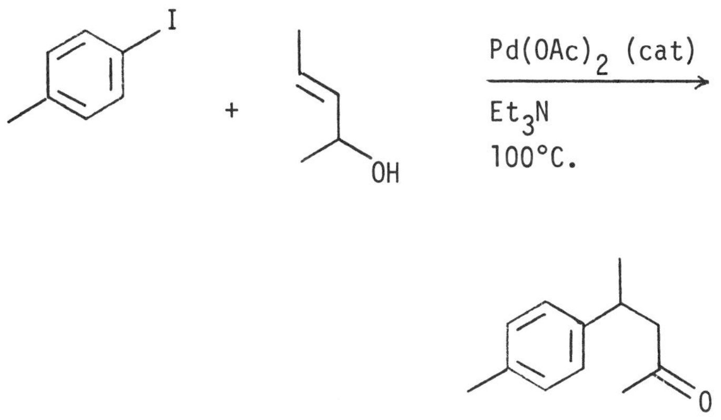

42%

I.F.2-2 A. N. Kashin et al, J. Org. Chem. (USSR), 17, 18
(1981); J. Setsune et al, Chem. Lett., 367 (1981); K. Matsui
et al, ibid, 1719.

$$RSnMe_3 \quad \xrightarrow[\substack{ArPdI(P\phi_3)_2 \ (cat) \\ ClCH_2CH_2Cl/130°C.}]{ArI} \quad \substack{R-Ar \\ \\ 76-96\%}$$

I.F.2-3 K. Sato, S. Inoue and K. Watanabe, J. Chem. Soc.,
Perkin I, 2411 (1981).

$$\xrightarrow[DMF]{\phi I}$$

51% (85% E)

I.F.2-4 M. Kumada et al, Tetrahedron Lett., 22, 4449 (1981).

$$Ar-O\overset{O}{\overset{\|}{P}}(OEt)_2 \quad \xrightarrow[Et_2O, \ 25°C.]{RM/Ni(acac)_2} \quad \substack{Ar-R \\ \\ 52-99\%}$$

RM = RMgX, ArMgX, R_3Al, $RCH=CHAl(R)_2$

I.F.2-5 M. Ochiai, M. Arimoto and E. Fujita, Tetrahedron Lett.,
22, 4491 (1981); T. Umemoto, Y. Kuriu and H. Shuyama, Chem.
Lett., 1663 (1981).

$$ArH \ + \ Me_3Si \diagup\diagdown\diagup \quad \xrightarrow[CH_2Cl_2, \ 0°C.]{Tl(OCOCF_3)_3} \quad Ar\diagup\diagdown\diagup$$

28-68%

I.F.2-6   T. Hirao et al, <u>Chem. Lett.</u>, 403 (1981).

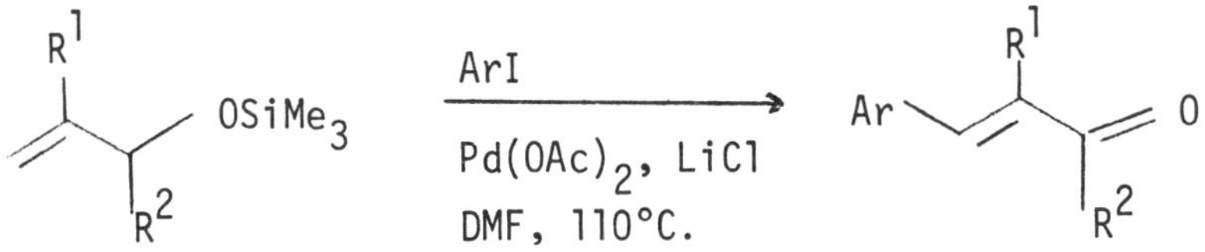

5-70% (GC)

I.F.2-7   N. Miyaura and A. Suzuki, <u>J. Organometal Chem.</u>, <u>213</u>, C53 (1981).

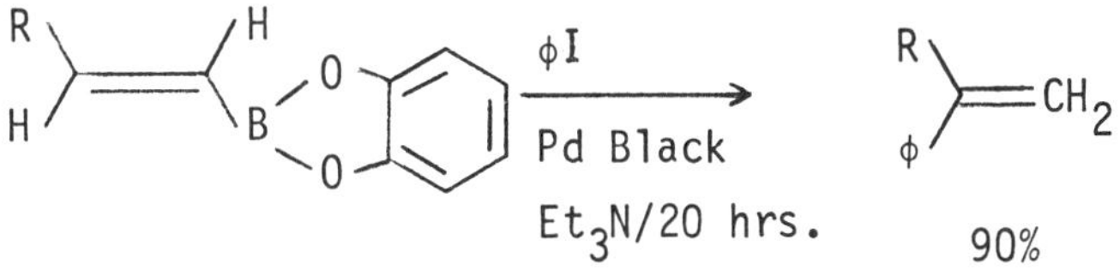

90%

I.F.2-8   P. Y. Johnson and J. Q. Wen, <u>J. Org. Chem.</u>, <u>46</u>, 2767 (1981);   T. Fuchikami, M. Yatabe and I. Ojima, <u>Synthesis</u>, 365 (1981).

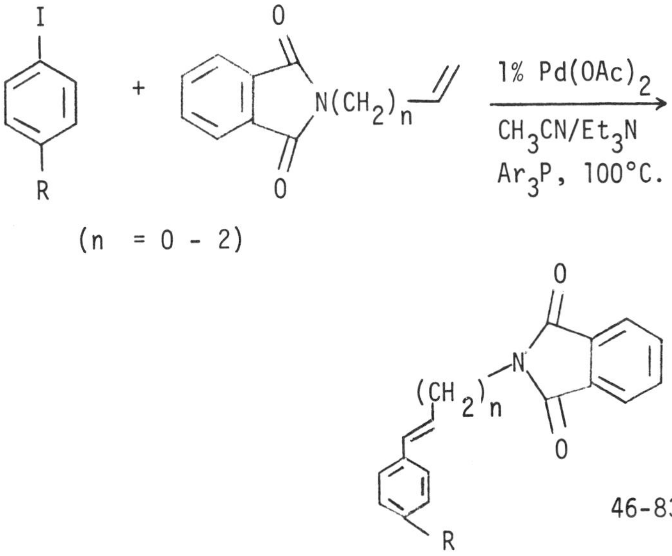

(n = 0 - 2)

46-83%

I.F.2-9   N. Jabri, A. Alexakis and J. F. Normant, <u>Tetrahedron</u>
<u>Lett.</u>, <u>22</u>, 3851 (1981);   S. Nunomoto, Y. Kawakami and Y.
Yamashita, <u>Bull. Chem. Soc. Jpn.</u>, <u>54</u>, 2831 (1981).

$$\left(\underset{R}{\overset{H}{\diagdown}} = \underset{}{\overset{}{\diagup}} \underset{}{\overset{H}{}} \right)_2 CuLi \quad \xrightarrow[\substack{2)\ 5\%\ (\phi_3P)_4Pd \\ 3)\ ArI}]{1)\ ZnBr_2/THF} \quad \underset{R}{\overset{H}{\diagdown}} = \underset{Ar}{\overset{H}{\diagup}}$$

65-80%

I.F.2-10   T. Matsuda et al, <u>J. Org. Chem.</u>, <u>46</u>, 4885 (1981);
T. Matsuda et al, <u>Tetrahedron</u>, <u>37</u>, 31 (1981).

$$ArNH_2 \quad + \quad CH_2=CH\phi \quad \xrightarrow[\substack{t_{BuONO} \\ HOAc/ClCH_2CO_2H \\ 50°C.}]{Pd(dba)_2} \quad ArCH=CH\phi$$

46-84%

dba = dibenzylideneacetone.

I.F.2-11   B. J. Brisdon, P. Nair and S. F. Dyke, <u>Tetrahedron</u>,
<u>37</u>, 173 (1981);   H. Horino and N. Inoue, <u>J. Org. Chem.</u>, <u>46</u>,
4416 (1981).

1) AcOH/CH$_2$Cl$_2$
→
Et$_3$N, 0°C.

2) ϕCH=CH$_2$

94%

I.F.2-12 T. Itahara, <u>Che. Commun.</u>, 859 (1981).

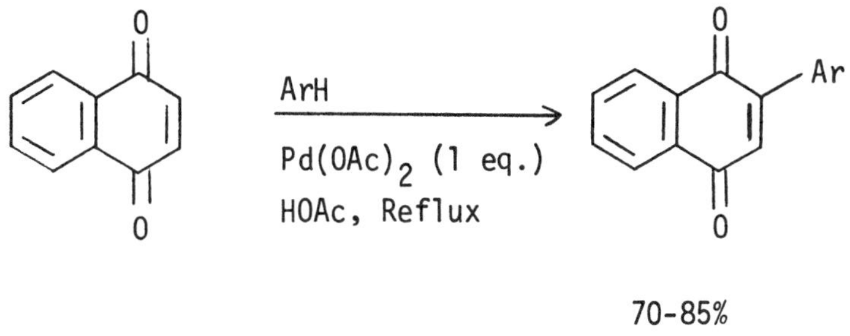

70-85%

I.F.2-13 L. L. Miller et al, <u>J. Org. Chem.</u>, <u>46</u>, 4545 (1981); B. Feringa and H. Wynberg, <u>ibid</u>, 2547; I. V. Berezin et al, <u>Tetrahedron Lett.</u>, <u>22</u>, 3793 (1981); S. Miyano, M. Tobita and H. Hashimoto, <u>Bull. Chem. Soc. Jpn.</u>, <u>54</u>, 3522 (1981).

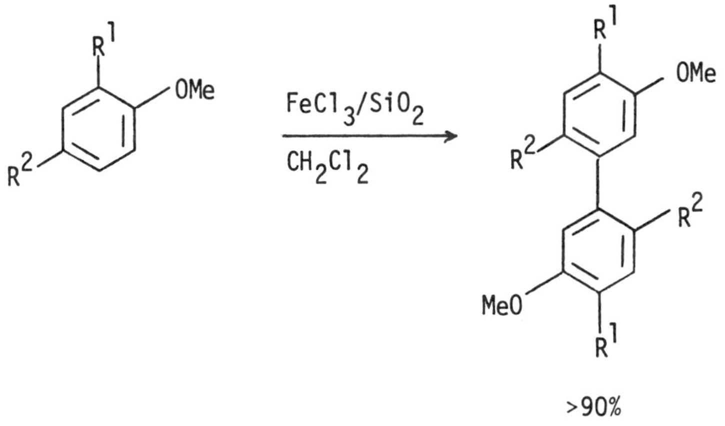

>90%

I.F.2-14 N. Miyaura, T. Yanagi and A. Suzuki, <u>Synth. Commun.</u>, <u>11</u>, 513 (1981).

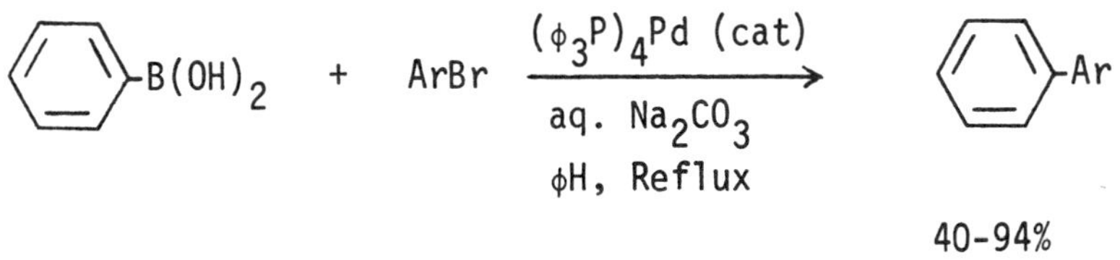

40-94%

I.F.2-15   S. K. Taylor et al, J. Org. Chem., 46, 2194 (1981).

$$2 \ ArMgX \quad \xrightarrow[\text{Et}_2\text{O, Reflux}]{\text{ClCH}_2\text{CH=CHCH}_2\text{Cl}} \quad Ar\text{-}Ar$$

I.F.2-16   R. A. Holton and K. J. Natalie, Jr., Tetrahedron Lett., 22, 267 (1981); A. Hallberg, L. Westfelt and B. Holm, J. Org. Chem., 46, 5414 (1981).

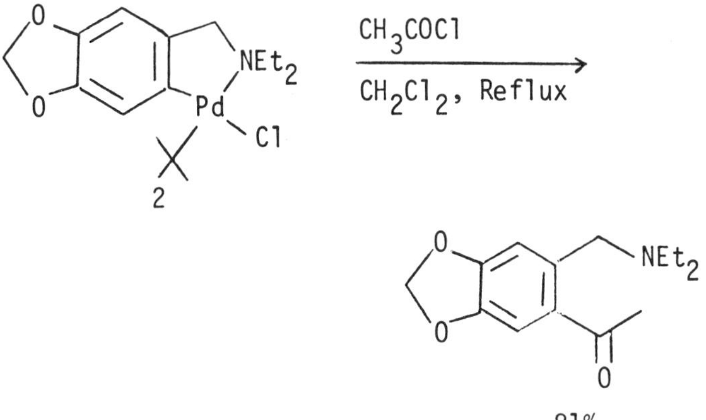

I.F.2-17   K. S. Y. Lau et al, J. Org. Chem., 46, 2280 (1981); D. E. Ames, D. Bull and C. Takundwa, Synthesis, 364 (1981).

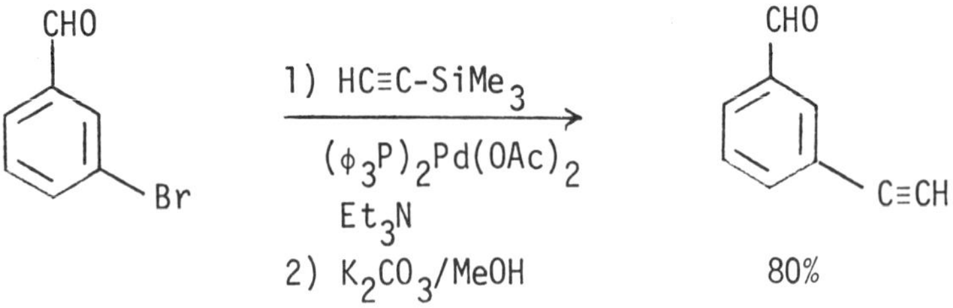

I.F.3. Other Aromatic Substitutions.

I.F.3-1 P. A. Wender, J. M. Erhardt and L. J. Letendre, <u>J.</u>
<u>Amer. Chem. Soc.</u>, <u>103</u>, 2114 (1981).

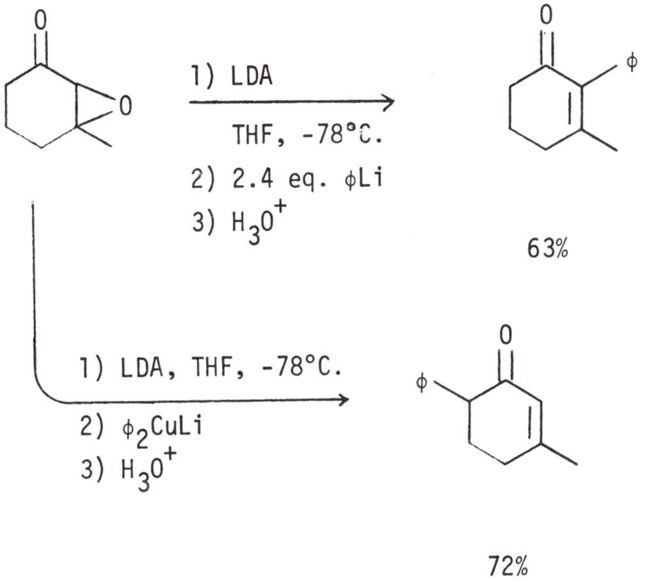

1) LDA

THF, -78°C.

2) 2.4 eq. ϕLi

3) H_3O^+

63%

1) LDA, THF, -78°C.

2) ϕ_2CuLi

3) H_3O^+

72%

I.F.3-2 G. Hanson and D. S. Kemp, <u>J. Org. Chem.</u>, <u>46</u>, 5441
(1981); T. Gungor, F. Marsais and G. Quenguiner, <u>J. Organo-</u>
<u>metal. Chem.</u>, <u>215</u>, 139 (1981); G. Queguiner et al, ibid, <u>216</u>,
139; R. Perez-Ossorio et al, <u>J. Chem. Soc.</u>, Perkin <u>II</u>, 597
(1981).

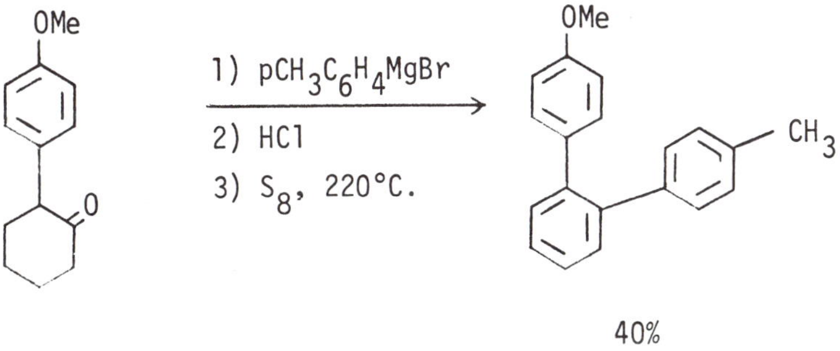

1) $pCH_3C_6H_4MgBr$

2) HCl

3) S_8, 220°C.

40%

I.F.3-3 V. N. Kashinskii, V. A. Demanov and I. I. Lapkin, J. Org. Chem. (USSR), 17, 72 (1981).

$$\underset{RCH_2\overset{O}{\overset{\|}{C}}-\overset{O}{\overset{\|}{C}}OCH_2R}{} \quad \xrightarrow[\text{2) ClCH}_2\text{OCH}_3]{\text{1) ArMgX}} \quad \underset{\substack{ArCHCO_2CH_2R \\ 36\text{-}58\%}}{\overset{OCH_2OCH_3}{|}}$$

I.F.3-4 M. Kumada et al, Chem. Commun., 313 (1981).

$$A = \text{(structure)}$$

$$B = R\diagup\!\!\diagdown\!\!\diagup\phi$$

I.F.3-5 J. C. Saddler and P. L. Fuchs, J. Amer. Chem. Soc., 103, 2112 (1981); J. S. R. Zilenovski and S. S. Hall, J. Org. Chem., 46, 4139 (1981); K. Koga et al, Tetrahedron, 37, 3951 (1981).

1) 2 eq. φLi

THF, -78°C.

2) NH₄Cl

3) H₂Cr₂O₇/Et₂O

4) NaOH/CH₂Cl₂

43%

I.F.3-6 W. Oppolzer and H. J. Loher, Helv. Chim. Acta, 64, 2808 (1981).

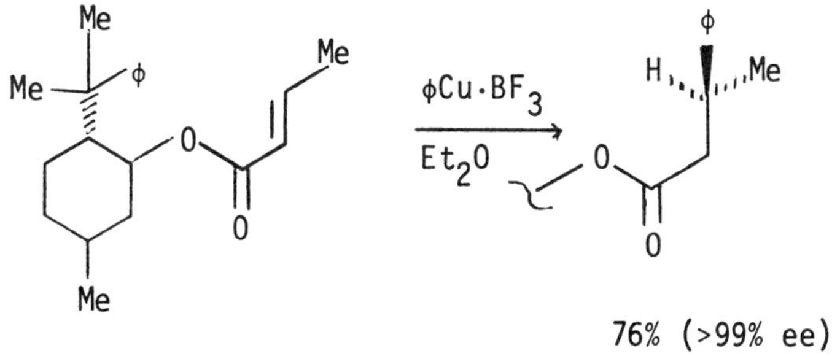

76% (>99% ee)

I.F.3-7 J. K. Sutherland, et al, Chem. Commun., 740, 1075 (1981); D. J. Mincher and G. Shaw, ibid, 508; K. Hirota, Y. Kitade and S. Senda, J. Org. Chem., 46, 3949 (1981).

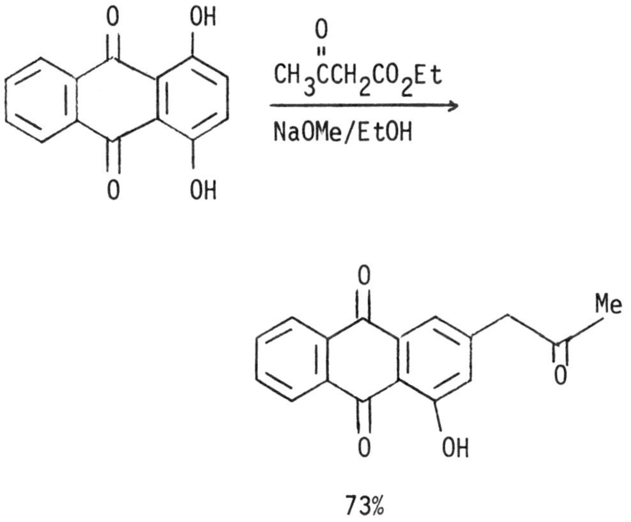

73%

I.F.3-8   C. A. Panetta and A. S. Dixit, Synthesis, 59 (1981);
H. Hommes, H. D. Verkruijsse and L. Brandsma, Chem. Commun.,
366 (1981);   C. A. Townsend and L. M. Bloom, Tetrahedron Lett.,
22, 3923 (1981).

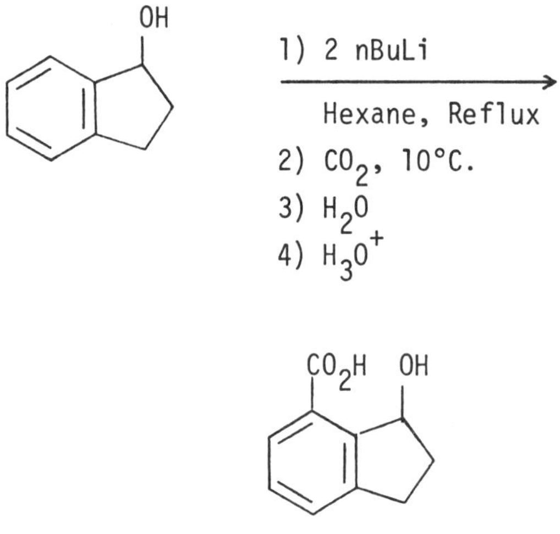

1) 2 nBuLi
────────────────→
   Hexane, Reflux
2) CO_2, 10°C.
3) H_2O
4) H_3O^+

28%

───────────────────────

I.F.3-9   V. Snieckus et al, Tetrahedron Lett., 22, 2349 (1981);
A. I. Meyers and W. B. Avila, J. Org. Chem., 46, 3881 (1981).

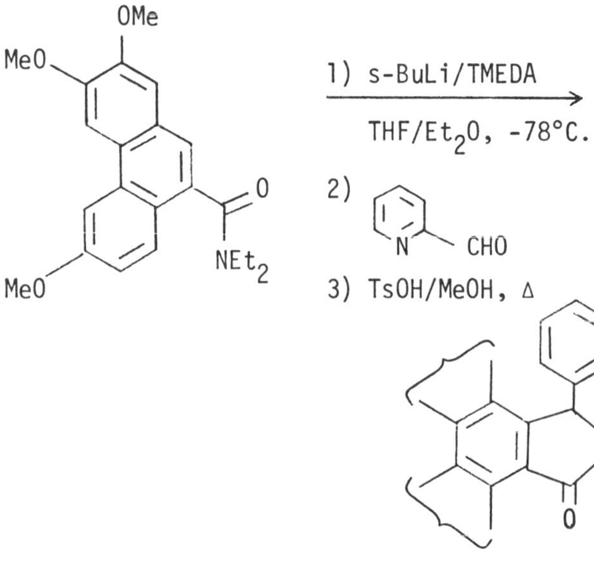

1) s-BuLi/TMEDA
────────────────────→
   THF/Et_2O, -78°C.
2)

3) TsOH/MeOH, Δ

I.F.3-10 M. S. Newman and V. G. Lee, Org. Prep. Proc. Int., 13, 426 (1981).

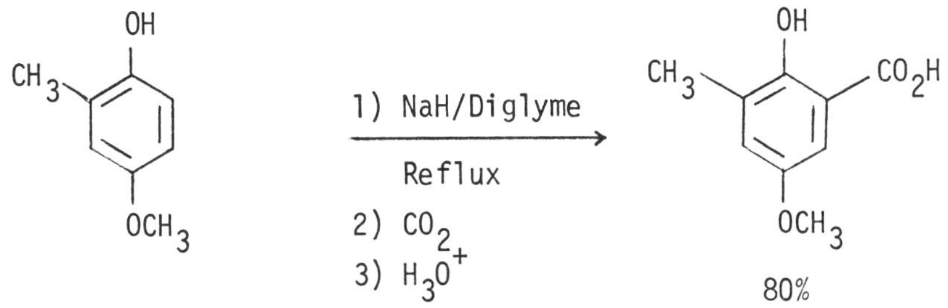
OH
CH3
OCH3
1) NaH/Diglyme
Reflux
2) CO2
3) H3O+
OH
CH3
CO2H
OCH3
80%

I.F.3-11 A. S. Kende et al, Tetrahedron Lett., 22, 1779 (1981); J. Org. Chem., 46, 2799 (1981); J. Amer. Chem. Soc., 103, 4247 (1981).

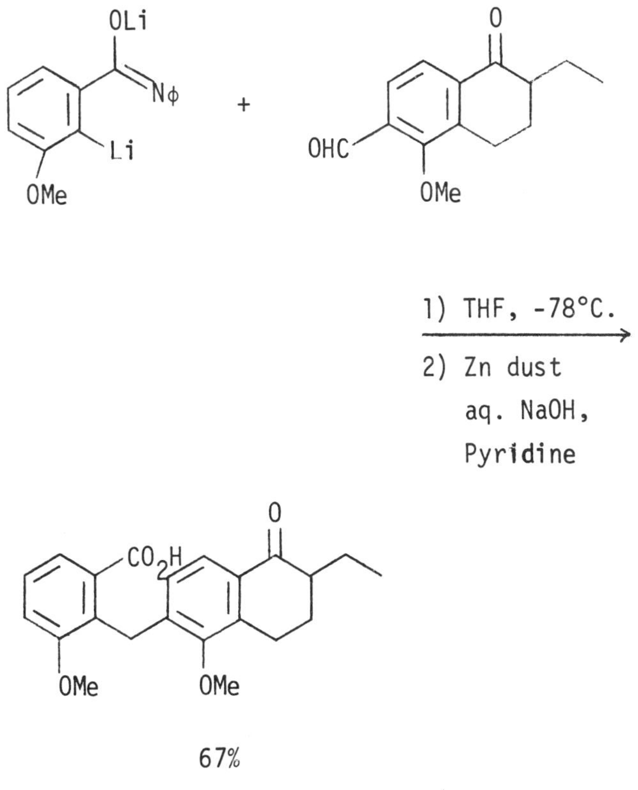
OLi
Nϕ
Li
OMe
+
O
OHC
OMe
1) THF, -78°C.
2) Zn dust
aq. NaOH,
Pyridine
O
CO2H
OMe
OMe
67%

I.F.3-12   J. S. Swenton et al, J. Org. Chem., 46, 4825 (1981);
M. Braun, Justus Liebigs Ann. Chem., 2247 (1981).

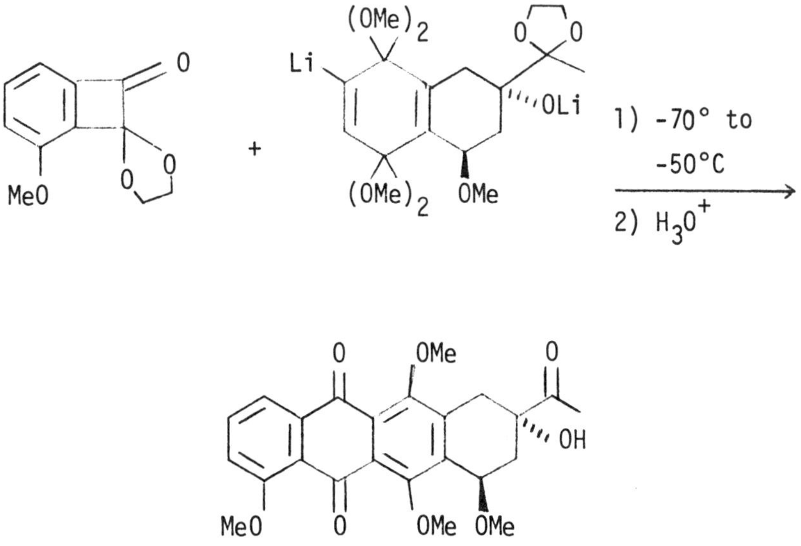

1) -70° to
   -50°C
2) H₃O⁺

56%

I.F.3-13   P. Helquist, P. L. Fuchs et al, J. Org. Chem., 46,
118 (1981); W. E. Parham, C. K. Bradsher and K. J. Edgar,
ibid, 1057.

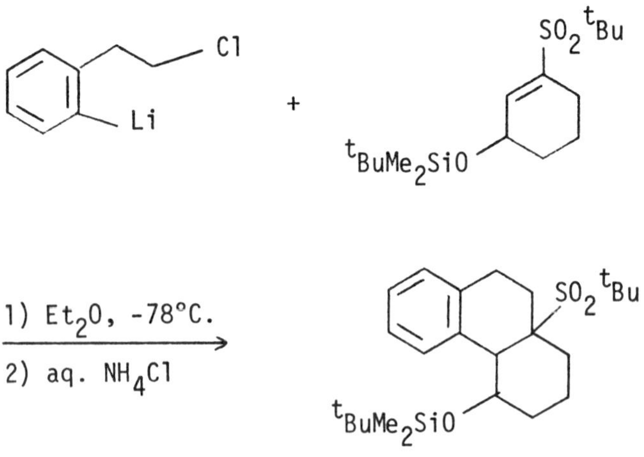

1) Et₂O, -78°C.
2) aq. NH₄Cl

78%

I.F.3-14   S. A. Jacobs, C. Cortez and R. G. Harvey, <u>Chem.</u>
<u>Commun.</u>, 1215 (1981).

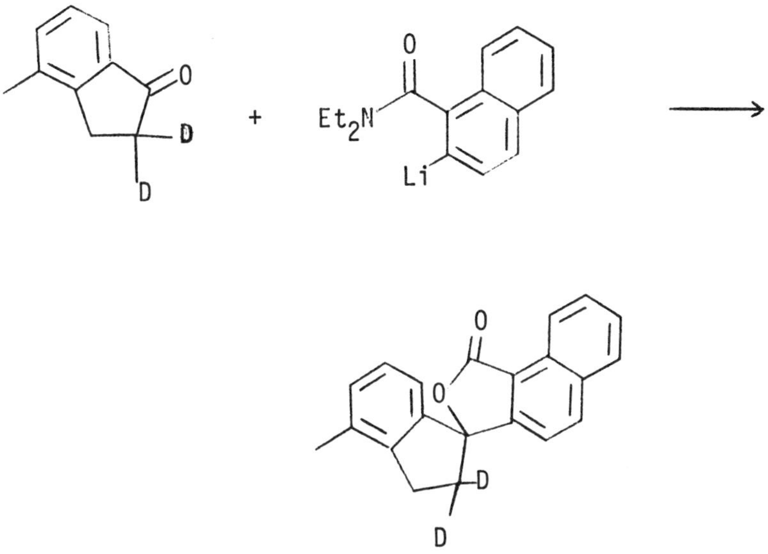

52%
(30% with α-Hydrogens)

I.F.3-15   G. R. Martinez, P. A. Grieco and C. V. Srinivasan,
J. Org. Chem., 46, 3760 (1981);   D. R. Schumacher and S. S.
Hall, ibid, 5060.

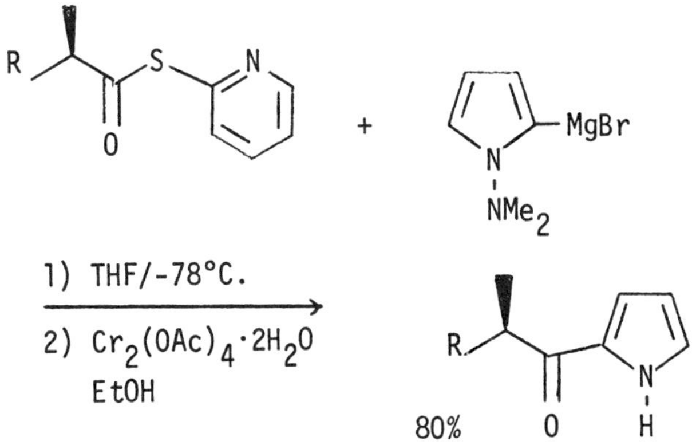

1) THF/-78°C.

2) Cr₂(OAc)₄·2H₂O
   EtOH

80%

I.F.3-16 A. I. Meyers, M. Reuman and R. A. Gabel, J. Org.
Chem., 46, 783 (1981); W. E. Parham, C. K. Bradsher and D. C.
Reames, ibid, 4804; T. Durst et al, ibid, 2730; C. K.
Bradsher et al, ibid, 4600, 4608.

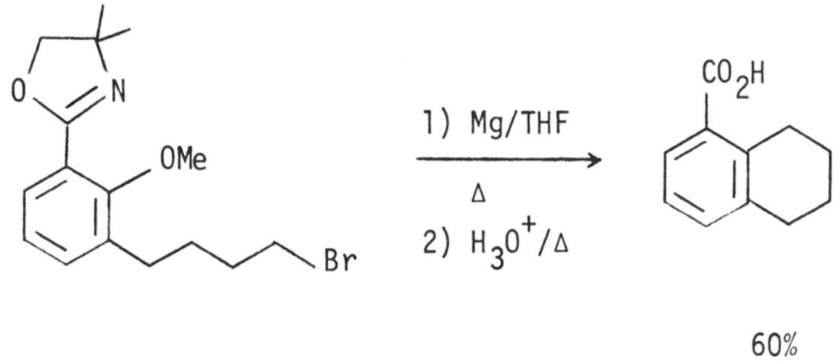

1) Mg/THF
$\xrightarrow{\Delta}$
2) H_3O^+/Δ

60%

Also, annulations to give benzo-fused heterocyclic
systems.

I.F.3-17 A. P. S. Narula and D. I. Schuster, Tetrahedron
Lett., 22, 3707 (1981); A. S. Kende, M. L. King and D. P.
Curran, J. Org. Chem., 46, 2826 (1981).

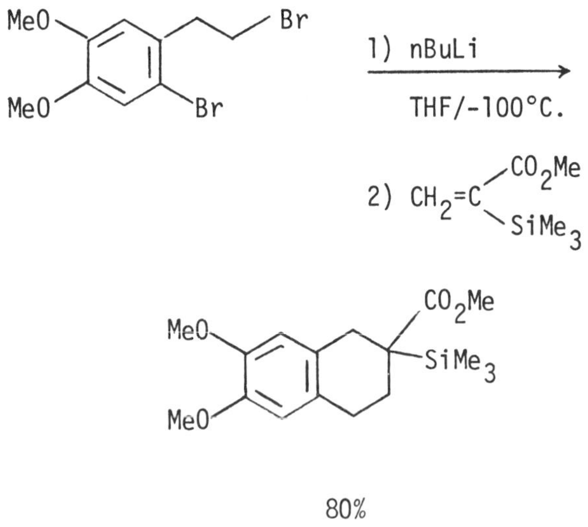

1) nBuLi
$\xrightarrow{\text{THF}/-100°C.}$

2) $CH_2=C\begin{smallmatrix}CO_2Me\\SiMe_3\end{smallmatrix}$

80%

I.F.3-18   D. L. Comins and J. D. Brown, Tetrahedron Lett., 22, 4213 (1981); M. E. K. Cartoon and G. W. H. Cheeseman, J. Organometal. Chem., 212, 1 (1981).

CHO

1) Li-N O

THF, -78°C.

2) nBuLi, -78°C.

3) CH₃I

4) H₃O⁺

CHO

CH₃

90%

I.F.3-19   R. Arshady, Chem. Ind., 250 (1981); L. S. Chen, G. J. Chen and C. Tamborski, J. Organometal. Chem., 215, 281 (1981).

Br

Br

1) Mg (1 eq.)

Et₂O

2) CH₃CHO

3) aq. NH₄Cl

4) KHSO₄, 185°C.

   t-Butyl Catechol

Br

50%

I.F.3-20   M. J. Gunter and L. N. Mander, Austr. J. Chem., 34, 675 (1981); L. N. Mander and J. V. Turner, Tetrahedron Lett., 22, 3683 (1981).

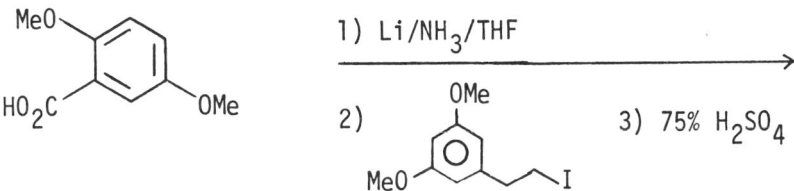

1) Li/NH₃/THF

2)

3) 75% H₂SO₄

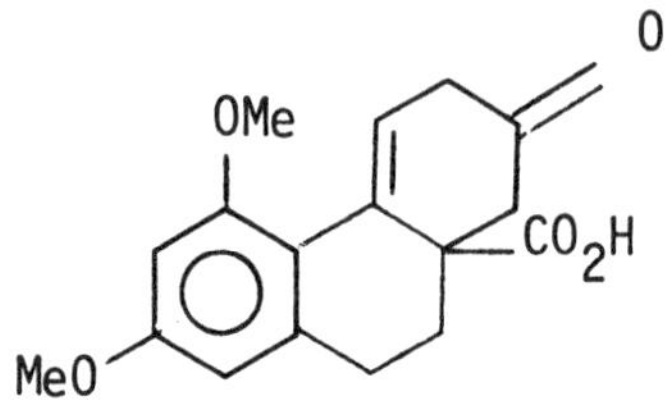

I.F.3-21   J. Cacchi, D. Misiti and G. Palmieri, Tetrahedron,
37, 2941 (1981).

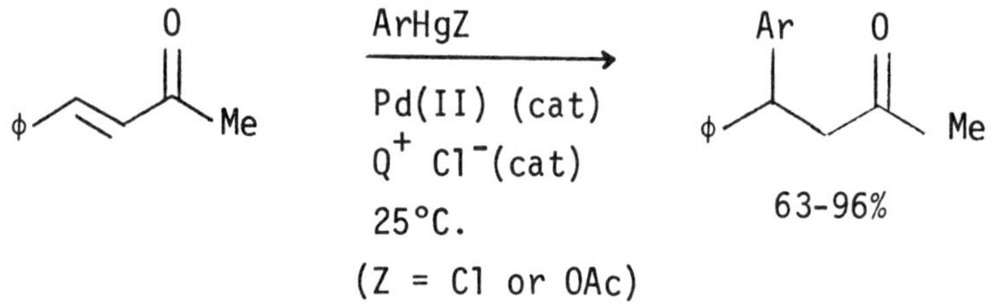

I.F.3-22   H. G. Adolph et al, J. Chem. Soc., Perkin I, 1815
(1981);   R. P. Kozyrod and J. T. Pinhey, Tetrahedron Lett., 22,
783 (1981).

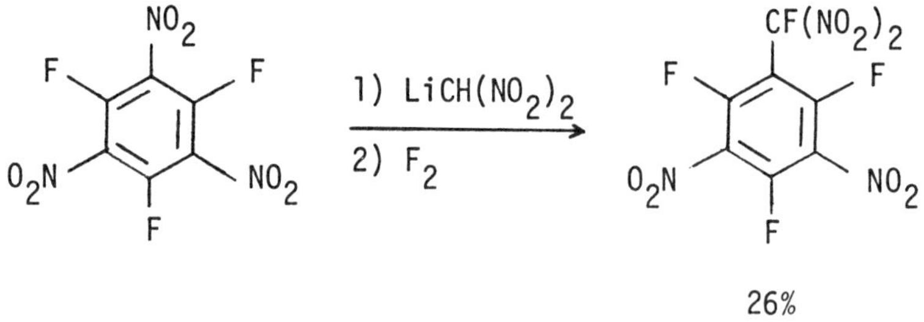

I.F.3-23 R. A. Russell and R. N. Warrener, Chem. Commun., 108 (1981); F. M. Hauser and S. Prasanna, J. Amer. Chem. Soc., 103, 6378 (1981); T. Li and Y. L. Wu, ibid, 7007; C. A. Townsend et al, ibid, 6885; D. J. Dodsworth, P. G. Sammes et al, J. Chem. Soc., Perkin I, 2120 (1981).

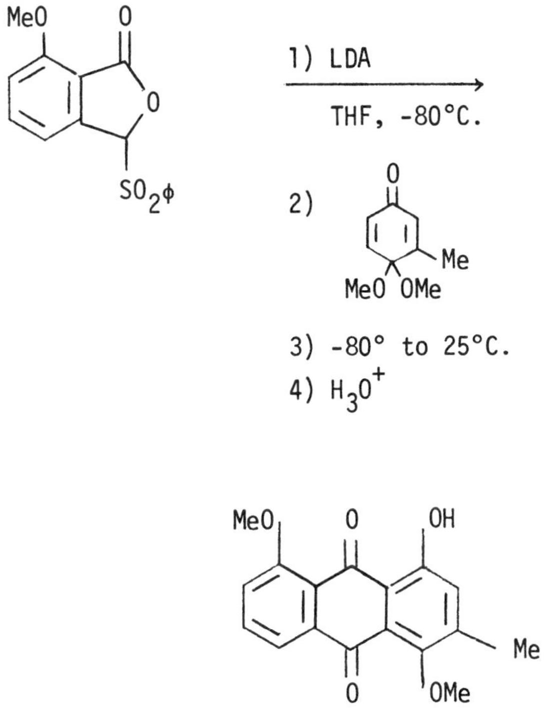

MeO
O
O
SO2ϕ
1) LDA
THF, -80°C.
2)
O
Me
MeO OMe
3) -80° to 25°C.
4) H3O+
MeO
O
OH
Me
O
OMe

55%

I.F.3-24 K. Maruyama et al, Chem. Lett., 47 (1981); Y. Naruta, H. Uno and K. Maruyama, Chem. Commun., 1277 (1981); Y. Naruta, H. Uno and K. Maruyama, Tetrahedron Lett., 22, 5221 (1981).

Me
O
O
1) Bu3Sn
BF3·Et2O/CH2Cl2
-78°C.
2) H3O+

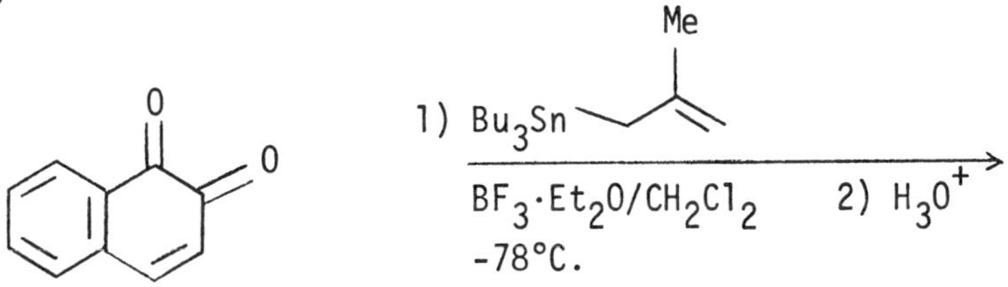

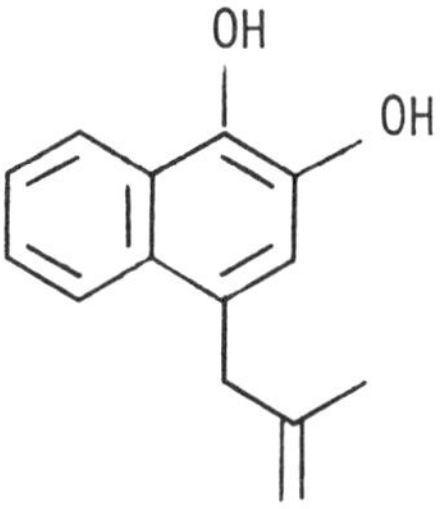

78%

I.F.3-25 N. Arumugam and A. Kumaraswamy, Synthesis, 367
(1981); K. B. Sukumaran and R. G. Harvey, J. Org. Chem., 46,
2740 (1981).

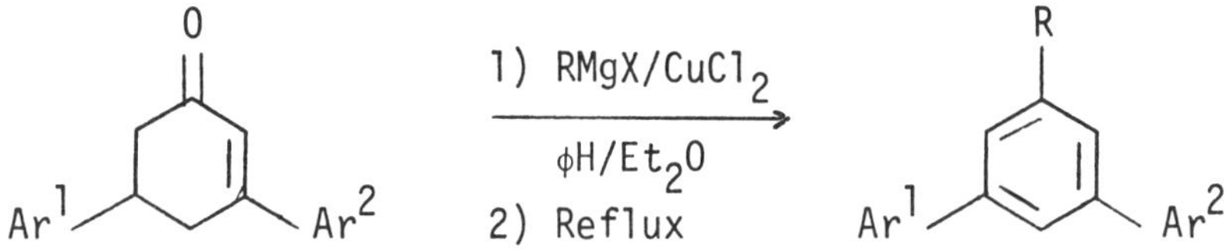

70-79%

I.F.3-26 K. Steinbeck, T. Schenke and J. Runsink, Chem. Ber.,
114, 1836 (1981); G. W. Gribble et al, J. Org. Chem., 46, 1025
(1981).

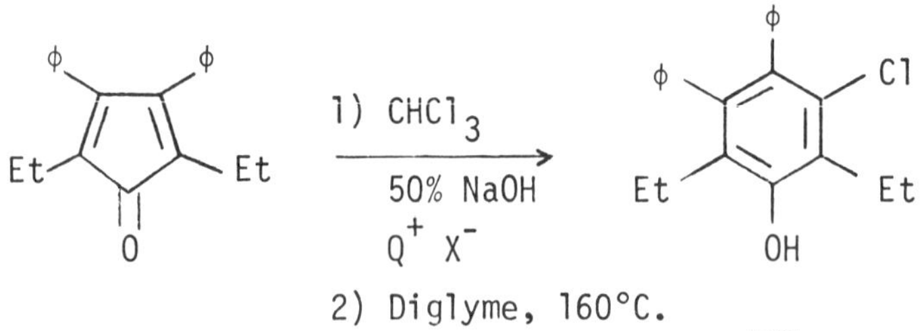

56%

I.F.3-27 H. Lee, N. Shyamasundar and R. G. Harvey, <u>Tetrahe-</u>
<u>dron</u>, <u>37</u>, 2563 (1981); M. P. Reddy and G. S. K. Rao, <u>J. Org.</u>
<u>Chem.</u>, <u>46</u>, 5371 (1981).

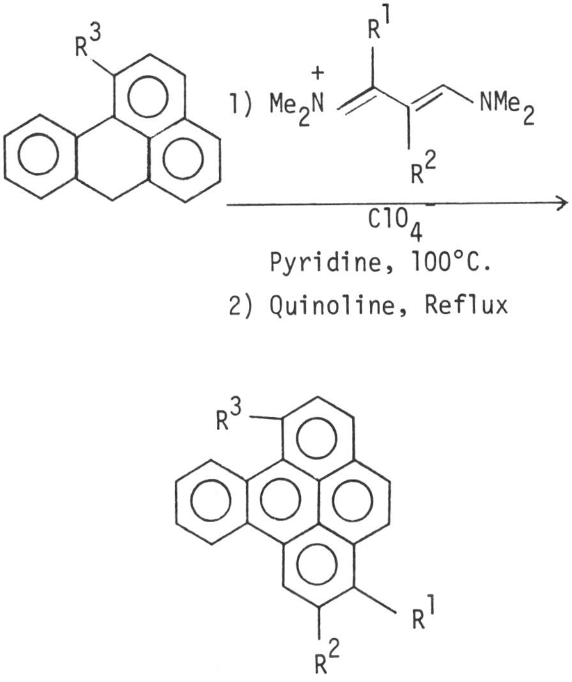

1) $Me_2N \overset{+}{\diagup}$ ... NMe_2
   R^1
   R^2

   ClO_4^-
   Pyridine, 100°C.
2) Quinoline, Reflux

25-40%

I.F.3-28 G. A. Taylor, <u>J. Chem. Soc., Perkin I</u>, 3132 (1981).

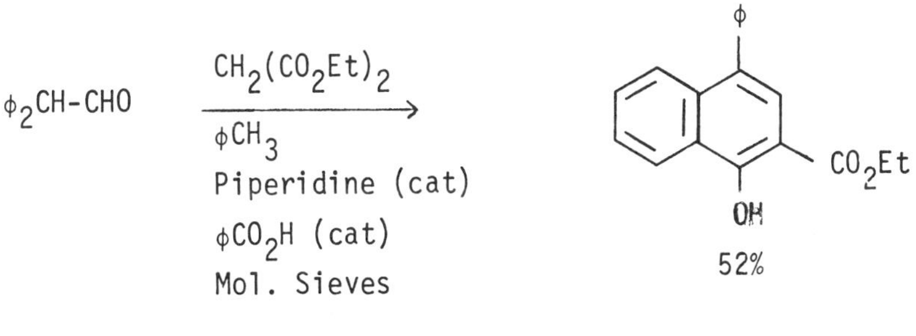

ϕ_2CH-CHO $\xrightarrow[\substack{\phi CH_3 \\ \text{Piperidine (cat)} \\ \phi CO_2H \text{ (cat)} \\ \text{Mol. Sieves}}]{CH_2(CO_2Et)_2}$

52%

I.F.3-29   Y. Tamura et al, <u>Chem. Pharm. Bull.</u>, <u>29</u>, 1312 (1981).

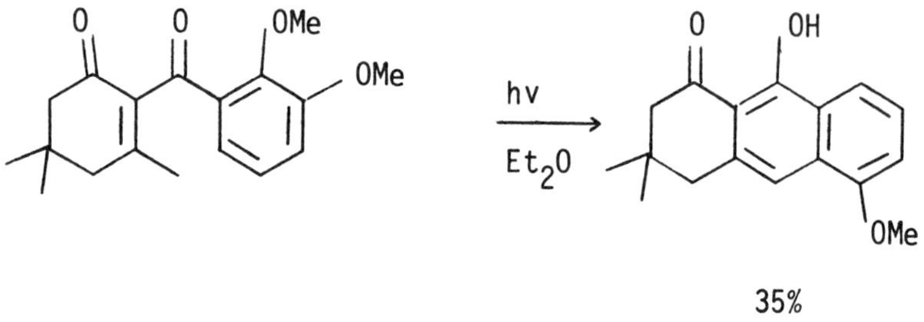

35%

I.F.3-30   S. V. Kessar et al, <u>Ind. J. Chem.</u>, <u>20B</u>, 1, 4, 7 (1981).

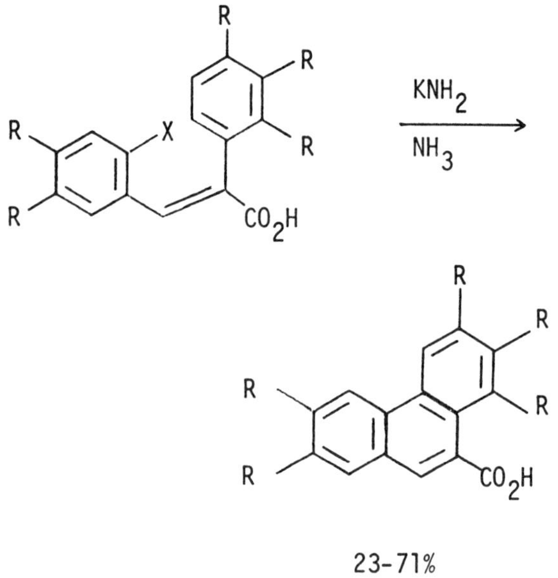

23-71%

I.F.3-31   P. Camps, C. Jaime and J. Molas, <u>Tetrahedron Lett.</u>, <u>22</u>, 2487 (1981).

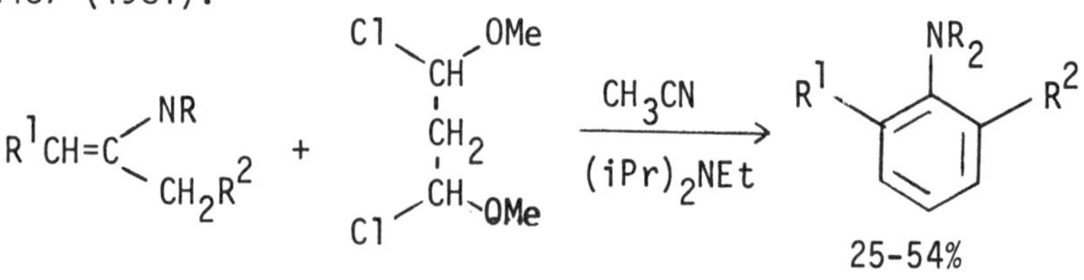

25-54%

I.F.3-32   K. Peseke and J. Q. Suarez, Zeit. Chem., 21, 405 (1981).

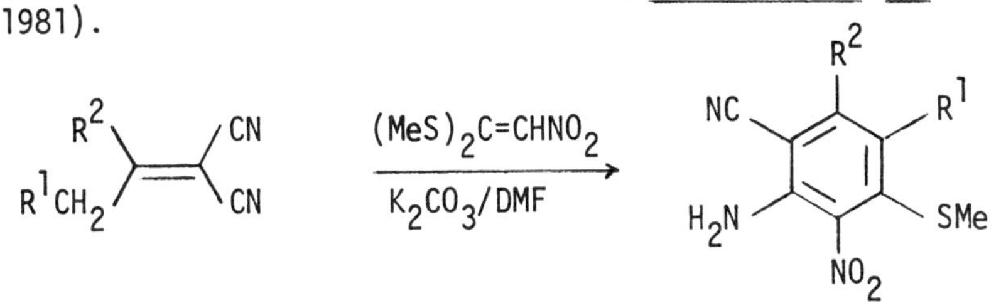

$$R^2\text{-}CH=C(CN)_2 \text{ with } R^1CH_2 \xrightarrow[K_2CO_3/DMF]{(MeS)_2C=CHNO_2}$$

69-90%

I.F.3-33   T. H. Chan and P. Brownbridge, Chem. Commun., 20 (1981); Tetrahedron, 37 (Suppl. 1), 387 (1981); M. A. Tius, Tetrahedron Lett., 22, 3335 (1981).

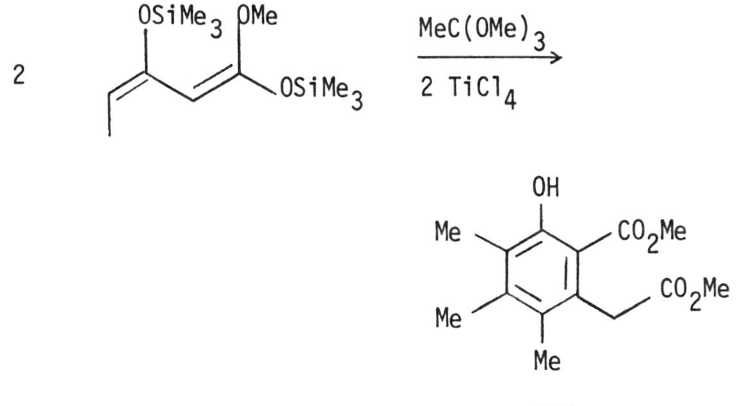

2

$$\xrightarrow[2\ TiCl_4]{MeC(OMe)_3}$$

53%

I.F.3-34   M. F. Semmelhack et al, Tetrahedron, 37, 3957 (1981).

Review: "Addition of Carbon Nucleophiles to Arene-Chromium Complexes."

I.F.3-35 E. Rose et al, J. Organometal. Chem., 221, 147, 157
(1981).

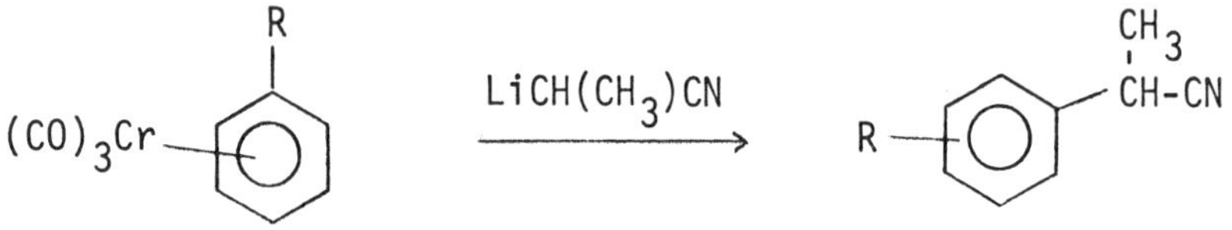

Study of Product
Distribution

I.F.3-36 W. D. Wulff, P. C. Tang and J. S. McCallum, J. Amer.
Chem. Soc., 103, 7677 (1981).

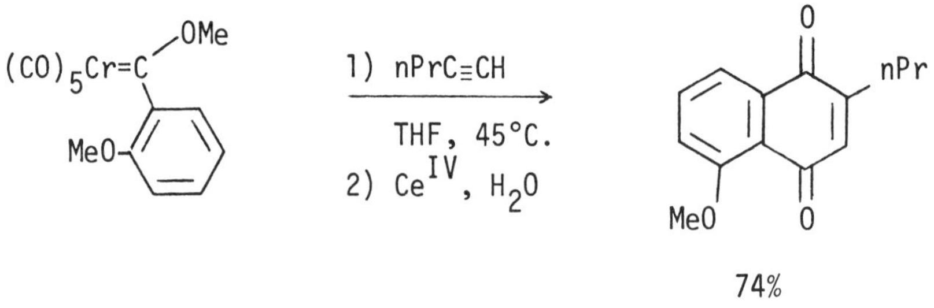

74%

I.F.3-37 F. Effenberger et al, Angew. Chem., Int. Ed. Engl.,
20, 265, 266 (1981).

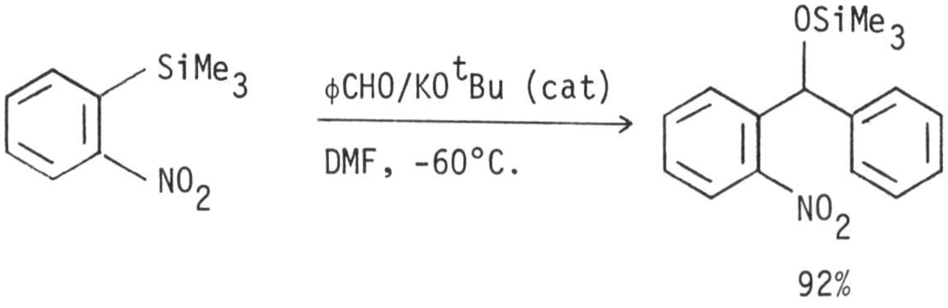

92%

Also, Tricarbonylchromium complexes reactions with
electrophiles.

I.F.3-38 J. Grimshaw and A. P. de Silva, Chem. Soc. Rev., 10, 181 (1981).

Review: "Photochemistry and Photocyclization of Aryl Halides."

I.F.3-39 D. J. Crouse, S. L. Hurlbut and D. M. S. Wheeler, J. Org. Chem., 46, 374 (1981); V. P. Pathak and R. N. Khanna, Synthesis, 882 (1981).

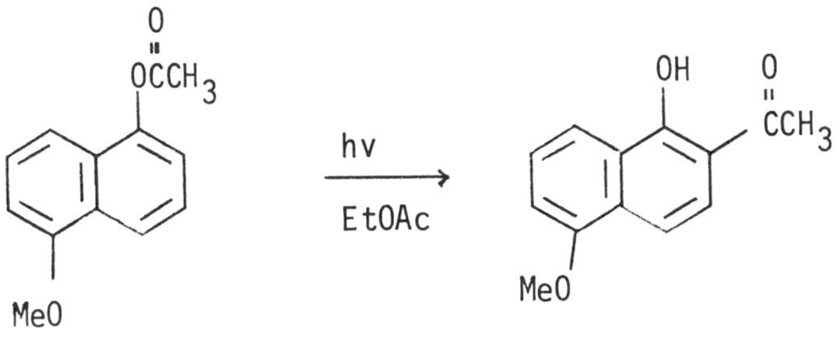

71%

I.F.3-40 A. Sugimori, M. Nishijima and T. Yashima, Chem. Lett., 303 (1981); T. Caronna et al, Tetrahedron Lett., 22, 155 (1981).

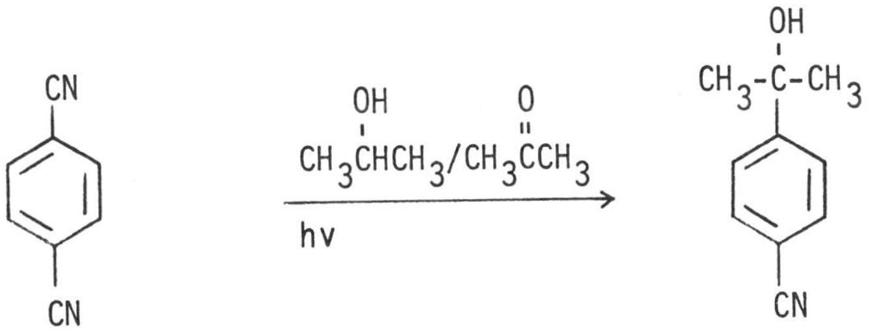

55%

I.F.3-41 L. L. Miller et al, J. Amer. Chem. Soc., 103, 4204, 4632 (1981).

Cyanation of Aromatic Compounds in a Gaseous Plasma.

I.F.3-42 M. Yamazaki et al, Chem. Pharm. Bull., 29, 1292, 1328 (1981); J. Krepelka, I. Vancurova and J. Holubek, Coll. Czech. Chem. Commun., 46, 1523 (1981).

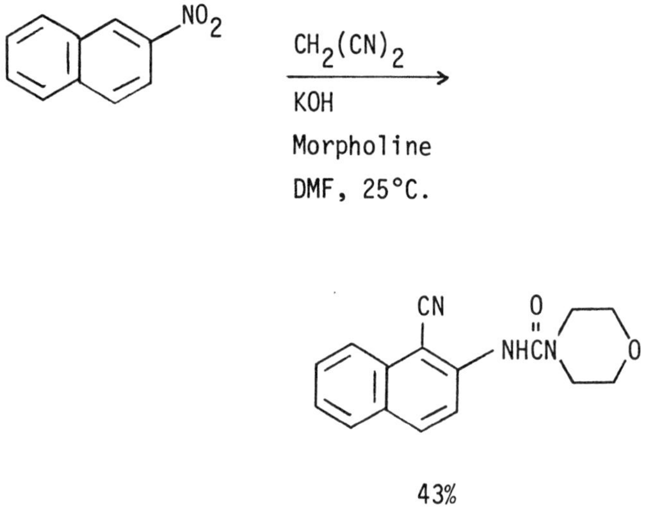

NO_2
$CH_2(CN)_2$
KOH
Morpholine
DMF, 25°C.
CN
O
NHCN
O
43%

I.F.3-43 J. Hocker and H. Giesecke, Org. Syn., 60, 49 (1981); P. Molina et al, Synthesis, 711 (1981).

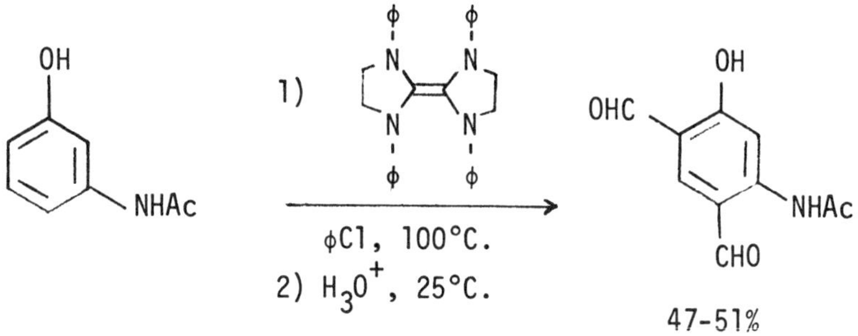

OH
NHAc
1)
ϕ
N
ϕCl, 100°C.
2) H_3O^+, 25°C.
OHC
OH
NHAc
CHO
47-51%

I.F.3-44 R. T. Borchardt and A. K. Sinhababu, J. Org. Chem.,
46, 5021 (1981).

1) Br$_2$/CHCl$_3$

2) 2 eq. HCHO
   2 eq. Me$_2$NH
   EtOH/H$_2$O/HOAc

3) Ni(R)
   3 N NaOH

68%

I.F.3-45 G. Tsuchihashi et al, Tetrahedron Lett., 22, 4305
(1981).

$$\begin{array}{c} \text{OMe} \quad \text{OSO}_2\text{R} \\ \text{Ar-C}\!-\!\!-\!\text{CHCH}_3 \\ \text{OMe} \end{array} \xrightarrow[\substack{\text{MeOH/H}_2\text{O} \\ \text{Reflux}}]{\text{CaCO}_3} \begin{array}{c} \text{CH}_3 \\ \text{Ar-CH-CO}_2\text{Me} \\ 75\text{-}93\% \end{array}$$

I.F.3-46 P. G. Gassman and T. Miura, Tetrahedron Lett., 22,
4787 (1981).

ϕSCl

1) MeSCH$_2$SiMe$_3$

CCl$_4$, -20°C.

2) 25°C.

13%

I.F.3-47   A. N. Kost, R. S. Sagitullin and A. A. Fadda, Org.
Prep. Proc. Int., 13, 203 (1981).

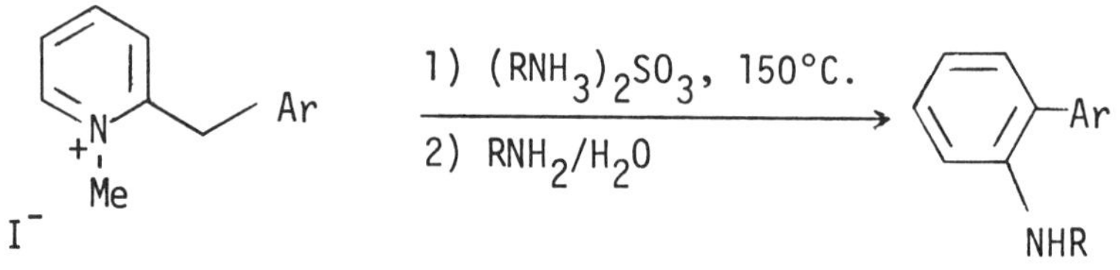

$$\text{pyridinium} \xrightarrow[\text{2) } RNH_2/H_2O]{\text{1) } (RNH_3)_2SO_3, \ 150°C.} \text{arene-Ar-NHR}$$

50-73%

I.F.3-48   M. E. Kurz, P. Ngoviwatchai and T. Tantrarant, J.
Org. Chem., 46, 4668, 4672 (1981).

$$ArH \ + \ CH_3NO_2 \xrightarrow[\text{HOAc}]{Mn(OAc)_3} ArCH_2NO_2$$

I.F.3-49   J. M. Bruce et al, Chem. Commun., 166, 169, 171
(1981).

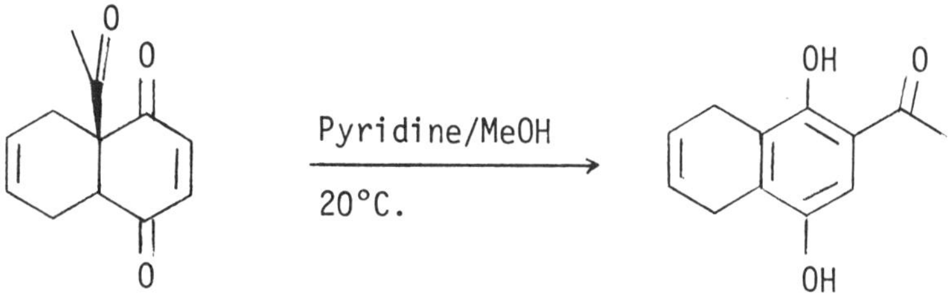

$$\xrightarrow[\text{20°C.}]{\text{Pyridine/MeOH}}$$

80%

Also, acyl migrations under acidic conditions.

I.F.3-50 M. Tiecco, Pure Appl. Chem., 53, 239 (1981).

Review: "Formation and Fate of Radical Ipso
         Intermediates in the Reactions of
         Carbon Radicals with Aromatics.Radical
         Ipso Substitution."

I.G. Synthesis via Organometallics.

I.G.1. Synthesis via Organoboranes.

I.G.1-1 H. C. Brown and J. B. Campbell, Aldrichimica Acta, 14, 3 (1981).

Review: "Synthesis and Applications of Vinylic
         Organoboranes."

I.G.1-2 H. C. Brown, P. K. Jadhav and A. K. Mandal, Tetrahedron, 37, 3547 (1981).

Review: "Asymmetric Syntheses via Chiral Organo-
         borane Reagents."

I.G.1-3 H. J. Bestmann, K. Suhs and T. Roder, Angew. Chem.,
Int. Ed. Engl., 20, 1038 (1981); D. Uguen, Bull. Soc. Chim.
Fr. II, 99 (1981); C. F. Reichert, W. E. Pye and T. A. Bryson,
Tetrahedron, 37, 2441 (1981).

$$\phi CH_2BH_2 \cdot P\phi_3 \quad \xrightarrow[\text{MeI}]{\text{1) } RCH=CH_2} \quad \phi CH_2C \underset{OH}{\overset{CH_2CH_2R}{\langle}} CH_2CH_2R$$

2) Cl_2CHOCH_3
3) LiO^tBu 30-52%
4) $H_2O_2/NaOH$

I.G.1-4 A. Pelter and J. M. Rao, Chem. Commun., 1149 (1981);
T. Yogo, J. Koshino and A. Suzuki, Chem. Lett., 1059 (1981).

$$R_3B \quad \xrightarrow[\text{2) } NaOH/H_2O_2]{\text{1) } (\phi S)_3CLi} \quad \overset{O}{\underset{}{\overset{\|}{R-C-R}}}$$

72-82%

Tertiary alcohols (R_3COH) also prepared.

I.G.1-5 M. M. Midland and Y. C. Kwon, J. Org. Chem., 46, 229
(1981).

1) 9-BBN

2) Base*/THF

3) $ClCH_2CN$

4) EtOH   60-70%

*Base = Potassium 2,6-di-t-butyl-4-methylphenoxide.

I.G.1-6 Y. Yamamoto, H. Yatagai and K. Maruyama, *J. Amer. Chem. Soc.*, 103, 1969 (1981).

$$R^1 \diagdown \diagup \overset{-}{BR_3} \quad \underset{Li^+}{} \quad \xrightarrow[Et_2O, -78°C.]{R^2 \diagdown \diagup X} \quad \underset{53-82\%}{R^1 \diagdown \diagup \diagdown \diagup R^2}$$

Also, reaction with Aldehydes.

I.G.1-7 A. Pelter and J. M. Rao, *Tetrahedron Lett.*, 22, 797 (1981).

$$(C_6H_{13})_3\overset{-}{B}C{\equiv}C-R^1 \quad \underset{Li^+}{} \quad \xrightarrow[2) \, MCPBA]{1) \, R^2CH=C(CO_2Et)_2}$$

$$\underset{R^1 \ R^2}{C_6H_{13}-\overset{O}{\overset{\|}{C}}-CH-CHCH(CO_2Et)_2}$$

$$67-91\%$$

I.G.1-8 H. C. Brown and T. M. Ford, *J. Org. Chem.*, 46, 647 (1981).

1) CO/Li(MeO)$_3$AlD
2) LiAlD$_4$
3) φCHO

52%

I.G.1-9   N. Miyaura and A. Suzuki, Chem. Lett., 879 (1981).

$$R^1, H \text{ C=C } R^2, B(O_2C_6H_4) \xrightarrow[\text{NaOAc}]{\substack{CO/MeOH \\ PdCl_2 \text{ (cat)}}} R^1, H \text{ C=C } R^2, CO_2Me$$

41-95% (GC)

I.G.2. Carbonylation Reactions

I.B.2-1   N. S. Nudelman and A. A. Vitale, J. Org. Chem., 46, 4625 (1981); Org. Prep. Proc. Int., 13, 144 (1981).

$$\phi Li + RBr \xrightarrow[\substack{CO \\ 2) \text{ aq. } NH_4Cl}]{1) \text{ THF}/-78°C.} \phi\text{-}\overset{\phi}{\underset{R}{C}}\text{-OH}$$

20-80%

I.G.2-2   N. Yoneda, A. Suzuki and Y. Takahashi, Chem. Lett., 767 (1981).

$$\text{Me, Me } \overset{O}{\underset{O}{\bigcirc}} \xrightarrow[\substack{CO, 0°C. \\ 2) H_3O^+}]{1) \text{ HF/SbF}_5} HO_2C\text{-}\underset{\text{Me Me}}{C}\text{-}CO_2H$$

100% (GC)

I.G.2-3 H. Alper and D. E. Laycock, Tetrahedron Lett., 22, 33 (1981); M. Tanaka, Synthesis, 47 (1981); T. Kobayashi and M. Tanaka, J. Organometal. Chem., 205, C27 (1981).

I.G.2-4 J. J. Brunet, C. Sidot and P. Caubere, Tetrahedron Lett., 22, 1013 (1981); D. Valentine, Jr., J. W. Tilley and R. A. LeMahieu, J. Org. Chem., 46, 4614 (1981); S. Gambarotta and H. Alper, J. Organometal. Chem., 212, C23 (1981).

$$ArX \xrightarrow[\substack{Co_2(CO)_8, \ Q^+ \ Br^- \\ aq. \ NaOH/\phi H \\ hv \\ 2) \ H_3O^+}]{1) \ CO} ArCO_2H$$

47-98% (GC)

I.G.2-5 M. Tanaka, Bull. Chem. Soc. Jpn., 54, 637 (1981); T. Kobayashi and M. Tanaka, Chem. Commun., 333 (1981); T. Matsuda et al, J. Org. Chem., 46, 4413 (1981); S. Komiya, A. Yamamoto and T. Yamamoto, Chem. Lett., 193 (1981).

$$ArX \xrightarrow[\substack{(\phi_3P)_2Pd(I)\phi \ (cat) \\ THF}]{CO/KCN} Ar-\overset{\displaystyle O}{\underset{\displaystyle \|}{C}}-CN$$

64-88%

I.G.2-6 S. Gambarotta and H. Alper, J. Org. Chem., 46, 2142 (1981).

$$R^1R^2C=C=C(R^3)H \xrightarrow[\substack{Co_2(CO)_8 \\ aq.\ NaOH \\ Q^+ X^-}]{CH_3I/CO}$$

(structure of product with OH, R^1, R^2, R^3, H, CCH_3, O, 23-43%)

Dienones also formed.

I.G.2-7 C. F. Hobbs and W. S. Knowles, J. Org. Chem., 46, 4422 (1981); M. J. Mirbach et al, J. Amer. Chem. Soc., 103, 7590, 7594 (1981).

$$CH_2=CHOAc \xrightarrow[\substack{80°C. \\ Rh(COD)acac + 6L}]{CO/H_2} CH_3CH{\substack{OAc \\ CHO}}$$

51% ee

$$L = \underset{H}{\overset{H}{\underset{O}{\overset{O}{\bigtimes}}}}\substack{CH_2PAr_2 \\ CH_2PAr_2}$$

I.G.2-8 P. Haelg, G. Consiglio and P. Pino, Helv. Chim. Acta, 64, 1865 (1981); H. C. Clark and J. A. Davies, J. Organometal. Chem., 213, 503 (1981).

$$\substack{Me \\ Me}C=C\substack{H \\ H} \xrightarrow[\substack{Pt(DIOP)(SnCl_3)Cl \\ \phi Et,\ 80°C. \\ 2)\ Ag_2O \\ 3)\ CH_2N_2}]{1)\ CO/D_2}$$

(product: CO_2Me, Me—H, Me—H, D)

I.G.3. Other Syntheses via Organometallics.

I.G.3-1 B. M. Trost and T. A. Runge, J. Amer. Chem. Soc., 103, 2485, 7550, 7559 (1981).

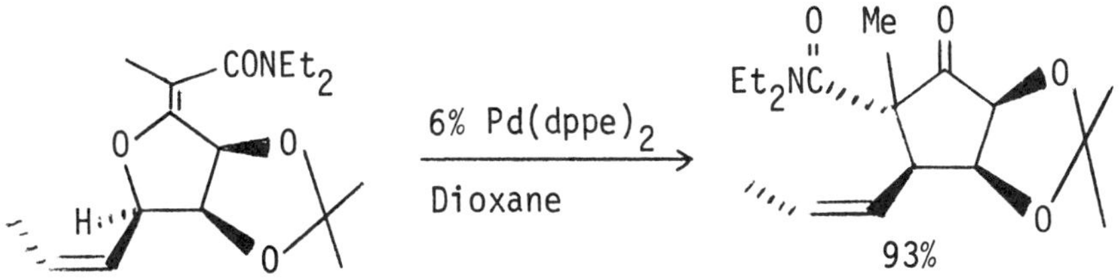

dppe = 1,2-bis(diphenylphosphino)ethane.

I.G.3-2 M. Moreno-Manas and A. Trius, Tetrahedron Lett., 22, 3109 (1981).

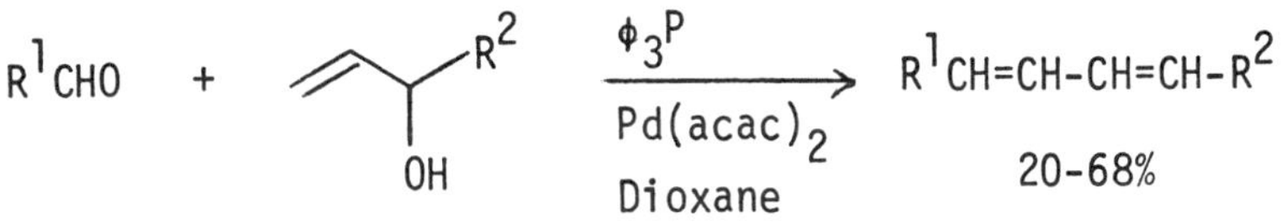

I.G.3-3 H. Suzuki, Y. Moro-Oka et al, Chem. Lett., 1361, 1435 (1981).

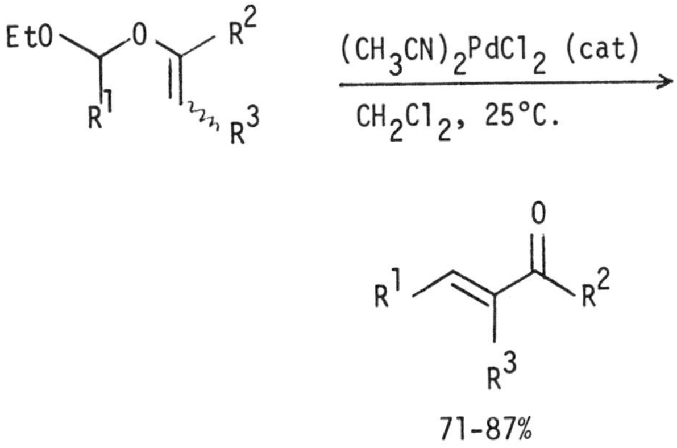

I.G.3-4   C. Moberg, Tetrahedron Lett., 22, 4827 (1981).

$$\xrightarrow[\text{Et}_3\text{Al}/\phi\text{CH}_3]{(\phi_3\text{P})_2\text{PdCl}_2 \text{ (cat)}}$$

81% (GC)

I.G.3-5   M. P. Doyle and D. Van Leusen, J. Amer. Chem. Soc., 103, 5917 (1981).

$$\xrightarrow[\substack{[\text{Rh(CO)}_2\text{Cl}]_2 \text{ or} \\ [\text{Ru(CO)}_3\text{Cl}_2]_2 \\ \text{(cat)} \\ 70°\text{C.}}]{\text{PtCl}_2 \cdot 2\phi\text{CN or}}$$

$$\underset{\text{MeO}}{\overset{\phi}{\diagdown}}\text{C=CHCH}_2\text{CO}_2\text{Et}$$

98%

I.G.3-6   B. Ganem et al, Tetrahedron Lett., 22, 4163 (1981).

$$\text{Me}_2\text{CHCH}_2\text{-}\underset{\underset{\text{N}_2}{\|}}{\text{C}}\text{-CO}_2\text{Me} \qquad \xrightarrow[\phi\text{H}, 25°\text{C.}]{\text{Rh(OAc)}_2 \text{ (cat)}} \qquad \underset{\text{H}}{\overset{\text{Me}_2\text{CH}}{\diagdown}}\text{C=C}\underset{\text{H}}{\overset{\text{CO}_2\text{Me}}{\diagup}}$$

39-99%

I.G.3-7 M. F. Semmelhack and S. J. Brickner, <u>J. Amer. Chem.</u>
<u>Soc.</u>, <u>103</u>, 3945, 6460 (1981).

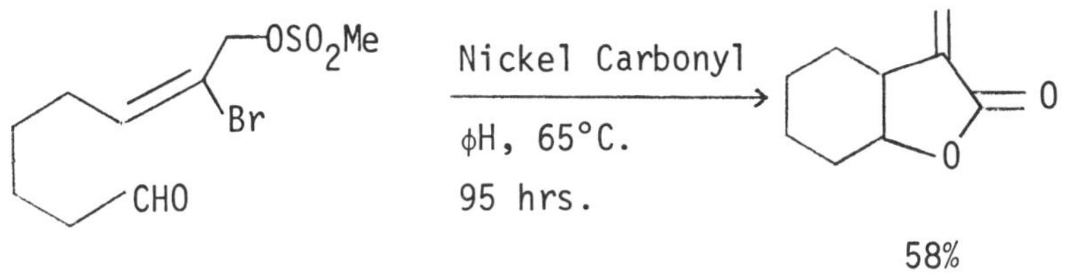

58%

I.G.3-8 A. Herrera and H. Hoberg, <u>Synthesis,</u> 831 (1981).

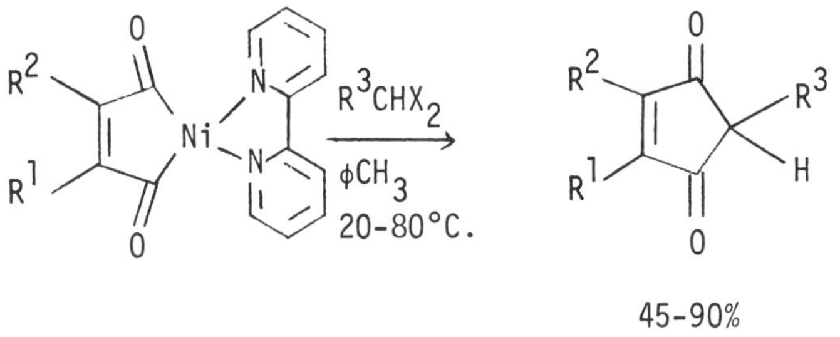

45-90%

I.G.3-9 N. E. Schore and M. C. Croudace, <u>J. Org. Chem.</u>, <u>46</u>,
5436 (1981); T. R. Gadek and K. P. C. Vollhardt, <u>Angew. Chem.</u>,
<u>Int. Ed. Engl.</u>, <u>20</u>, 802 (1981); Y. Yamamoto and H. Yamazaki,
<u>Bull. Chem. Soc. Jpn.</u>, <u>54</u>, 787 (1981).

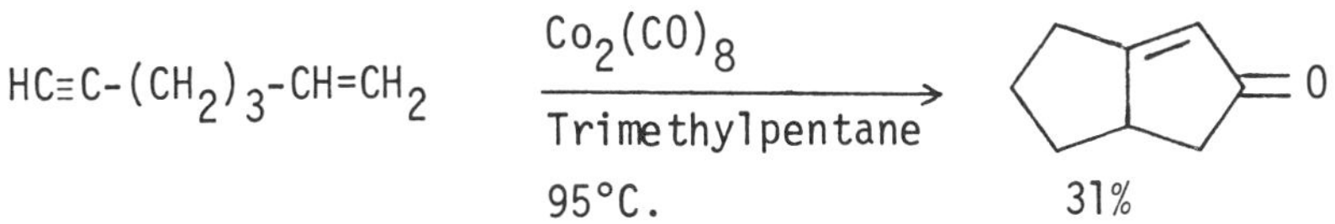

$HC{\equiv}C-(CH_2)_3-CH{=}CH_2$ $\xrightarrow[\substack{\text{Trimethylpentane} \\ 95°C.}]{Co_2(CO)_8}$

31%

I.G.3-10   M. Okano et al, Chem. Ind., 96 (1981).

1) Mn(OAc)$_3$
HOAc, Ac$_2$O
KOAc
2) LAH/Et$_2$O

8%

I.G.3-11   M. Nishizawa and R. Noyori, Bull. Chem. Soc. Jpn., 54, 2233 (1981).

Fe$_2$(CO)$_9$
ϕH, Reflux

80% (GC)

I.G.3-12   B. Myrboh, H. Ila and H. Junjappa, Synthesis, 126 (1981).

$$\underset{R-C-CH_3}{\overset{O}{\parallel}}$$

1) BF$_3$·Et$_2$O/MeOH
Pb(OAc)$_4$/ϕH
2) H$_2$O

RCH$_2$CO$_2$Me

I.G.3-13   J. Tsuji and S. Hashiguchi, J. Organometal. Chem., 218, 69 (1981);   T. Gibson and L. Tulich, J. Org. Chem., 46, 1821 (1981).

(CH$_2$)$_7$CO$_2$Me

WCl$_6$
———→
Cp$_2$TiMe$_2$
ϕH

MeO$_2$C(CH$_2$)$_7$-CH=CH-(CH$_2$)$_7$CO$_2$Me

34%

I.G.3-14 R. Nakajima, K. Morita and T. Hara, <u>Bull. Chem. Soc.</u>
<u>Jpn.</u>, <u>54</u>, 3599 (1981); F. S. Pinault and A. L. Crumbliss, <u>J.</u>
<u>Organometal.</u>, <u>Chem.</u>, <u>215</u>, 229 (1981); Y. Yamada and D. I.
Momose, <u>Chem. Lett.</u>, <u>1277</u> (1981).

$$2 \text{ RI} \xrightarrow[\substack{\text{NaOH/MeOH} \\ \text{NH}_2\text{NH}_2 \cdot \text{H}_2\text{O}}]{\text{PdCl}_2 \text{ (cat)}} \text{R-R}$$

1-74% (GC)

I.G.3-15 P. Girard, R. Couffignal and H. B. Kagan, <u>Tetrahedron</u>
<u>Lett.</u>, <u>22</u>, 3959 (1981); R. G. H. Kirrstetter and U. Vagt,
<u>Chem. Ber.</u>, <u>114</u>, 630 (1981).

$$\underset{\substack{\text{R-C-Cl}}}{\overset{\text{O}}{\|}} \xrightarrow[\substack{\text{THF, 25°C.} \\ \text{2) H}_3\text{O}^+}]{\text{1) 2 eq. SmI}_2} \underset{\substack{\text{R-C-C-R}}}{\overset{\text{O O}}{\|\ \|}}$$

40-80%

I.G.3-16 J. M. Pons, J. P. Zahra and M. Santelli, <u>Tetrahedron</u>
<u>Lett.</u>, <u>22</u>, 3965 (1981).

25%

(d,l:meso = 70:30)

I.G.3-17 T. Y. Luh, W. H. So and S. W. Tam, <u>J. Organometal.</u>
<u>Chem.</u>, <u>218</u>, 261 (1981).

37%

I.G.4. Organometallic Reviews.

I.G.4-1 P. Heimbach and H. Schenkluhn, Pure Appl. Chem., 53,
2419 (1981).

Review: "Control in Transition Metal Catalyzed
Organic Synthesis."

I.G.4-2 M. Pereyre and J. P. Quintard, Pure Appl. Chem., 53,
2401 (1981).

Review: "Organotin Chemistry for Synthesis Applications."

I.G.4-3 J. Tsuji, Pure Appl Chem., 53, 2371 (1981).

Review: "Palladium Catalysis in Natural Product
Synthesis."

I.G.4-4 E. Negishi, Pure Appl. Chem., 53, 2333 (1981).

Review: "Bimetallic Catalytic Systems Containing Ti, Zr,
Ni and Pd. Their Applications to Selective
Organic Synthesis."

I.G.4-5   R. F. Heck, Pure Appl. Chem., 53, 2323 (1981).

    Review: "Palladium-Catalyzed Synthesis of Conjugated
              Polyenes."

I.G.4-6   B. M. Trost, Pure Appl. Chem., 53, 2357 (1981).

    Review: "Transition Metal Templates for Selectivity
              in Organic Synthesis."

I.G.4-7   B. M. Trost, Aldrichimica Acta, 14, 43 (1981).

    Review: "Transition-Metal Templates for Selectivity
              in Organic Synthesis."

I.G.4-8   M. Orchin, Acct. Chem. Res., 14, 259 (1981).

    Review: "HCo(CO)$_4$, the Quintessential Catalyst."

I.G.4-9   J. F. Knifton, CHEMTECH, 609 (1981).

    Review: "Big Acids from Little Ones" (by Homogeneous
              Catalysis with Ru).

I.G.4-10   H. Lehmkuhl, Bull. Soc. Chim. Fr. II, 87 (1981).

Review: "The Insertion of Olefins into Metal Carbon
          Bonds."

I.G.4-11   G. van Koten and K. Vrieze, Rec. Trav. Chim., 100,
129 (1981).

Review: "Interaction of Metal Centres with the 1,4-
          Diaza-1,3-Butadiene(α-diimine) Ligand.
          Versatile Coordination Chemistry and
          Applications in Organic Synthesis and
          Catalysis."

I.G.4-12   B. B. Snider et al, Tetrahedron, 37, 3927 (1981).

Review: "Alkylaluminum Halides. Lewis Acid
          Catalysts which are Bronsted Bases."

II

OXIDATIONS

II.A. C—O Oxidations

1. Alcohol⟶ Ketone, Aldehyde

II.A.1-1   K. Oshima et al., Tetrahedron Lett., 22, 1605 (1981).

Oxidation of primary alcohols in the presence of secondary alcohols, e.g.:

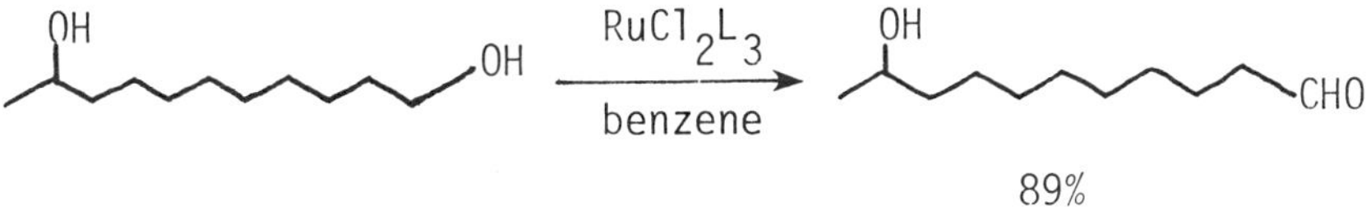

89%

II.A.1-2   J.M.J. Fréchet, P. Darling, and M.J. Farrall, J. Org. Chem., 46, 1728 (1981).

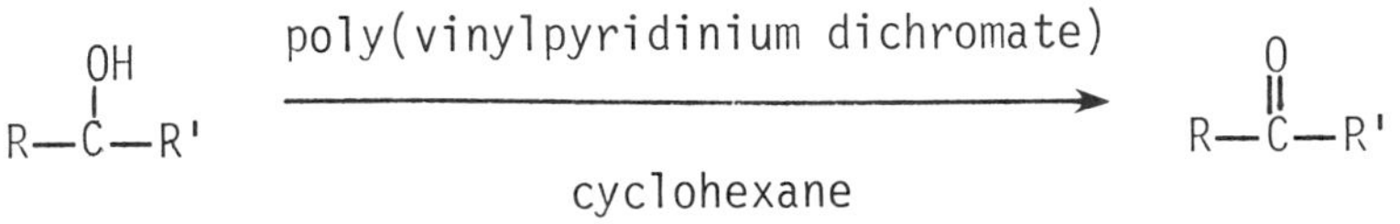

R, R' = H, Ph, alkyl, cyclic, Bz   ∿70-100%

II.A.1-3   S. Hanessian, D.H. Wong, and M. Therien, Synthesis, 394 (1981).

$$R-\underset{\underset{OH}{|}}{CH}-R' \quad \xrightarrow[\substack{Bu_4N\ Cl \\ \oplus\ominus}]{NCS \quad CH_2Cl_2} \quad R-\underset{\overset{O}{\|}}{C}-R'$$

$\sim 80\text{-}98\%$

R, R' = H, alkyl, cyclic, subst. Ph

II.A.1-4   O. Mitsunobu and N. Yoshida, Tetrahedron Lett., 22, 2295 (1981).

$$R-\underset{\underset{OH}{|}}{CH}\diagdown_{R'} \quad \xrightarrow[\substack{2) \quad Ph_3P \\ 3) \quad EtOCCH_2NO_2 \\ \quad\quad\overset{\|}{O}}]{1) \quad EtOOC-N{=}N-COOEt} \quad R\diagup\overset{\overset{O}{\|}}{C}\diagdown_{R'}$$

R = alkyl, Ph, Bz
R' = H, Me

$\sim 52\text{-}85\%$

II.A.1-5   B.S. Bal, K.S. Kockhar, and H.W. Pinnick, J. Org. Chem., 46, 1492 (1981).

$$\underset{R}{\overset{H}{\diagdown}}C\underset{R'}{\overset{OBz}{\diagup}} \quad \xrightarrow{\text{Jones reagent}} \quad R\diagup\overset{\overset{O}{\|}}{C}\diagdown_{R'}$$

R, R' = alkyl, cyclic, Ph

$\sim 16\text{-}79\%$

II.A.1-6   P. Müller and J. Godoy, Tetrahedron Lett., 22, 2361 (1981).

$$\underset{\underset{R}{\overset{\overset{OH}{|}}{CH}}\diagdown R'}{} \xrightarrow[\text{PhIO or PhI(OAc)}_2]{\text{RuCl}_2L_3, \text{ CH}_2\text{Cl}_2} \underset{R}{\overset{\overset{O}{\parallel}}{C}}\diagdown R'$$

$\sim$ 80-90%

R = alkyl, aryl, Bz, cyclic
R' = H, alkyl, cyclic with R

II.A.1-7/II.A.2-1   F.M. Menger and C. Lee, Tetrahedron Lett., 22, 1655 (1981).

$$R-CHO \xrightarrow{\text{solid NaMnO}_4} R-COOH$$
77-80%

$$R-CH_2OH \xrightarrow{\text{solid NaMnO}_4} R-COOH$$
67-81%

$$\underset{R-\overset{\overset{OH}{|}}{CH}-R'}{} \xrightarrow{\text{solid NaMnO}_4} R-\overset{\overset{O}{\parallel}}{C}-R'$$
84-100%

II.A.2.   Alcohol, Aldehyde ⟶ Acid, Acid Derivative

II.A.2-2   B.S. Bal, W.E. Childers, Jr., and H.W. Pinnick, Tetrahedron, 37, 2091 (1981).

$$R\diagup\overset{\parallel}{\diagup}\diagdown CHO \xrightarrow[\text{NaH}_2\text{PO}_4, \text{ H}_2\text{O}]{\text{sodium chlorite, t-butanol}} R\diagup\overset{\parallel}{\diagup}\diagdown COOH$$

87-95%

R = alkyl, Ph, etc.

II.A.2-3   F. Camps et al., Tetrahedron Lett., 22, 3895 (1981).

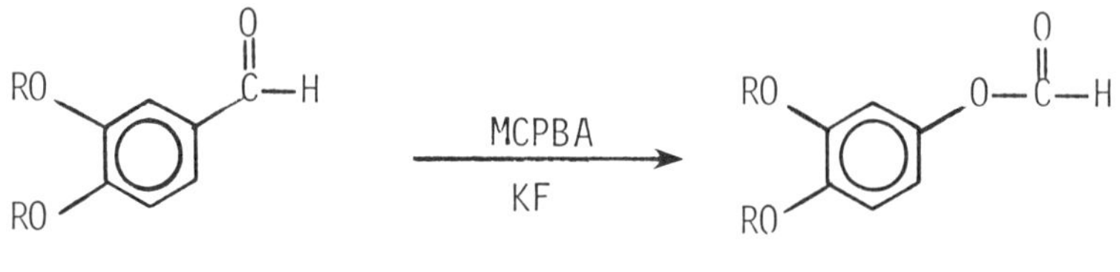

R = alkyl   77-95%

II.B.   C—H Oxidations

1.   C—H ⟶ C—O

II.B.1-1   S.N. Suryawanshi and P.L. Fuchs, Tetrahedron Lett., 22, 4201 (1981).

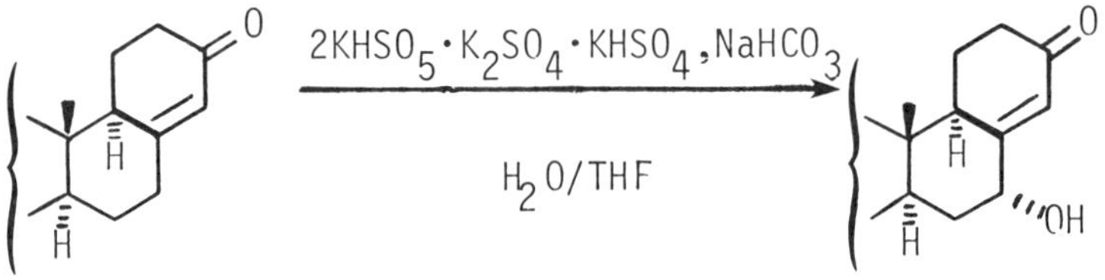

97%

II.B.1-2   R.M. Moriarty, H. Hu, and S.C. Gupta, Tetrahedron Lett., 22, 1283 (1981).

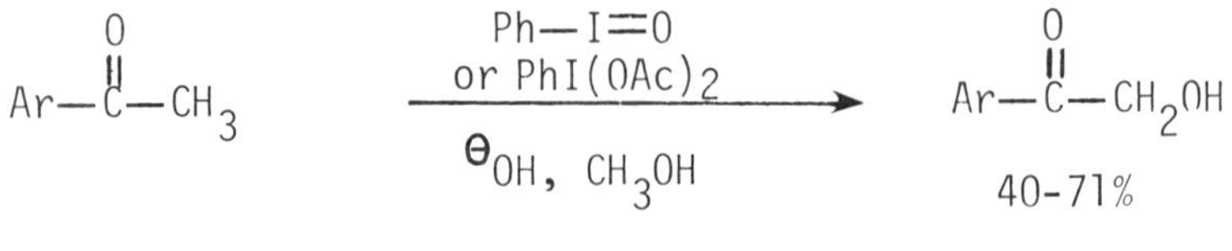

$$Ar\!-\!\overset{O}{\overset{\|}{C}}\!-\!CH_3 \quad \xrightarrow[\underset{OH,\ CH_3OH}{\ominus}]{\begin{array}{c}Ph\!-\!I\!=\!O\\ or\ PhI(OAc)_2\end{array}} \quad Ar\!-\!\overset{O}{\overset{\|}{C}}\!-\!CH_2OH$$

40-71%

Ar = subst. Ph, ferrocenyl, etc.

II.B.1-3   D.V. Rao and F.A. Stuber, Tetrahedron Lett., 22, 2337 (1981).

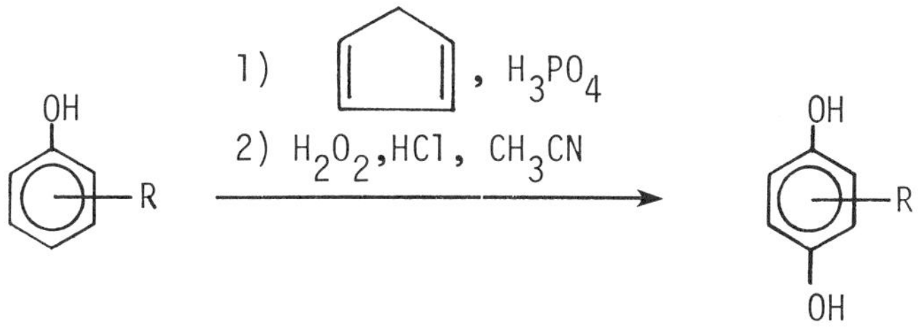

R = H,Me

~ 70-80%
overall

II.B.1-4   A.M. Klibanov, Z. Berman, and B.N. Alberti, J. Am. Chem. Soc., 103, 6263 (1981).

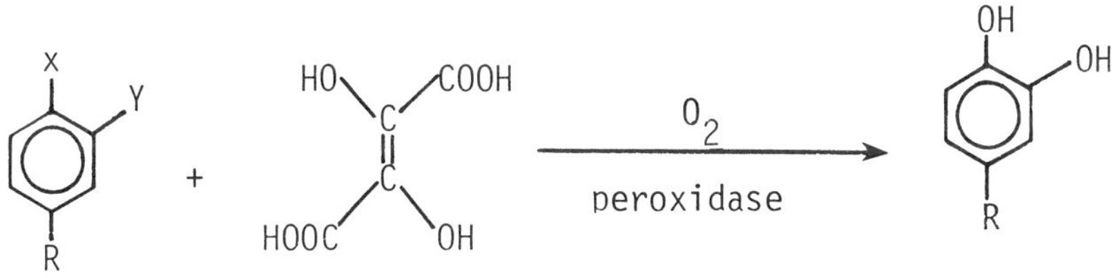

R   amino acid
X, Y = one is H, the other OH

up to 70%,
no racemization

II.B.1-5   H.L. Holland, U. Daum, and E. Riemland, Tetrahedron Lett., 22, 5127 (1981).

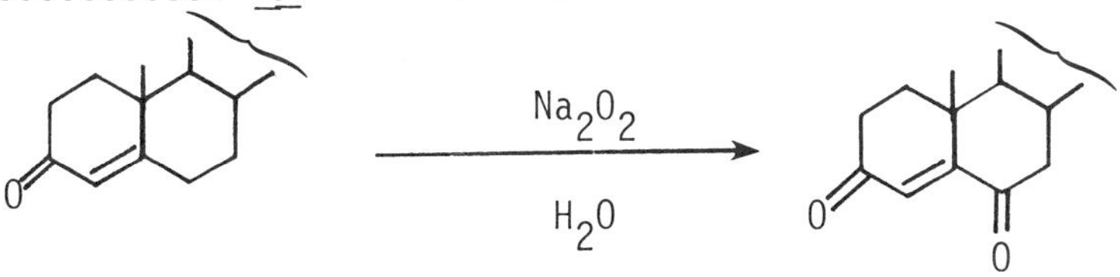

43-63%

II.B.1-6   J.P. McCormick, W. Tomasik, and M.W. Johnson,
Tetrahedron Lett., 22, 607 (1981).

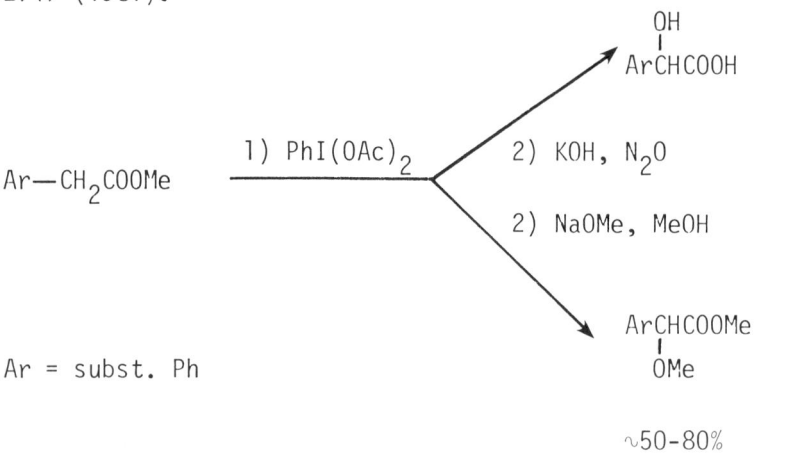

54-98%

NMMO = N-methylmorpholine-N-

R = alkyl, cycloalkyl, subst. Ph   oxide
R', R" = H, alkyl, cyclic with R

II.B.1-7   R.M. Moriarty and H. Hu, Tetrahedron Lett., 22,
2747 (1981).

$$Ar-CH_2COOMe \xrightarrow{\text{1) PhI(OAc)}_2}$$

OH
|
ArCHCOOH

2) KOH, N_2O

2) NaOMe, MeOH

ArCHCOOMe
|
OMe

Ar = subst. Ph

∿50-80%

II.B.1-8   P.T. Perumal and M.V. Bhatt, Indian J. Chem., 20B,
153 (1981); Tetrahedron Lett., 22, 2605 (1981).

$$Ar-CH_2-R \xrightarrow[\text{or } CuSO_4/peroxydisulfate]{S_2O_8^{-2}, Cu^{+2}} Ar-\overset{O}{\overset{\|}{C}}-R$$

widely varying yields

Ar = subst. Ph
R = H, alkyl

II.B.2.   C—H ⟶   C—Hal

II.B.2-1   D. Masilamani and M.M. Rogic, J. Org. Chem., 46, 4486 (1981).

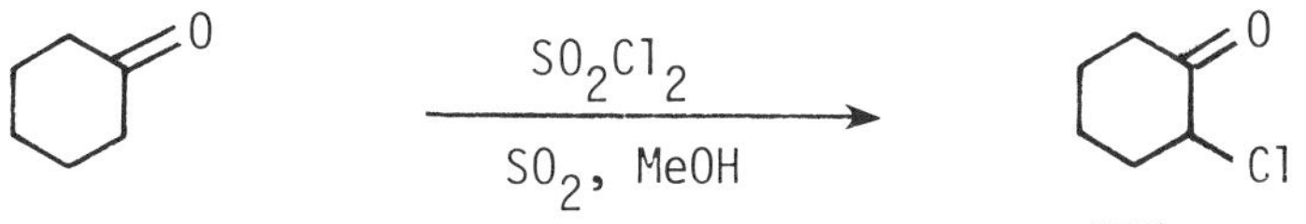

$$\xrightarrow[\text{SO}_2, \text{ MeOH}]{\text{SO}_2\text{Cl}_2}$$

95%

II.B.2-2   C.A. Horiuchi and J.Y. Satoh, Synthesis, 312 (1981).

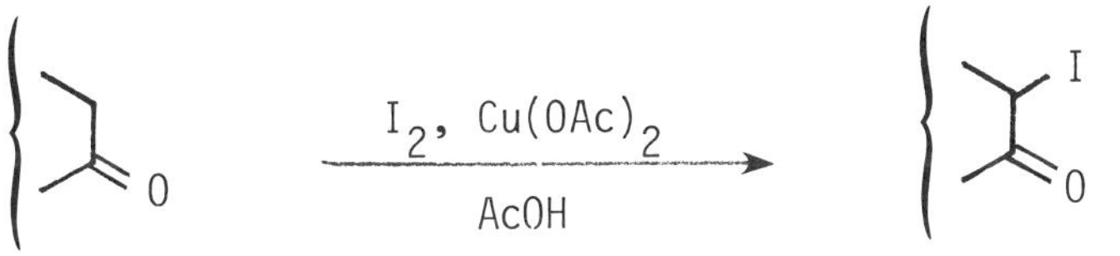

$$\xrightarrow[\text{AcOH}]{\text{I}_2, \text{ Cu(OAc)}_2}$$

∿ 70-100%

II.B.2-3   S.V. Ley and A.J. Whittle, Tetrahedron Lett., 22, 3301 (1981).

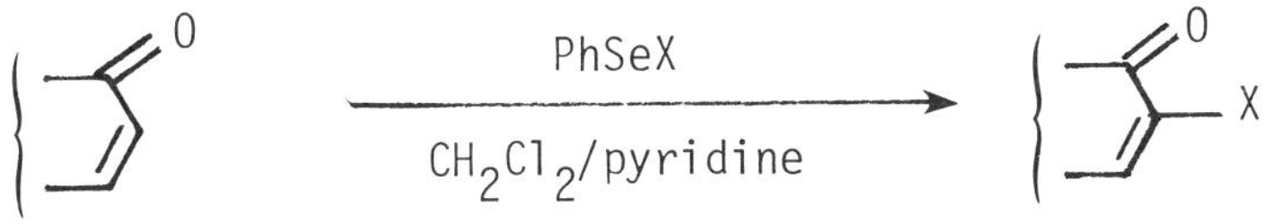

$$\xrightarrow[\text{CH}_2\text{Cl}_2/\text{pyridine}]{\text{PhSeX}}$$

X = Cl, Br   ∿ 50-100%

II.B.2-4   A. Chaintreau, G. Adrian, and D. Couturier,
Synth. Comm., 11, 669 (1981).

$$Ar—CH_3 \xrightarrow[\text{AcOH, } Ac_2O]{\text{CuBr}_2, \underline{t}\text{-BuOOH}} Ar—CH_2Br$$

Ar = subst. Ph   43-95%

II.B.2-5   S. Stavber and M. Zupau, J.C.S. Chem. Comm., 148
(1981).

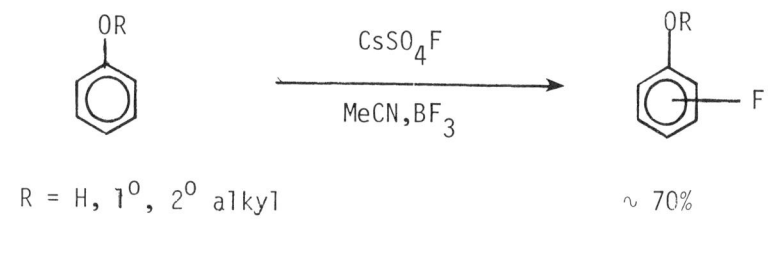

R = H, 1°, 2° alkyl   ∿ 70%

II.B.2-6   J.J. Harrison, J.P. Pellegrini, and C.M. Selwitz,
J. Org. Chem., 46, 2169 (1981).

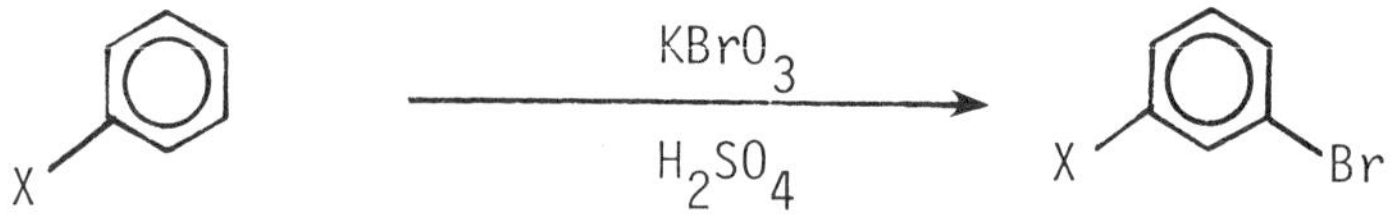

X = H, NO_2, COOH, $COCH_3$ 50-97%

II.B.2-7 T. Sugiyama, Bull. Chem. Soc. Japan, 54, 2847 (1981).

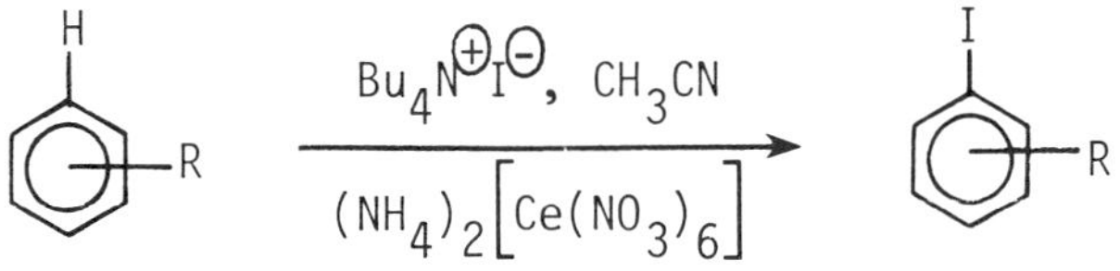
H
R
$Bu_4N^{\oplus}I^{\ominus}$, CH_3CN
$(NH_4)_2[Ce(NO_3)_6]$
I
R

R = 1, 2, 3, or 4 methyl groups ~50-70%

II.B.2-8 S. Torii et al., Tetrahedron Lett., 22, 3193 (1981).

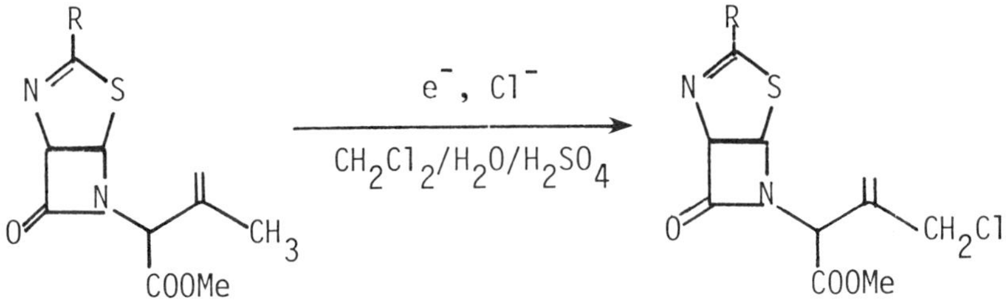
R
N
S
N
O
CH_3
COOMe
e^-, Cl^-
$CH_2Cl_2/H_2O/H_2SO_4$
R
N
S
N
O
CH_2Cl
COOMe

R = Bz, PhCO, $PhOCH_2$, $PhCCl_2$ ~80%

II.B.2-9 E.K. Ryu and M.MacCoss, J. Org. Chem., 46, 2819 (1981).

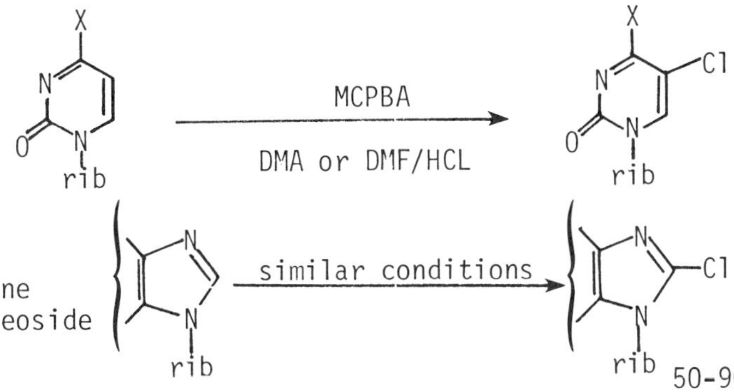
X
N
O
N
rib
MCPBA
DMA or DMF/HCL
X
Cl
N
O
N
rib
purine
nucleoside
N
N
rib
similar conditions
N
Cl
N
rib

50-90%

II.B.3.   Other C—H Oxidations

II.B.3-1   P. Bakuzis and M.L.F. Bakuzis, _J. Org. Chem._, 46,
235 (1981).

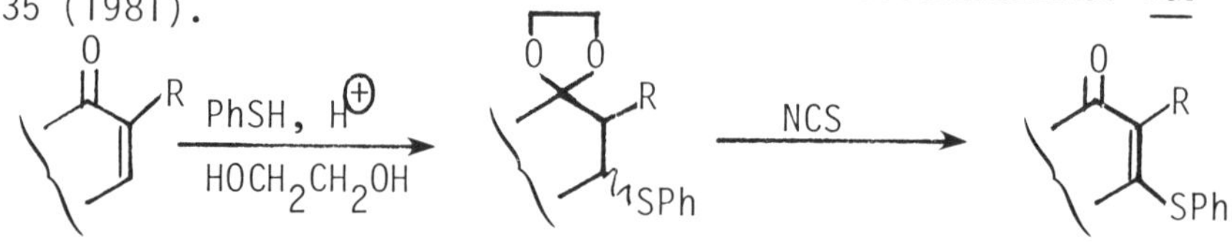

R = H, alkyl   ∿ 60% overall

II.C.   C—N Oxidations

II.C-1   J. H. Babler and B.J. Invergo, _J. Org. Chem._, 46,
1937 (1981).

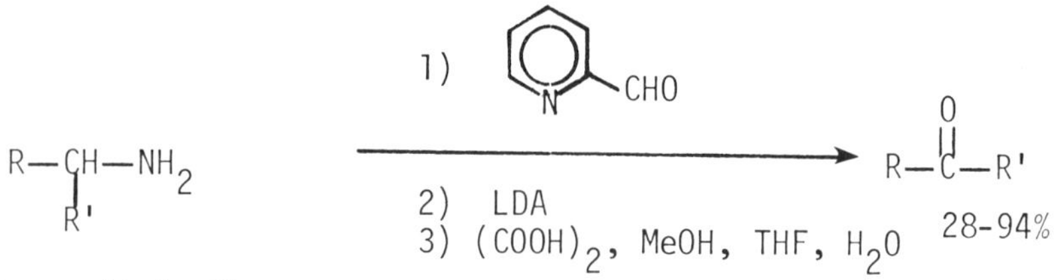

R = alkyl, Ph
R' = H, Me, cyclic with R

II.C-2   B.M. Trost and G. Liu, _J. Org. Chem._, 46, 4617 (1981).

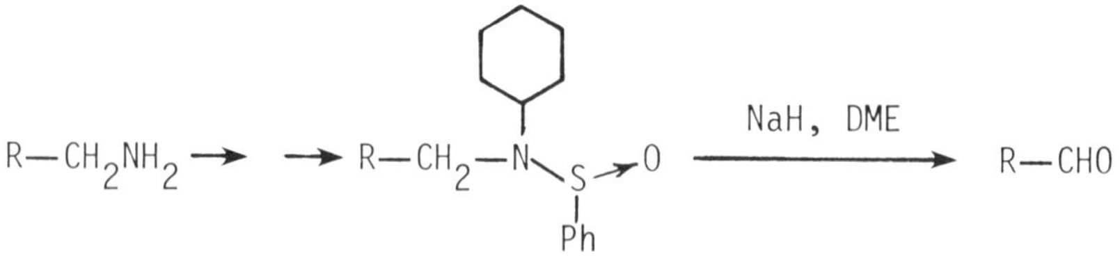

R = naphthyl, vinyl   66-82%

II.C-3   M.R. Galobardes and H.W. Pinnick, <u>Tetrahedron Lett.</u>, 22, 5235 (1981).

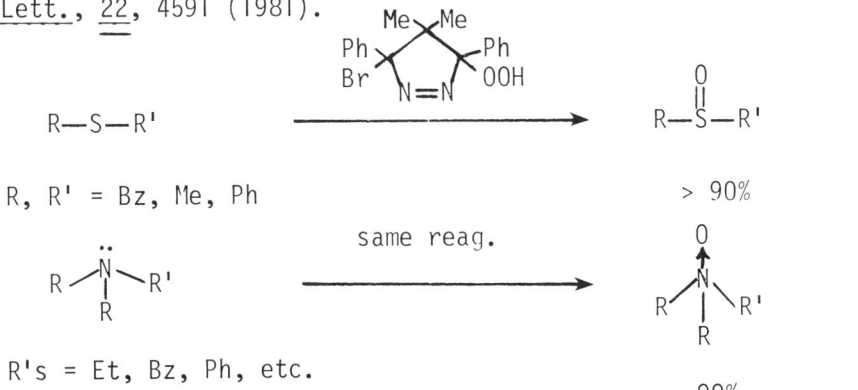

$$\begin{array}{c} H \\ \diagdown \\ \quad C \diagup NO_2 \\ R \diagup \quad \diagdown R' \end{array} \xrightarrow[\text{2) MoO}_5\text{·pyr·HMPA}]{\text{1) LDA}} \begin{array}{c} O \\ \| \\ C \\ R \diagup \quad \diagdown R' \end{array}$$

R, R' = alkyl, Bz   same conditions   ~ 70-90%

$$R-CH_2-NO_2 \longrightarrow R-COOH$$

R = Me, Ph   69-74%

II.C-4   H. Ku and J.R. Barrio, <u>J. Org. Chem.</u>, 46, 5239 (1981)

$$R-\!\!\left\langle\!\bigcirc\!\right\rangle\!\!-NH_2 \xrightarrow[\substack{\text{3) Me}_3\text{SiCl} \\ \text{LiBr or NaI}}]{\substack{\text{1) NaNO}_2\text{, HCl} \\ \text{2) K}_2\text{CO}_3\text{, Et}_2\text{NH}}} R-\!\!\left\langle\!\bigcirc\!\right\rangle\!\!-Br$$
(or I)

R = H, Me, Ac, CN   60-95%

II.D.   Amine Oxidations

II.D-1/II.E-1   A.L. Baumstark and D.R. Chrisope, <u>Tetrahedron Lett.</u>, 22, 4591 (1981).

$$R-S-R' \xrightarrow{\substack{Me \diagup Me \\ Ph \diagdown \diagup Ph \\ Br \diagdown_{N=N} \diagup OOH}} \begin{array}{c} O \\ \| \\ R-S-R' \end{array}$$

R, R' = Bz, Me, Ph   > 90%

$$R-\overset{\displaystyle ..}{\underset{\displaystyle R}{N}}-R' \xrightarrow{\text{same reag.}} R-\overset{\displaystyle \uparrow O}{\underset{\displaystyle R}{N}}-R'$$

R's = Et, Bz, Ph, etc.   > 90%

II.D-2/II.E-2   F.M. Menger and C. Lee, Tetrahedron Lett., 22, 1655 (1981).

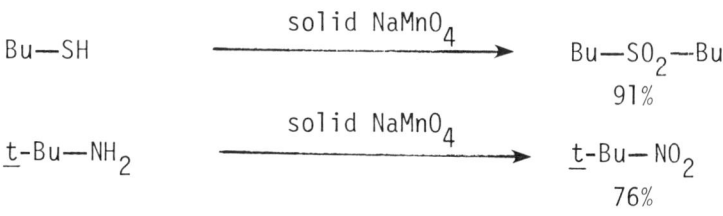

II.E. Sulfur Oxidations

II.E-3   J.Drabowicz and M. Mikolajczyk, Synth. Comm., 11, 1025 (1981).

$$R—S—R' \xrightarrow[\text{MeOH}]{H_2O_2} R—\overset{\overset{\text{O}}{\|}}{S}—R'$$

R, R' = alkyl, Ph, Bz, etc. $\sim$ 90%

II.E-4   Y. Watanabe, T. Numata, and S. Oae, Synthesis, 204 (1981).

$$R—S—R' \xrightarrow[\text{MeOH/H}_2\text{O/MeCN}]{TiCl_3,\ H_2O_2} R—\overset{\overset{\text{O}}{\|}}{S}—R'$$

R, R' = alkyl, aryl, Bz, cyclic, etc.   93-100%

II.E-5   B.M. Trost and D.P. Curran, Tetrahedron Lett., 22, 1287 (1981).

$$R—S—R' \xrightarrow[\text{CH}_3\text{OH}]{KHSO_5} R—\overset{\overset{\text{O}}{\|}}{\underset{\underset{\text{O}}{\|}}{S}}—R'$$

R= alkyl,aryl, vinyl, etc.
R' = Me, Ph   77-100%

II.E-6   D. Scholz, _Monats. Chem._, 112, 241 (1981).

$$R-S-R'$$
or
$$\underset{\overset{\|}{O}}{R-S-R'}$$

$$\xrightarrow[\text{CH}_2\text{Cl}_2,\ \text{AcOH}]{\text{Et}_3\overset{\overset{\text{Bz}}{|}}{N}{}^{\oplus}\ \text{MnO}_4{}^{\ominus}}$$

$$\underset{\overset{\|}{O}}{\overset{\overset{O}{\|}}{R-S-R'}}$$

R, R' = alkyl, aryl, etc.

II.F.   Oxidative Additions to C—C Multiple Bonds

1. Epoxidations

II.F.1-1   Y. Sawaki and Y. Ogata, _J. Am. Chem. Soc._, 103, 2049 (1981).

$$\text{C}=\text{C} \xrightarrow[\underset{\overset{\|}{O}\ \ \overset{|}{O}H}{Ph-C-CH-Ph}]{O_2,\ h\upsilon} \text{C}-\text{C}$$

widely varying yields

II.F.1-2   E.D. Mihelich, K. Daniels, and D.J. Eickhoff, _J. Am. Chem. Soc._, 103, 7690 (1981).

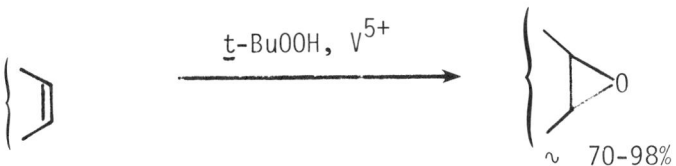

$$\xrightarrow{\underline{t}\text{-BuOOH},\ V^{5+}}$$

~ 70-98%

Generally highly stereo-
selective

This paper discusses how to predict the direction of stereoselectivity of this reaction.

II.F.1-3   B.E. Rossiter, T. Katsuki, and K.B. Sharpless,
J. Am. Chem. Soc., 103, 464 (1981).

Use of t-Bu-OOH, Ti(O-i-Pr)$_4$, and dialkyl tartrate esters to
accomplish asymmetric epoxidations of allylic alcohols
leading to key synthetic intermediates.   Yields are ∿50-80%,
>90% ee.

II.F.1-4   R.D. Bach and J.W. Knight, Org. Syn., 60, 63
(1981).

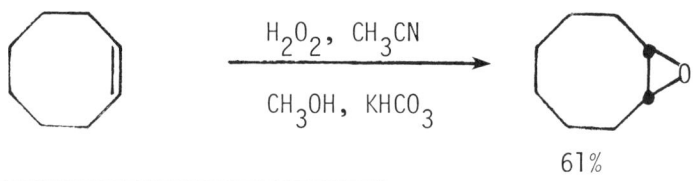

H$_2$O$_2$, CH$_3$CN

CH$_3$OH, KHCO$_3$

61%

II.F.1-5   C.J. Stark, Tetrahedron Lett., 22, 2089 (1981).

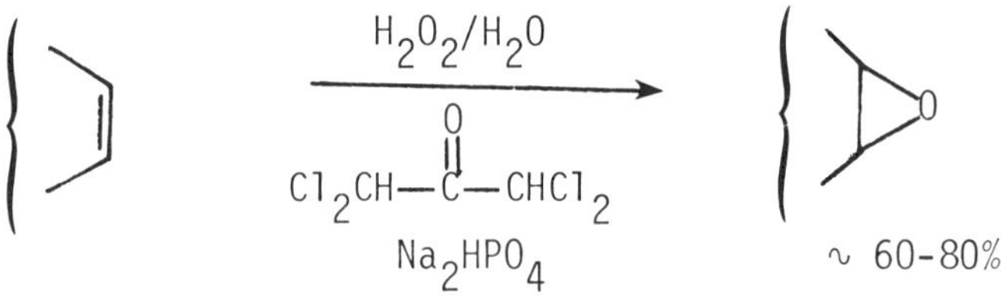

H$_2$O$_2$/H$_2$O

Cl$_2$CH—C—CHCl$_2$

Na$_2$HPO$_4$

∿ 60-80%

II.F.1-6   F. Camps et al., Tetrahedron Lett., 22, 3895
(1981).

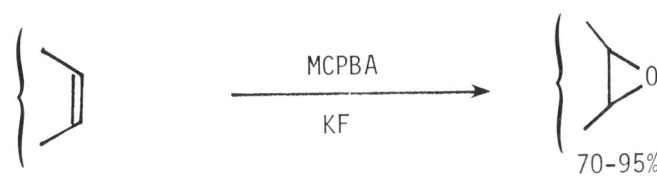

$$\begin{array}{c} \text{MCPBA} \\ \hline \text{KF} \end{array}$$

70-95%

II.F.1-7   A. Mizuno, Y. Hamada, and T. Shioiri, Chem.
Pharm. Bull., 29, 1774 (1981).

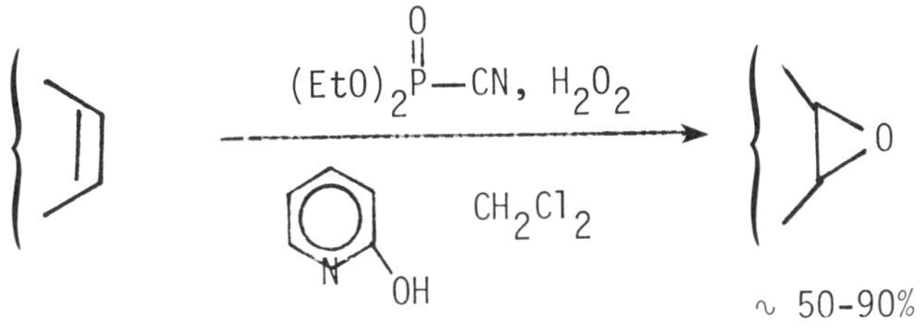

$$(EtO)_2\overset{\overset{O}{\|}}{P}-CN, H_2O_2$$

$$CH_2Cl_2$$

~ 50-90%

II.F.1-8   S. Torii et al., J. Org. Chem., 46, 3312 (1981).

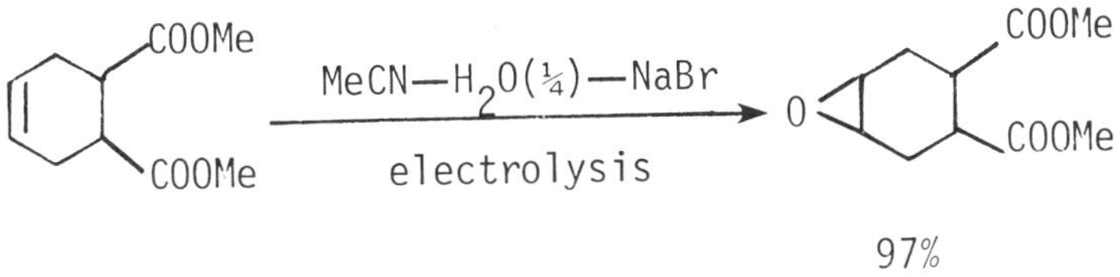

$$\begin{array}{c} MeCN-H_2O(\frac{1}{4})-NaBr \\ \hline \text{electrolysis} \end{array}$$

COOMe
COOMe

97%

II.F.2. Hydroxylation

II.F.2-1 A.G. Abatjoglou and D.R. Bryant, Tetrahedron
Lett., 22, 2051 (1981).

OsO_4-catalyzed oxidations of olefins to glycols may use

diphenyl selenoxide as the stoichiometric oxidant. Yields

are ∿ 80-90%.

II.F.2-2 G.A. Olah, A.P. Fung, and D. Meidar, Synthesis,
280 (1981).

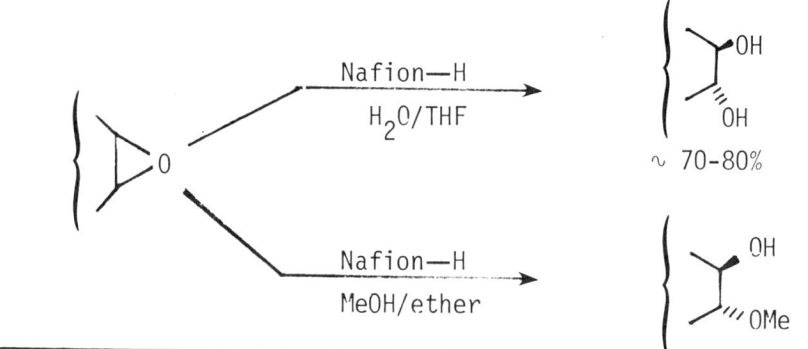

II.F.2-3 S. Torii et al., J. Am. Chem. Soc., 103, 4606
(1981).

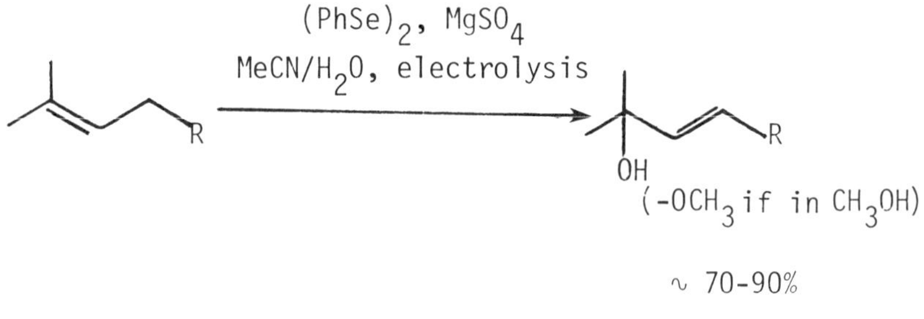

II.F.2-4/II.F.3-1   J. Tsuji et al., Tetrahedron Lett., 22, 131 (1981).

$$R\diagdown\diagup\diagdown COOMe \xrightarrow[C_5H_{11}ONO]{PdCl_2, KOAc} R\diagdown\diagup\diagup COOMe$$
$$\underset{OAc}{}$$

R = alkyl   ~ 50%

II.F.3.   Other Oxidative Additions to C-C Multiple Bonds

II.F.3-2   S. Torii et al., J. Org. Chem., 46, 3312 (1981).

$$\xrightarrow[\text{electrolysis}]{MeCN-H_2O(1/19)-NaBr-H_2SO_4}$$

Br, COOMe OH'' COOMe 72%

$$\xrightarrow[\text{electrolysis}]{MeCN-H_2O(¼)-NaBr-H_2SO_4}$$

Br, COOMe Br'' COOMe 92%

COOMe / COOMe

II.F.3-3   B. Damin, J. Garapon, and B. Sillion, Synthesis, 362 (1981).

$$R-CH{\equiv}CH-R' \xrightarrow[H_3O^+/acetone]{Ts-\overset{Cl}{\underset{}{N}}{}^{\ominus}Na^{\oplus}} \underset{Cl}{R}\diagdown CH-CH\underset{OH}{\diagup R'}$$

40-75%

R, R' = H, alkyl, cyclic

II.F.3-4   M.A. Loreto, L. Pellacani, and P.A. Tardella, Synth. Comm., 11, 287 (1981).

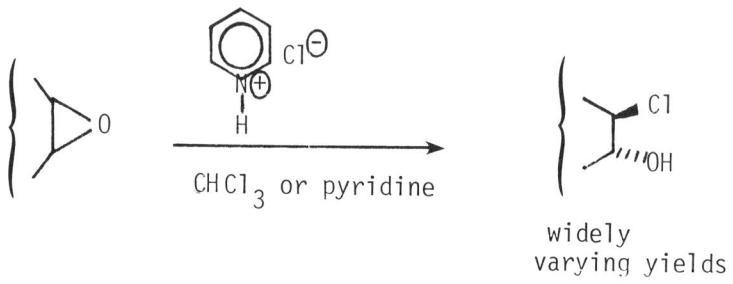

widely
varying yields

II.F.3-5   L.E. Overman and Lee A. Flippin, Tetrahedron Lett., 22, 195 (1981).

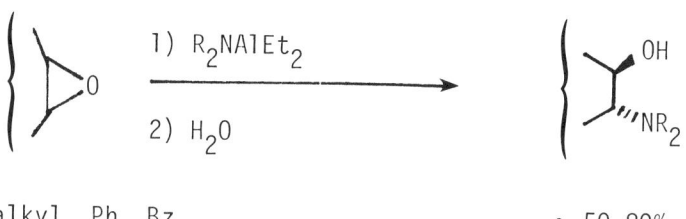

R = H, alkyl, Ph, Bz

∿ 50-80%

II.F.3-6   J. Barluenga, L.A. Cires, and G. Asensio, Synthesis, 376 (1981).

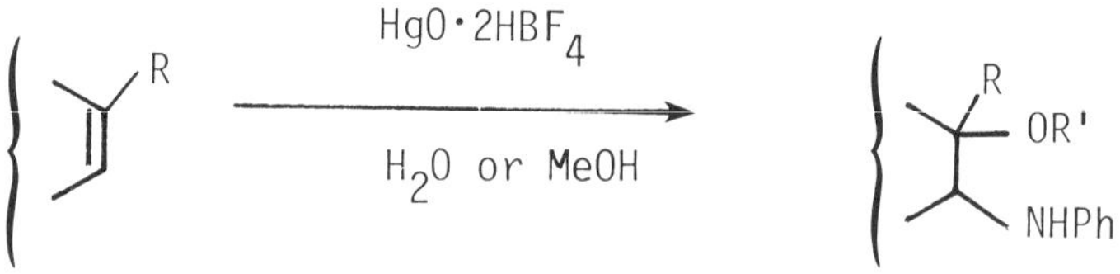

R = H, Me, Ph

(R' = H, Me)
54-80%

II.F.3-7   G. Cardillo et al., J.C.S. Chem. Comm., 465
(1981).

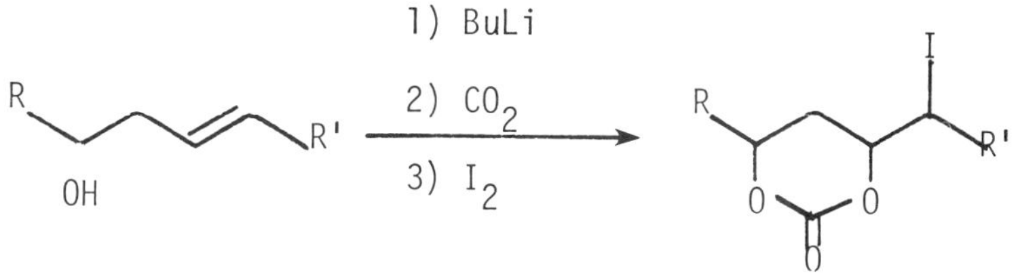

$$\text{1) BuLi}$$
$$\text{2) } CO_2$$
$$\text{3) } I_2$$

R, R' = H, alkyl, terpenoid, etc.   60-90%
Similar reaction for allylic alcohols.

II.F.3-8   A. Kohda, K. Ueda, and T. Sato, J. Org. Chem., 46,
509 (1981).

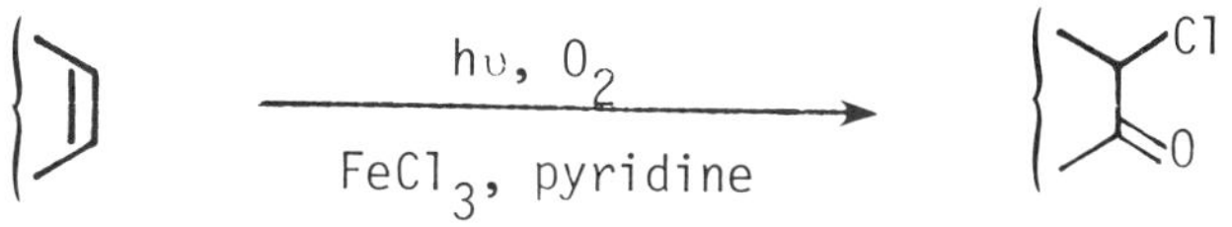

$$h\upsilon, O_2$$
$$FeCl_3, \text{ pyridine}$$

widely varying yields

II.F.3-9   K. Taya et al., J.C.S. Chem. Comm., 1274 (1981).

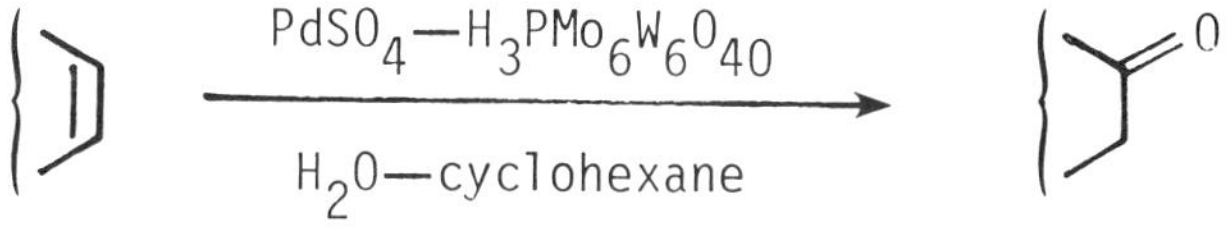

$$PdSO_4 - H_3PMo_6W_6O_{40}$$
$$H_2O - \text{cyclohexane}$$

simple cycloalkenes $\sim$ 80-100%

II.F.3-10   S. Uemura et al., Bull. Chem. Soc. Japan, 54,
2843 (1981).

$$R-C\equiv C-R' \xrightarrow{SO_2Cl_2} \begin{matrix} R \\ \diagdown \\ Cl \end{matrix} C=C \begin{matrix} Cl \\ \diagup \\ R' \end{matrix}$$

R, R' = Ph, H, alkyl

∿ 50-90%

II.F.3-11   P. Muller and J. Godoy, Helv. Chim. Acta, 64,
2531 (1981).

$$R-C\equiv C-R' \xrightarrow[RuCl_2L_3]{PhIO} \begin{matrix} R \\ \diagdown \\ O \end{matrix} C-C \begin{matrix} O \\ \diagup\diagdown \\ R' \end{matrix}$$

65-86%

$$R-C\equiv C-H \xrightarrow{\text{same conditions}} R-COOH$$

R = Ph, alkyl   69-81%

II.G. Phenol⟶Quinone Oxidations

II.G-1   D.N. Gupta, P. Hodge, and J.E. Davies, J.C.S.
Perkin I, 2970 (1981).

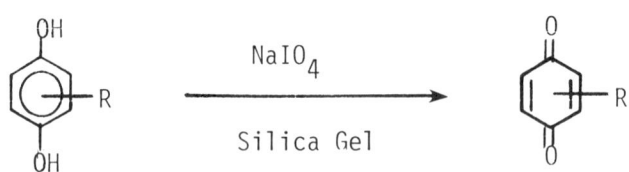

R = H, alkyl, benzo-, etc.   widely varying yields

II.G-2   D.H.R. Barton et al., J.C.S. Perkin I, 1473 (1981).

Benzeneseleninic anhydride converts phenols into o-quinones, e.g.:

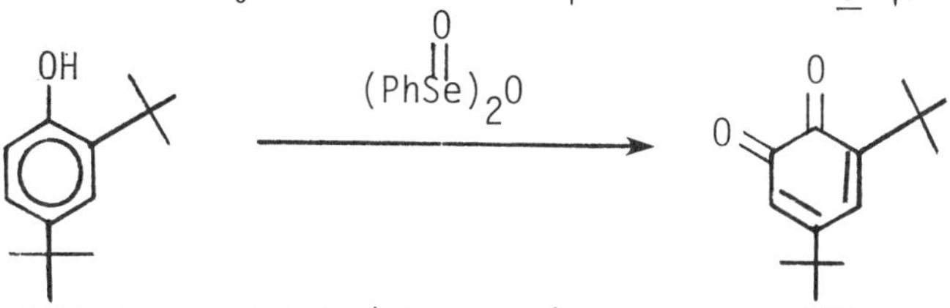

Hydroquinones are converted to p-quinones.   68%

II.H. Oxidative Cleavages

II.H-1   T.R. Beebe, P. Hii, and P. Reinking, J. Org. Chem., 46, 1927 (1981).

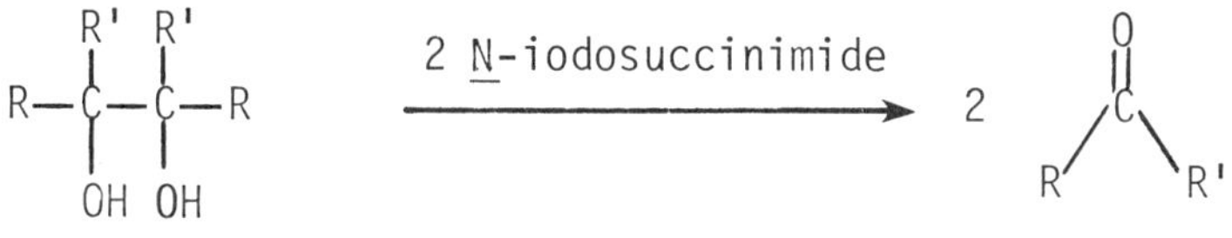

R,R' = alkyl, Ph, H

~ 90%

II.H-2   K. Kaneda et al., Tetrahedron Lett., 22, 2595 (1981).

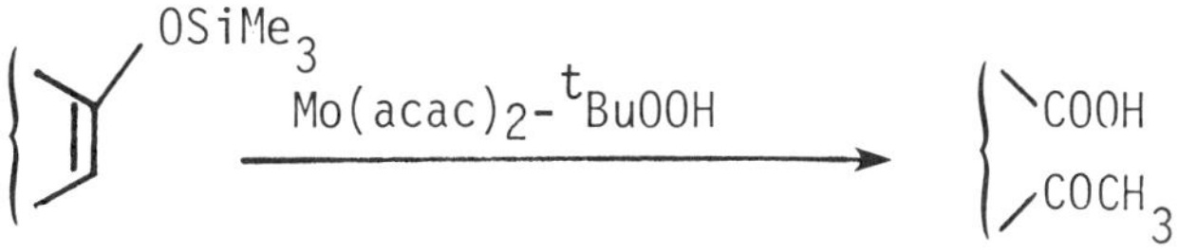

~ 90%

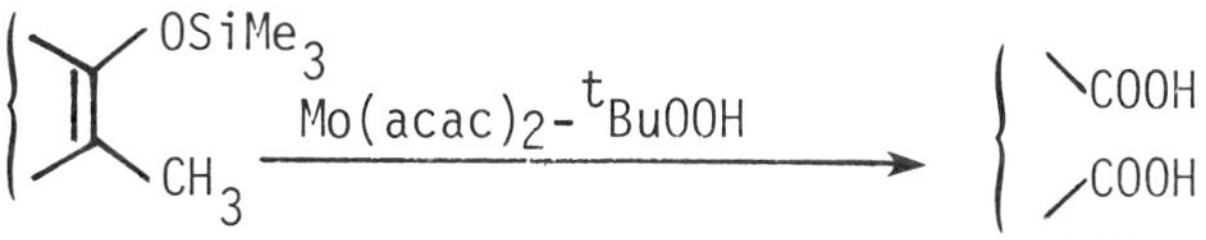

~ 90%

II.H-3   Y.A. Serguchev and I.P. Beletskaya, <u>Russ. Chem. Rev.</u>, <u>49</u>, 1119 (1981).

Review: "Oxidative Decarboxylation of Carboxylic Acids"

II.I. Photosensitized Oxygenations

II.I-1   A. Kohda, K. Ueda, and T. Sato, <u>J. Org. Chem.</u>, <u>46</u>, 509 (1981).

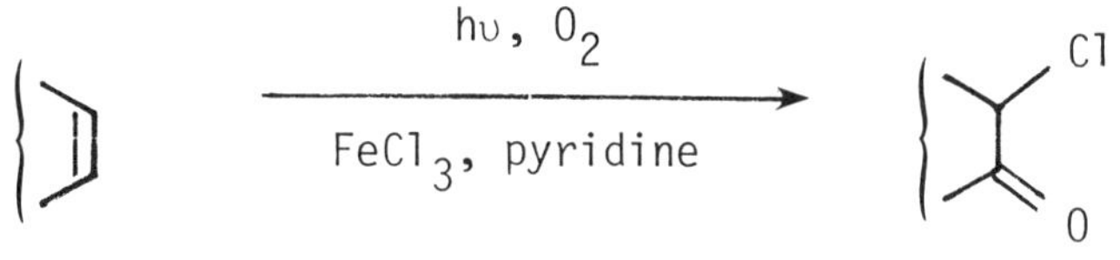

widely varying yields

II.I-2   Y. Sawaki and Y. Ogata, <u>J. Am. Chem. Soc.</u>, <u>103</u>, 2049 (1981).

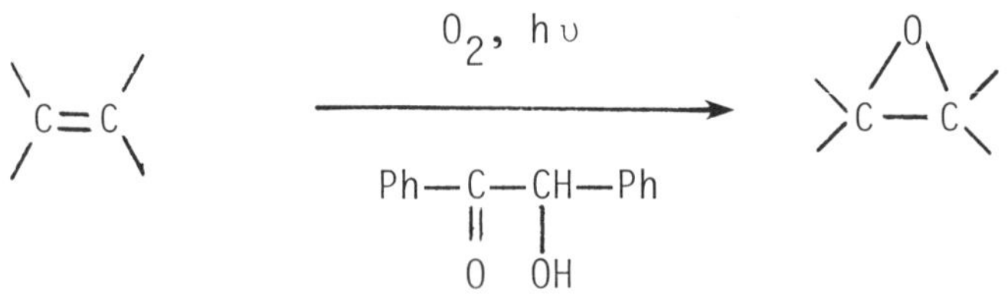

widely varying yields

II.I-3   A. Albini, Synthesis, 249 (1981).

Review: "Photosensitization in Organic Syntheses"

II.J. Dehydrogenation

II.J-1   D.H.R. Barton, J.W. Morzycki, and W.B. Motherwell, J.C.S. Chem. Comm., 1044 (1981).

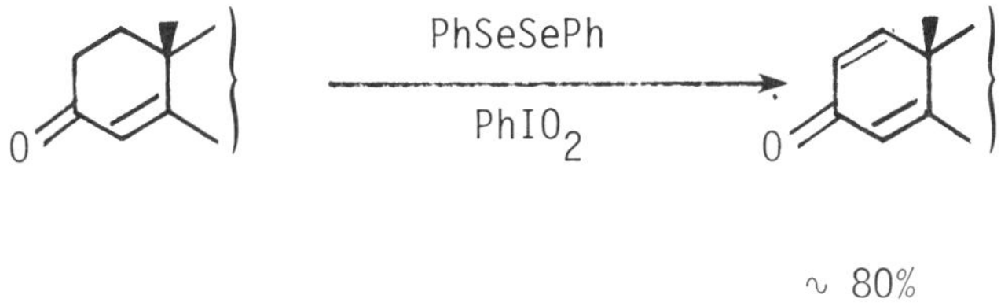

$\sim$ 80%

II.J-2   D. Liotta et al., J. Org. Chem., 46, 2920 (1981).

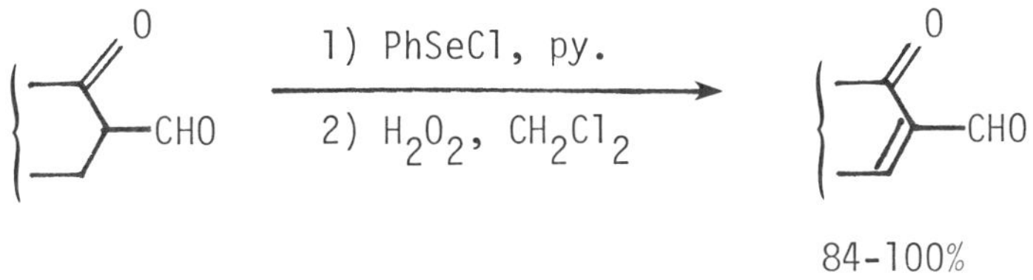

84-100%

II.J-3   J.P. Marino and R.D. Larsen, Jr., J. Am. Chem. Soc., 103, 4642 (1981).

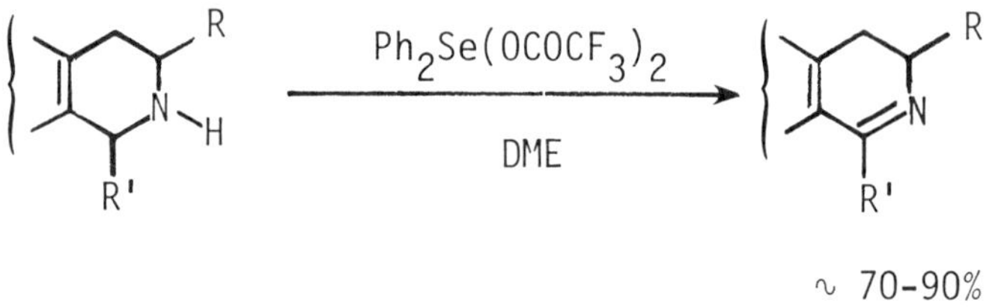

$\sim$ 70-90%

may also be used to form

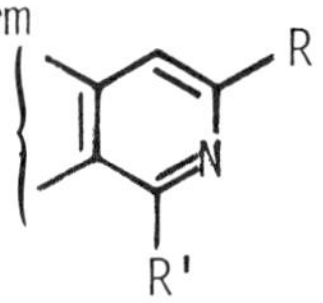

II.K. Other Oxidations and Reviews

II.K-1   D.T. Sawyer and J.S. Valentine, Accounts Chem. Res., 14, 393 (1981).

Review: "How Super is Superoxide?"

II.K-2   A.J. Mancusco and D. Swern, Synthesis, 165 (1981).

Review: "Activated Dimethyl Sulfoxide: Useful Reagents for Organic Synthesis"

II.K-3   H.H. Wasserman and J.L. Ives, Tetrahedron, 37, 1825 (1981).

Review: "Singlet Oxygen in Organic Synthesis"

II.K-4   H. Mimoun, Pure and Appl. Chem., 53, 2389 (1981).

    Review: "Activation of Molecular Oxygen and Selective

              Oxidation of Olefins Catalyzed by Group VIII

              Transition Metal Complexes"

II.K-5   D.H.R. Barton et al., Tetrahedron, 37, Supp. #1,
73 (1981).

    "Functional Group Oxidation by Pentavalent Organobismuth

    Reagents"

    A full paper summarizing the uses of pentavalent organo-
    bismuth reagents in oxidizing allylic, benzylic, and
    other alcohols.

II.K-6   I.P. Beletskaya and D.I. Makhon'kov, Russ. Chem.
Rev., 50, 534 (1981).

    Review: "Oxidation of Alkyl Derivatives of Aromatic

              Hydrocarbons by Transition Metal Salts"

II.K-7   G.R. Krow, Tetrahedron, 37, 2697 (1981).

    Review: "Oxygen Insertion Reactions of Bridged Bicyclic

              Ketones"

III

REDUCTIONS

III.A.   C≡O Reductions

(Reductions of carboxylic acids to aldehydes and alcohols are included in section III.F.1.)

III.A-1   Y. Inouye et al., J. Am. Chem. Soc., 103, 4613 (1981).

$$\underset{R}{\overset{O}{\underset{R'}{\parallel}}} C \xrightarrow{\text{chiral dihydropyridine}} R-\underset{*}{\overset{OH}{\underset{|}{C}}}H-R'$$

R = Ph, 2-pyridyl
R' = -COOEt, -CF$_3$, Me, Ph

∿ 70%
up to 98% ee (R)

III.A-2   P. Jouin, C.B. Troostwijk, and R.M. Kellogg, J. Am. Chem. Soc., 103, 2091 (1981).

$$R-\overset{O}{\overset{\parallel}{C}}-R' \xrightarrow[\text{Mg}^{++}]{\text{chiral macrocyclic dihydropyridine}} R-\overset{OH}{\underset{*}{\underset{|}{C}}}H-R'$$

R = subst. Ph
R' = -COOR or -CONHR

∿ 60-80%
up to 90% ee
(predominantly S)

III.A-3   R. Noyori, <u>Pure and Appl. Chem.</u>, <u>53</u>, 2315 (1981).

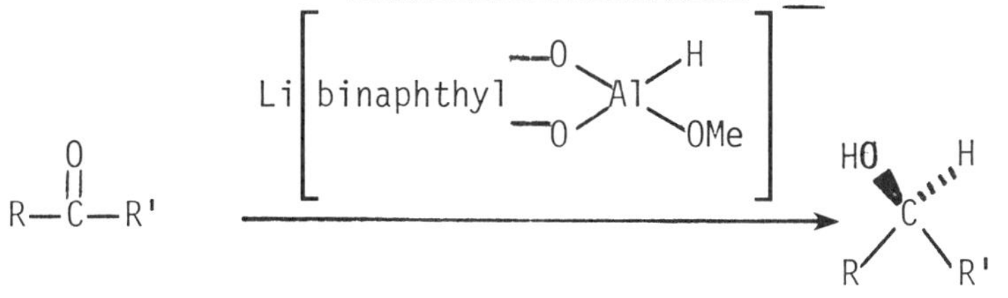

R = Ph, vinyl, acetylenic, etc.   ∿ 70-90%
R' = alkyl   ∿ 80-90% ee

III.A-4   M. Nishizawa, M. Yamada, and R. Noyori, <u>Tetrahedron</u>
<u>Lett.</u>, <u>22</u>, 247 (1981).

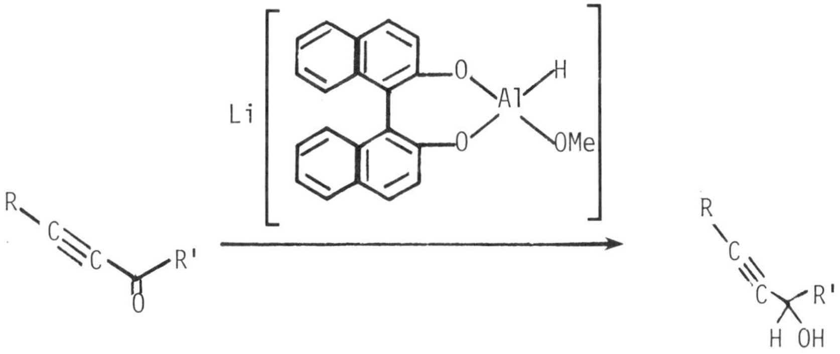

R = H, <u>n</u>-alkyl, -COOCH$_3$ ∿70-90%
R' = alkyl, -CH$_2$CH$_2$COOCH$_3$ ∿90% ee

III.A-5 A. Hirao et al., Bull. Chem. Soc. Japan, 54, 1424 (1981).

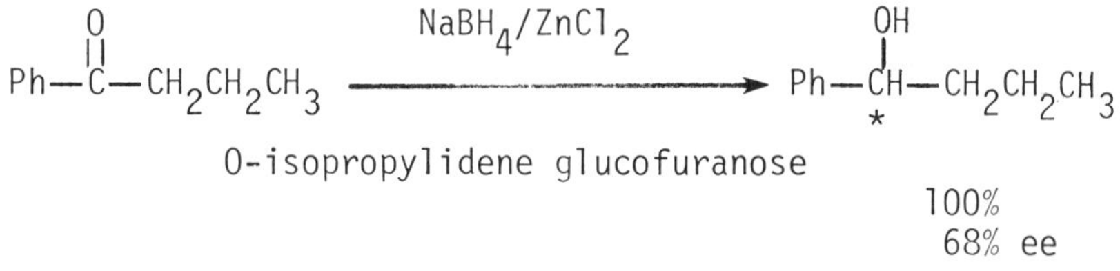
O
Ph—C—CH2CH2CH3
NaBH4/ZnCl2
O-isopropylidene glucofuranose
OH
Ph—CH—CH2CH2CH3
*
100%
68% ee

III.A-6 M. Hayashi et al., Bull. Chem. Soc. Japan, 54, 3033 (1981).

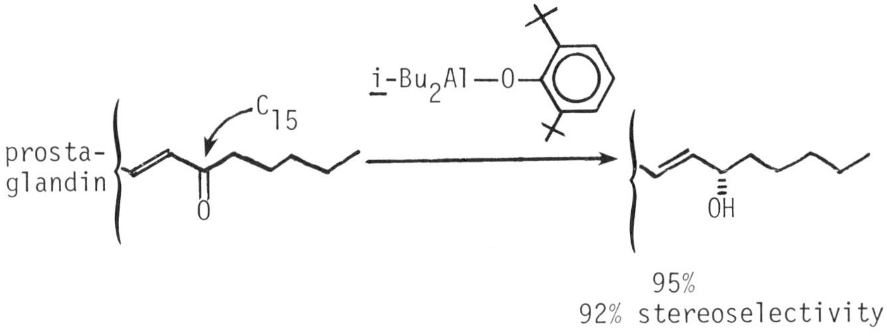
i-Bu2Al—O
C15
prosta-
glandin
O
OH
95%
92% stereoselectivity

III.A-7 J.H. Babler and S.J. Sarussi, J. Org. Chem., 46, 3367 (1981).

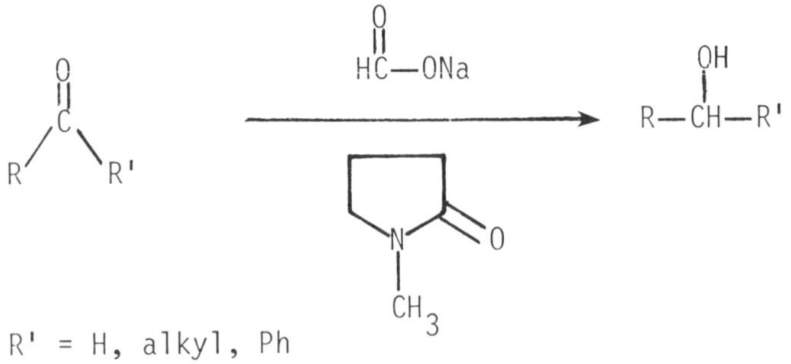
O
C
R R'
O
HC—ONa
N
O
CH3
OH
R—CH—R'

R, R' = H, alkyl, Ph

III.A-8   A.P. Krapcho and D.A. Seidman, Tetrahedron Lett., 22, 179 (1981).

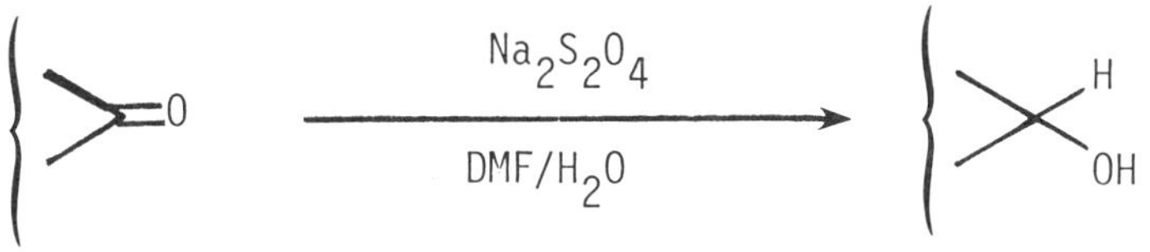

cyclic, bicyclic   ∿ 60-100%

Cyclohexanones yield mainly equatorial alcohols, while bicyclic ketones give mainly endo alcohols.

III.A-9   R. Contreras et al., Synthesis, 214 (1981).

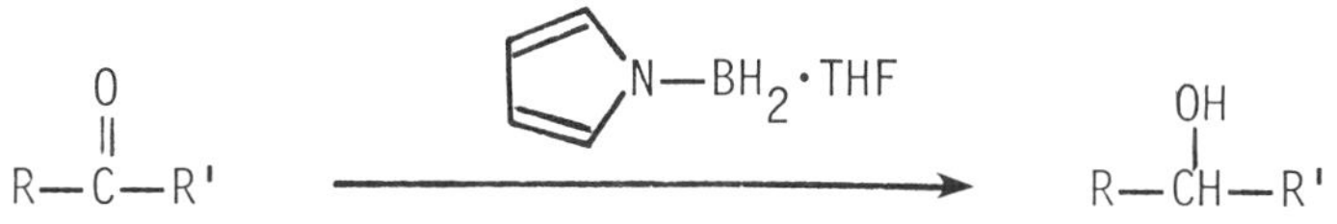

R, R' = H, alkyl, Ph   ∿ 90-100%

III.A-10   G. Bram, E. D'Incan, and A. Loupy, J.C.S. Chem. Comm., 1066 (1981).

NaBH$_4$ reductions of ketones and aldehydes may use "Fontainebleau sand" instead of alumina as a solid support. Most yields are high, above 80%.

III.A-11   R.J.P.Corriu et al., J.C.S. Chem. Comm., 121
(1981); Tetrahedron, 37, 2165 (1981).

$$
R-\overset{\overset{\displaystyle O}{\|}}{C}-R' \quad \xrightarrow[\text{KF or CsF}]{\begin{array}{c}(EtO)_3SiH \\ \text{or } Me(EtO)_2SiH\end{array}} \quad R-\overset{\overset{\displaystyle OH}{|}}{CH}-R'
$$

R,R' = H, Ph, styryl, alkyl   ∿ 70-90%

Aldehydes may be reduced in the present of ketones.

III.A-12   J.H. Babler and B.J. Invergo, Tetrahedron Lett.,
22, 621 (1981).

$$
R-CHO \quad \xrightarrow[\text{2) } H_2O]{\text{1) } \left[HCOOH + 2EtMgBr \right]} \quad R-CH_2OH
$$

R = alkyl, aryl   ∿ 70-90%

Aldehydes are reduced in the presence of ketones.

III.A-13   A.L. Gemal and J.L. Luche, J. Am. Chem. Soc.,
103, 5454 (1981).

$$
\xrightarrow{NaBH_4-CeCl_3}
$$

∿ 90%

III.A-14   T. Satoh et al., Chem. Lett., 1029 (1981).

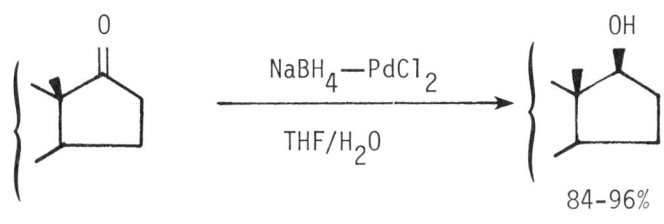

84-96%

III.A-15   T. Nakata, T. Tanaka, and T. Oishi, Tetrahedron
Lett., 22, 4723 (1981).

R's = Me, Et, Bu, H   ~ 80%
                                  predominately erythro

III.A-16   G.W.J. Fleet and P.J.C. Harding, Tetrahedron
Lett., 22, 675 (1981).

$$R—CHO \xrightarrow{(Ph_3P)_2CuBH_4} R—CH_2OH$$

R = alkyl, aryl, vinyl, etc.

$$R—\overset{\overset{\displaystyle O}{\|}}{C}—R' \xrightarrow{(Ph_3P)_2CuBH_4} R—\overset{\overset{\displaystyle OH}{|}}{CH}—R'$$

R, R' = alkyl, aryl, vinyl, etc.

Aldehydes may be reduced in the presence of ketones.

III.A-17   S. Yamaguchi, K. Kabuto, and F. Yasuhara, Chem. Lett., 461 (1981).

Aldehydes may be reduced in the presence of ketones by

$$NaHB\left(-O-\bigcirc\diagdown\diagup\right)_3$$

Yields are > 90%.

III.A-18   A.L. Gemal and J.L. Luche, Tetrahedron Lett., 22, 4077 (1981).

$NaBH_4$ in the presence of $ErCl_3$ reduces conjugated aldehydes in the presence of non-conjugated ones. $CrCl_3/NaBH_4$ is also effective in selective reduction of ketones in the presence of aldehydes.

III.A-19   S. Krishnamurthy, J. Org. Chem., 46, 4628 (1981).

$$R\!-\!CHO \xrightarrow{\text{LiAlH}\left[OCEt_3\right]_3} R\!-\!CH_2OH$$

100%

Reduces aldehydes in the presence of ketones.

III.B. Nitrile Reductions

III.B-1 H.C. Brown, Y.M. Choi, and S. Narasimhan, Synthesis, 605 (1981).

$$R—C\equiv N \xrightarrow[\text{3) NaOH}]{\begin{array}{c}\text{1) } Me_2S\cdot BH_3 \\ \text{2) HCl, } H_2O\end{array}} R—CH_2NH_2$$

R = 1°, 2°, 3° alkyl, subst. Ph   61-88%

III.C. Reduction of Sulfur Compounds

III.C-1   R. Marchelli et al., Synthesis, 141 (1981).

$$R—\overset{\overset{\text{O}}{\|}}{S}—R' \xrightarrow{\text{t-BuBr}} R—S—R'$$

R, R' = 1° alkyl, Bz, subst. Ph   80-100%

III.C-2   A.H. Schmidt and M. Russ, Chem. Ber., 114, 822 (1981).

$$R—\overset{\overset{\text{O}}{\|}}{S}—R' \xrightarrow[\text{Zn}]{Me_3SiCl} R—S—R'$$

77-96%

R = alkyl, Bz, subst. Ph

III.D.   N—O Reductions

III.D-1   A. Nose and T. Kudo, <u>Chem. Pharm. Bull.</u>, <u>29</u>, 1159 (1981).

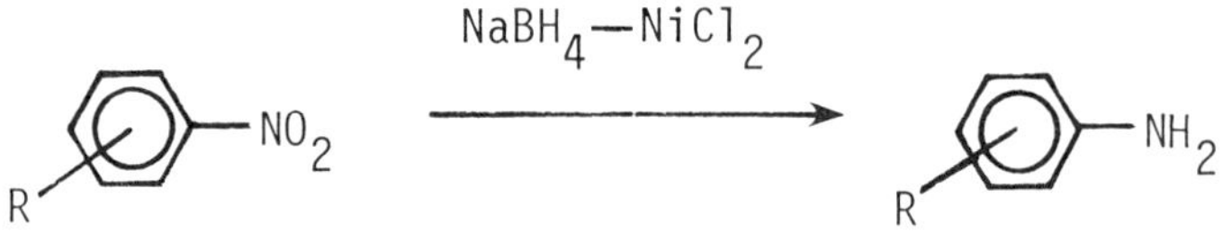

R = H, Me, OMe, Cl, OH, COOH   76-95%

III.D-2   T. Satoh <u>et al.</u>, <u>Chem. Pharm. Bull.</u>, <u>29</u>, 1443 (1981).

$$Ar-NO_2 \xrightarrow[\text{EtOH}]{NaBH_4-SnCl_2} Ar-NH_2$$

Ar = subst. Ph   65-95%

III.D-3   J.H. Babler and S.J. Sarussi, <u>Synth. Comm.</u>, <u>11</u>, 925 (1981).

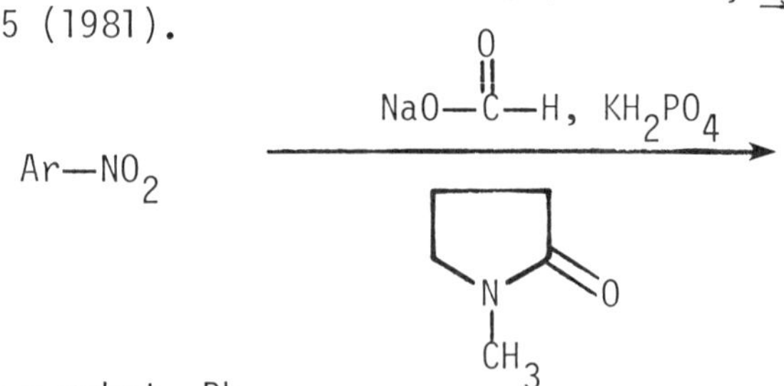

Ar = subst. Ph

III.D-4   H. Alper and M. Gopal, J. Org. Chem., 46, 2593
(1981).

$$\left\{ \begin{array}{c} \end{array} \right\} \!\! \diagup\!\! \searrow\!\!\! N\!\!\rightarrow\!\! 0 \quad \xrightarrow[\text{or } Fe_3(CO)_{12}/Al_2O_3]{Mo(CO)_6/Al_2O_3} \quad \left\{ \begin{array}{c} \end{array} \right\} \!\! \diagup\!\! \searrow\!\! N\colon$$

Also reduces azoxybenzenes to azobenzenes.

III.E.   C—C Multiple Bond Reductions

1.   C=C Reductions

(Reductions to form amino acids are included in section
VI.A.6.)

III.E.1-1   R. Bar and Y. Sasson, Tetrahedron Lett., 22,
1709 (1981).

Use of phase-transfer conditions to accelerate the reduction

of olefins by formate ion in the presence of $RuCl_2L_3$.

III.E.1-2   J.W. Suggs et al., Tetrahedron Lett., 22, 303
(1981).

widely varying yields

III.E.1-3   S. Antus, A. Gottsegen, and M. Nogradi, Synthesis, 574 (1981).

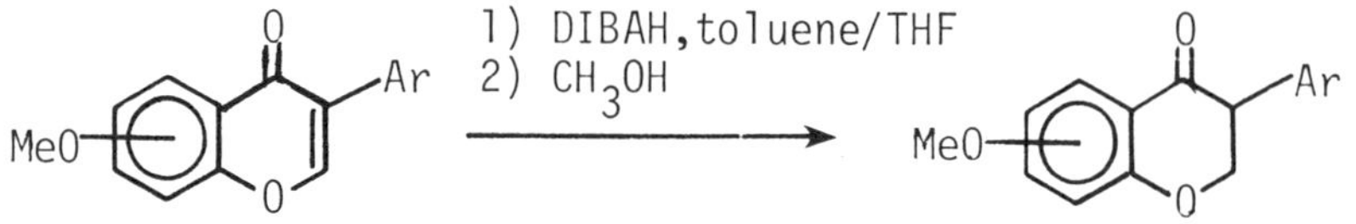

75-90%

III.E.1-4   R. Brettle and S.M. Shibib, J.C.S. Perkin I,
2912 (1981).

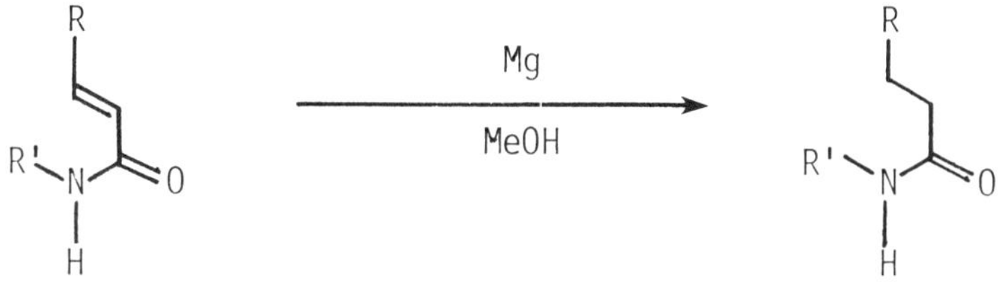

R = H, Me, Ph   50-92%
R' = H, Ph, Et

III.E.1-5   V. Caplar, G. Comisso, and V. Sunjic, Synthesis,
85 (1981).

    Review: "Homogeneous Asymmetric Hydrogenation"

III.E.2.   C≡C Reductions

III.E.2-1   A.U. Ronchi et al., J. Org. Chem., 46, 5340 and 5344 (1981).

$$R-C\equiv C-R' \xrightarrow[\text{potassium-graphite}]{\overset{H_2}{\text{nickel on}}} \underset{H}{\overset{R}{>}}C=C\underset{H}{\overset{R'}{<}}$$

70-98%

R, R' = H, alkyl, Ph, -COOR

III.E.2-2   D. Savoia et al., J.C.S. Chem. Comm., 540 (1981).

$$R-C\equiv C-R' \xrightarrow[\text{Pd-graphite}]{H_2,\ \text{MeOH}} \underset{R}{\overset{H}{>}}C=C\underset{R'}{\overset{H}{<}}$$

>90%

R, R' = H, alkyl, Ph, ester

III.E.3.   Reduction of Aromatic Rings

III.E.3-1   M. Yasuda, C. Pac, and H. Sakurai, J. Org. Chem., 46, 788 (1981).

R = Me, OMe   ∿60-80%

Products very similar to Birch reduction.

III.F.   Hydrogenolysis of Hetero Bonds

1. C—O ⟶ C—H

III.F.1-1   F. Sato, T. Jinbo, and M. Sato, Synthesis, 871, (1981).

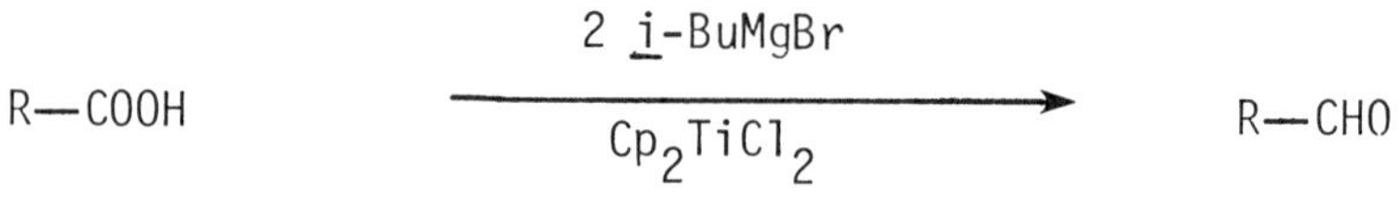

$$R—COOH \quad \xrightarrow[\text{Cp}_2\text{TiCl}_2]{2\ \underline{i}\text{-BuMgBr}} \quad R—CHO$$

48-73%

R = Ph, Bz, alkyl

III.F.1-2   J.C. Craig et al., Synthesis, 303 (1981).

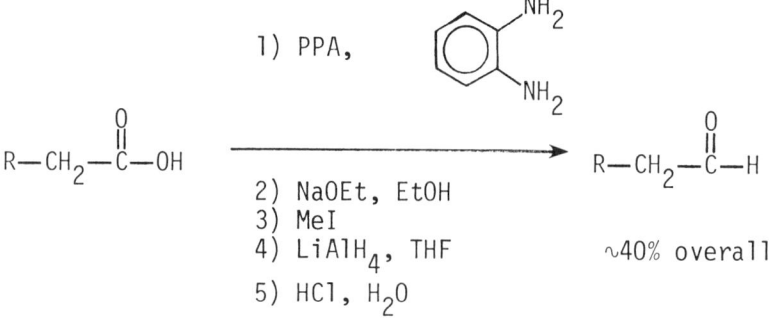

R = Me, Ph, furyl

III.F.1-3   R.J.P. Corriu et al., Synthesis, 558 (1981).

$$\underset{\substack{\| \\ \text{O}}}{\text{R}-\text{C}-\text{OR}'} \xrightarrow{\begin{array}{l}1)\ \ \text{HSi(OEt)}_3,\ \text{CsF} \\ 2)\ \ \text{H}_2\text{O}\end{array}} \text{R}-\text{CH}_2\text{OH}$$

∿60-90%

R = Ph, 1° alkyl; may contain olefinic double bonds.
R' = Me, Et, menthyl

_____

III.F.1-4   H.C. Brown and Y.M. Choi, Synthesis, 439 (1981).

$$\text{R}-\text{COOEt} \xrightarrow[\text{THF}]{\text{Me}_2\text{S}\cdot\text{BH}_3} \text{R}-\text{CH}_2\text{OH}$$

89-97%

R = 1°, 2°, 3° alkyl, Ph

$$\underset{\substack{\| \\ \text{O}}}{\text{R}-\text{C}-\text{NH}_2} \xrightarrow[\text{2)\ \ ether,\ HCl}]{\text{1)}\ \ \text{Me}_2\text{S}\cdot\text{BH}_3} \text{R}-\text{CH}_2\text{NH}_2\cdot\text{HCl}$$

∿80-90%

R = 1°, 2°, 3° alkyl, Ph

III.F.1-5   E. Santaniello, P. Ferraboschi, and P. Sozzani,
J. Org. Chem., 46, 4584 (1981).

$$R-\overset{O}{\underset{\|}{C}}-OR' \xrightarrow[\text{polyethylene glycol}]{NaBH_4} R-CH_2OH$$

73-90%

R = alkyl, haloalkyl, subst. Ph
R' = Me, Et

III.F.1-6   S.L. Baxter and J.S. Bradshaw, J. Org. Chem., 46,
832 (1981).

$$R-\overset{O}{\underset{\|}{C}}-OR' \xrightarrow{I} R-\overset{S}{\underset{\|}{C}}-OR' \xrightarrow{Ra-Ni} R-CH_2-OR'$$

R = subst. Ph, t-Bu, heterocyclic
R' = alkyl, Ph

I = MeO—⟨○⟩—P(=S)⟨S—S⟩P(=S)—⟨○⟩—OMe

May be used to form crown ethers.

III.F.1-7   J.R. Rasmussen et al., J. Org. Chem., 46,
4843 (1981).

1) N,N'-thiocarbonyldiimidazole

$$R-OH \xrightarrow[\text{2) } Bu_3SnH]{} R-H$$

(deoxy sugar)

R = protected sugar

∿30-80%

III.F. 1-8 G.A. Kraus *et al.*, *J. Org. Chem.*, **46**, 2417 (1981).

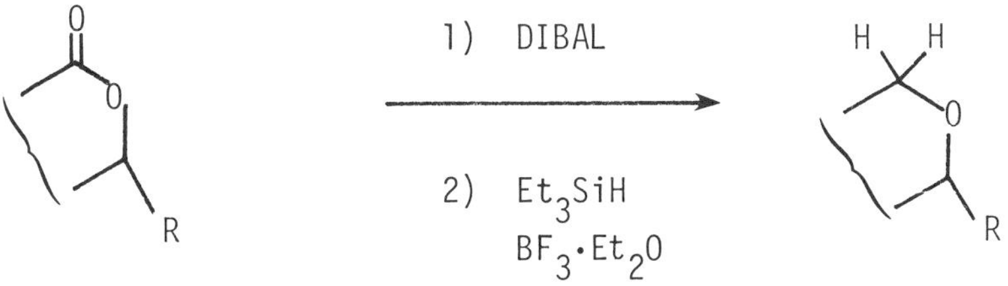
O
O
R
1) DIBAL
2) Et_3SiH
$BF_3 \cdot Et_2O$
H
H
O
R

50-88%

R = Ph, OH

α,β-unsaturated lactones react similarly; the double bond is unaffected.

III.F.1-9 S.R. Wann, P.T. Thorsen, and M.M. Kreevoy, *J. Org. Chem.*, **46**, 2579 (1981).

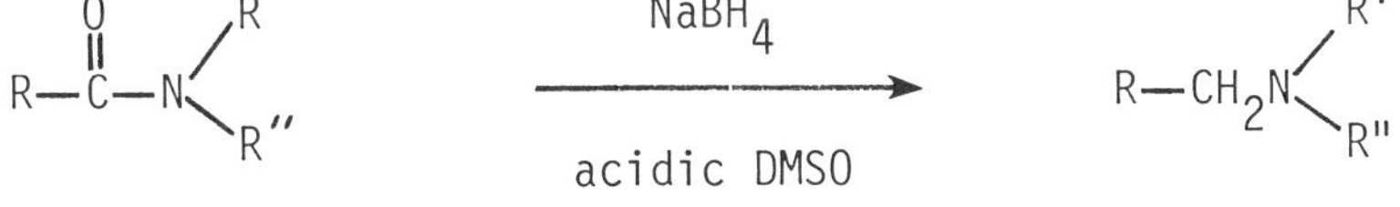
O
R—C—N
R'
R''
$NaBH_4$
acidic DMSO
R—CH_2N
R'
R''

R = alkyl, Ph,Bz
R', R'' = H, alkyl, Ph

III.F.1-10   H.C. Brown, S. Narasimhan, and Y.M. Choi,
Synthesis, 996 (1981).

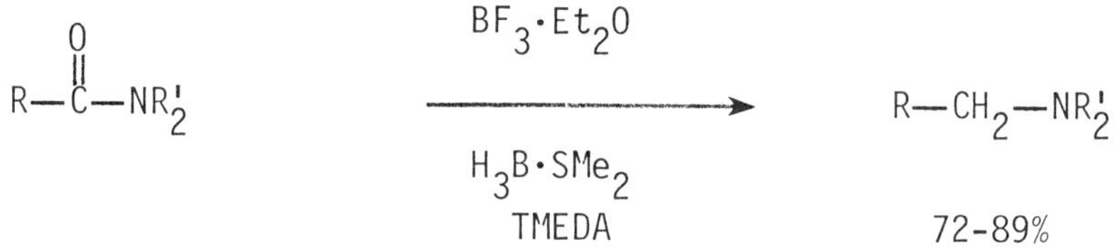

$$R-\overset{\overset{\displaystyle O}{\|}}{C}-NR'_2 \quad \xrightarrow[\substack{H_3B \cdot SMe_2 \\ TMEDA}]{BF_3 \cdot Et_2O} \quad R-CH_2-NR'_2$$

72-89%

R = alkyl, aryl, cycloalkyl
R' = H, Me, i-Pr

III.F.1-11   J.L. Fry, S.B. Silverman, and M. Orfanopoulos,
Org. Syn., 60, 108 (1981).

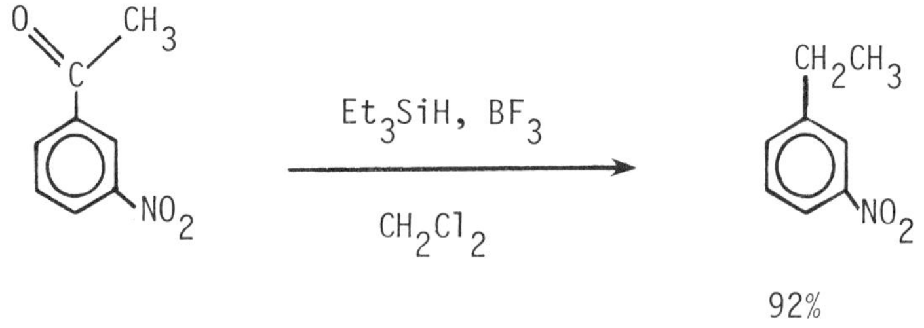

92%

III.F.1-12   T. Morita, Y. Okamoto, and H. Sakurai,
Synthesis, 32 (1981).

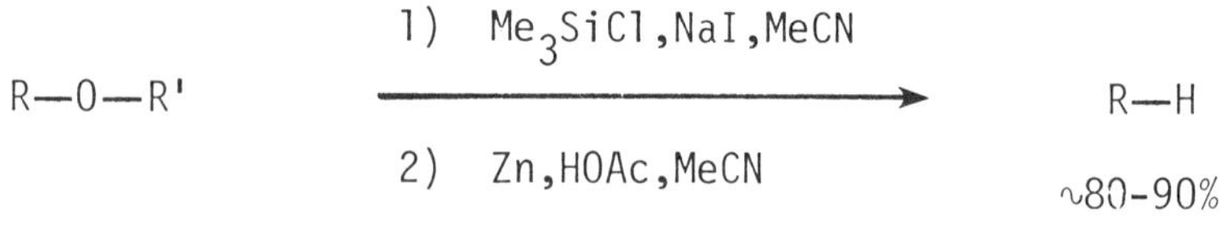

$$R-O-R' \quad \xrightarrow[\substack{2) \ Zn,HOAc,MeCN}]{1) \ Me_3SiCl,NaI,MeCN} \quad R-H$$

∼80-90%

R = 1°, 2° alkyl, Bz
R' = H, Me, SiMe_3

III.F.1-13   T. Satoh et al., Chem Lett., 1029 (1981).

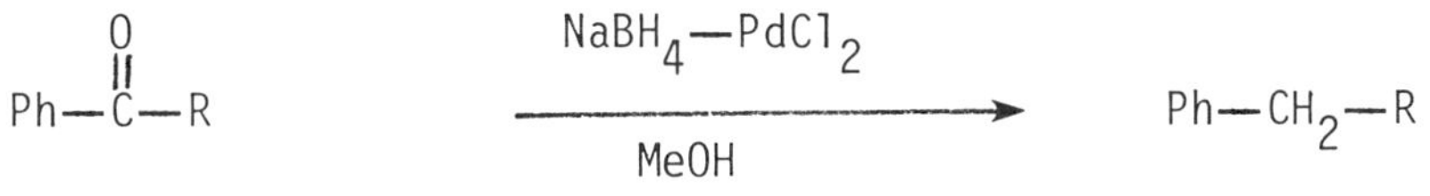

$$Ph-\overset{\overset{\displaystyle O}{\|}}{C}-R \quad \xrightarrow[\text{MeOH}]{NaBH_4-PdCl_2} \quad Ph-CH_2-R$$

53-83%

$$Ph-\overset{\overset{\displaystyle OH}{|}}{CH}-R \quad \xrightarrow[\text{MeOH}]{NaBH_4-PdCl_2} \quad Ph-CH_2-R$$

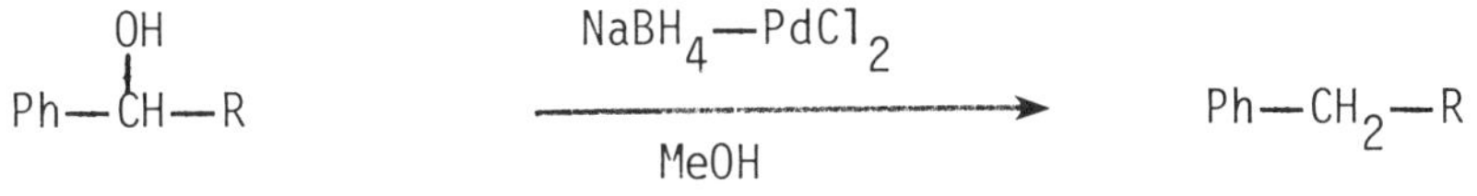

25-91%

R = Me, Ph

___

III.F.1-14   R.A. Lessor and N.J. Leonard, J. Org. Chem., 46, 4300 (1981).

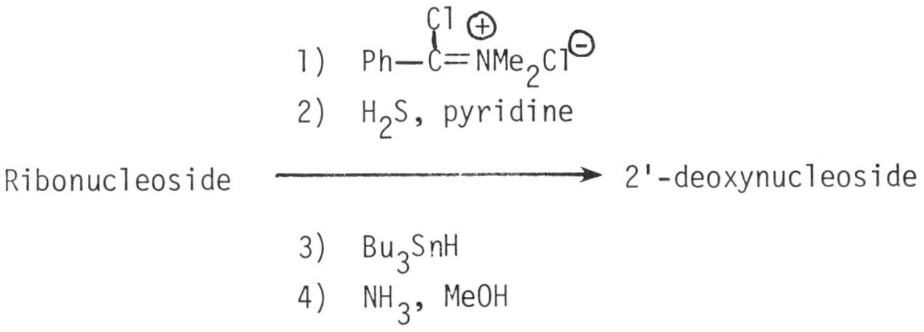

Ribonucleoside

1) $Ph-\overset{\overset{\displaystyle Cl}{|}}{C}=\overset{\oplus}{N}Me_2Cl^{\ominus}$

2) H_2S, pyridine

$\xrightarrow{\hspace{3cm}}$ 2'-deoxynucleoside

3) Bu_3SnH

4) NH_3, MeOH

∿50% overall

III.F.1-15 G.W. Kabalka and S.T. Summers, J. Org. Chem.,
46, 1217 (1981).

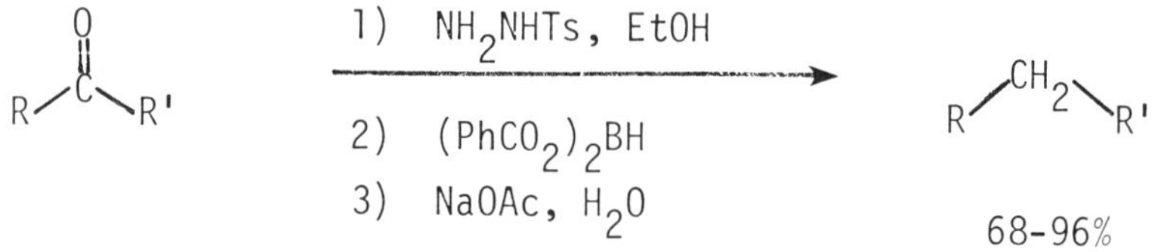

1) NH$_2$NHTs, EtOH

2) (PhCO$_2$)$_2$BH

3) NaOAc, H$_2$O

68-96%

R, R' = H, alkyl, Ph, cyclic

same conditions

R' = H, alkyl

III.F.1-16 M.J. Robins and J.S. Wilson, J. Am. Chem. Soc.,
103, 932 (1981).

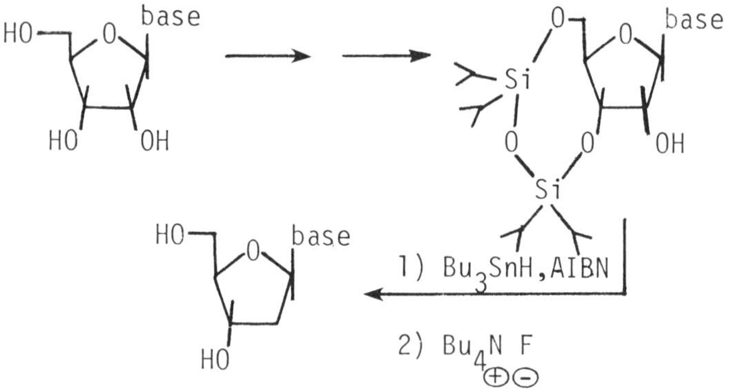

1) Bu$_3$SnH, AIBN

2) Bu$_4$N F
       ⊕⊖

58-85% overall

III.F.2.   C—Hal⟶   C—H

III.F.2-1   R. Vanderesse, J.-J. Brunet, and P. Caubere,
J. Org. Chem., 46, 1270 (1981).

$$R—X \xrightarrow[\text{Ni or Zn salts(some cases)}]{\text{NaH, } \underline{t}\text{-AmONa}} R—H$$

∿80-100%

R = 1⁰, 2⁰, 3⁰ alkyl, vinyl, allyl, benzyl

III.F.2-2   F. Rolla, J. Org. Chem., 46, 3909 (1981).

$$R—X \xrightarrow[\text{phase-transfer conditions}]{\text{aq. NaBH}_4} R—H$$

R = 1⁰, 2⁰ alkyl, Bz
X = Cl, Br, I, MeSO₃, OTs

III.F.2-3   H. Toi et al., Tetrahedron, 37, 2261 (1981).

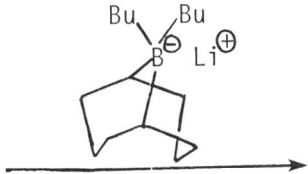

R—X   R—H

∿ 50-100%

R = alkyl, benzyl, allyl

III.F. 2-4   T. Ohsawa et al., Tetrahedron Lett., 22, 2583 (1981).

$$R—F \xrightarrow[\text{dicyclohexyl 18-crown-6}]{\text{K or Na-K, toluene}} R—H$$

R = alkyl, steroidal

~80%

III.F.2-5   T. Satoh et al., Chem. Lett., 1029 (1981).

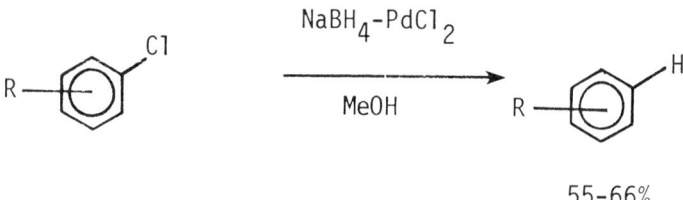

55-66%

R = H, -COOMe, Benzo-

III.F.2-6   P. Four and F. Guibe, J. Org. Chem., 46, 4439 (1981).

$$R—\overset{\displaystyle O}{\overset{\|}{C}}—Cl \xrightarrow[\text{PdL}_4]{\text{Bu}_3\text{SnH}} R—\overset{\displaystyle O}{\overset{\|}{C}}—H$$

~60-90%

R = alkyl, subst. Ph, styryl, etc.

III.F.2-7   J.H. Babler and B.J. Invergo, <u>Tetrahedron Lett.</u>,
<u>22</u>, 11 (1981).

$$R-\overset{\overset{\text{O}}{\|}}{C}-Cl \xrightarrow[\text{THF/DMF}]{\text{1)}\quad NaBH_4,\,-70^0} R-CHO$$

$$\text{2)}\quad CH_3CH_2COOH,\; HCl \qquad \sim70\text{-}80\%$$
$$EtOCH{=\!=}CH_2$$

R = alkyl, aryl

___

III.F.2-8   J.N. Denis and A. Krief, <u>Tetrahedron Lett.</u>,
<u>22</u>, 1431 (1981).

$$R-\overset{\overset{\text{X}}{|}}{C}H-\overset{\overset{\text{O}}{\|}}{C}-R' \xrightarrow{PI_3 \text{ or } P_2I_4} R-CH_2-\overset{\overset{\text{O}}{\|}}{C}-R'$$

$$\sim90\%$$

X = Br, I
R = alkyl, Ph,⎱
R' = H, alkyl, ⎰ cyclic

___

III.F.2-9   T. -L. Ho, <u>Synth. Comm.</u>, <u>11</u>, 101 (1981).

X = Br, Cl   85-96%

III.F.3.   C—S ⟶ C—H

III.F.3-1   H. -J. Liu, R.R. Bukownik, and P.R. Pednekar, Synth. Comm., 11, 599 (1981).

$$R-\overset{\overset{O}{\|}}{C}-SR' \quad \xrightarrow[\text{EtOH}]{NaBH_4} \quad R-CH_2OH$$

R = alkyl, aryl, cyclic, etc.
R' = Et, Ph, t-Bu, c-Hx

   Can be accomplished without affecting ester and nitrile groups.

III.F.3-2   R.J. Sundberg, C.P. Walters, and J.D. Bloom, J. Org. Chem., 46, 3730 (1981).

$$R-\overset{\overset{S}{\|}}{C}-N\overset{R'}{\underset{R'}{\Big\langle}} \quad \xrightarrow[\underline{or}\ NaBH_3CN]{NaBH_4} \quad R-CH_2-N\overset{R'}{\underset{R'}{\Big\langle}}$$

                                                        44-96%

R = Ph, Me, Bz
R' = alkyl, cyclic, Ph

III.F.3-3   B.M. Trost and P.L. Ornstein, Tetrahedron Lett., 22, 3463 (1981).

                                                        50-76%

R = acetals, benzo-, etc.
R' = alkyl, Ph

III.F.4. C—N ⟶ C—H

III.F.4-1   R.J. Lahoti, V. Parameswaran, and D. R. Wagle, Indian J. Chem., 20B, 767 (1981).

$$\text{Ar—N}{\equiv}\overset{\oplus}{\text{N}} \quad \xrightarrow{\text{DMF}} \quad \text{Ar—H}$$

$\sim$60-90%

Ar = subst. Ph

III.F.4-2   D.D. Tanner, E.V. Blackburn, and G.E. Diaz, J. Am. Chem. Soc., 103, 1557 (1981).

$$R{-}NO_2 \quad \xrightarrow[\text{initiator}]{Bu_3SnH} \quad R{-}H$$

$\sim$80-90%

R must be tertiary.

III.F.4-3   N. Ono et al., Tetrahedron Lett., 22, 1705(1981).

$$R{-}NO_2 \quad \xrightarrow[\text{AIBN}]{Bu_3SnH} \quad R{-}H$$

$\sim$60-90%

R = 2^0 or 3^0 alkyl, benzylic, etc.

Many examples.

III.G.   Reductive Eliminations

III.G-1   K. Fukunaga and H. Yamaguchi, Synthesis, 879
(1981).

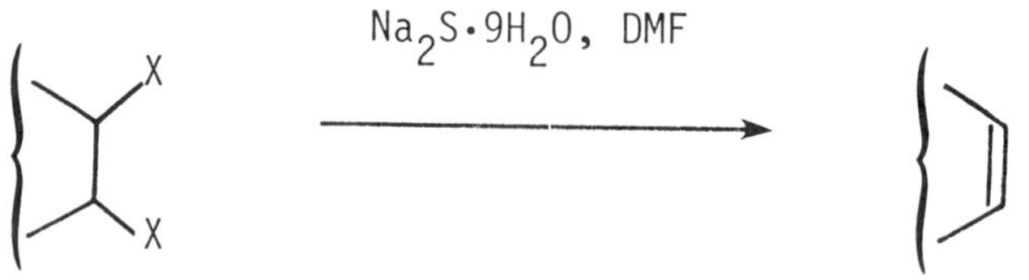

X = Cl, Br   ~70-90%

(alkyl, Ph, cycloalkyl, etc.)

III.G-2   R.N. Majumdar and H.J. Harwood, Synth. Comm., 11,
901 (1981).

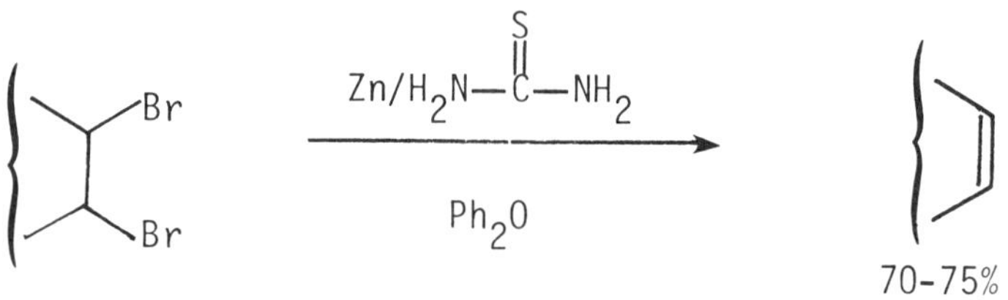

70-75%

III.G-3   M.A. Umbreit and K.B. Sharpless, <u>Org. Syn.</u>, <u>60</u>, 29 (1981).

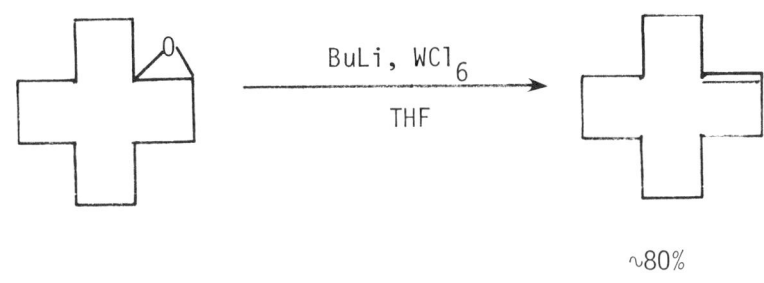

~80%

III.G-4   R. Caputo <u>et al.</u>, <u>Tetrahedron Lett.</u>, <u>22</u>, 3551 (1981).

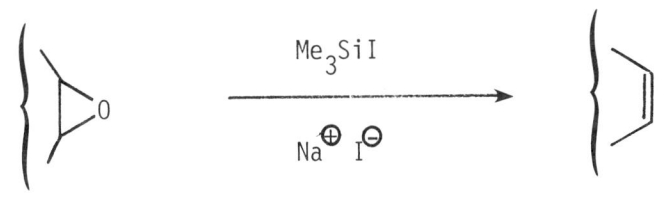

>90%

(retention of stereochemistry)

III.G-5   D.L.J. Clive and V.N. Kale, <u>J. Org. Chem.</u>, <u>46</u>, 231 (1981).

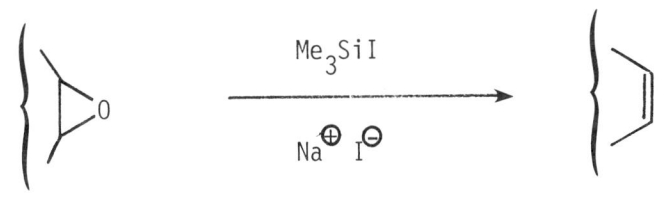

70-91%

III. H.   Reductive Cleavages

III.H-1   W.S. Murphy and S. Wattanasin, Tetrahedron Lett.,
22, 695 (1981).

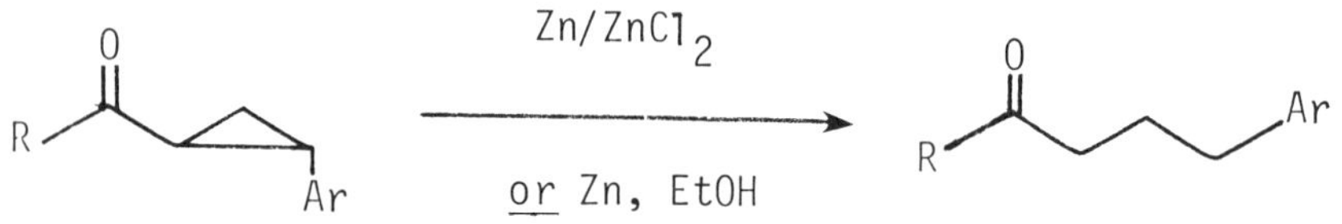

R = alkyl, aryl
Ar = Subst. Ph

III.H-2   Y. Tsuda and Y. Sakai, Synthesis, 119 (1981).

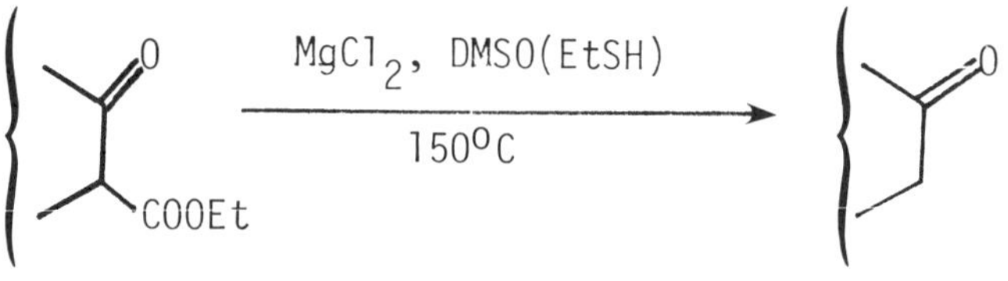

∿50-90%

III.I. Hydroboration (Reduction only)

III.I-1   R. Contreras et al., Synthesis, 214 (1981).

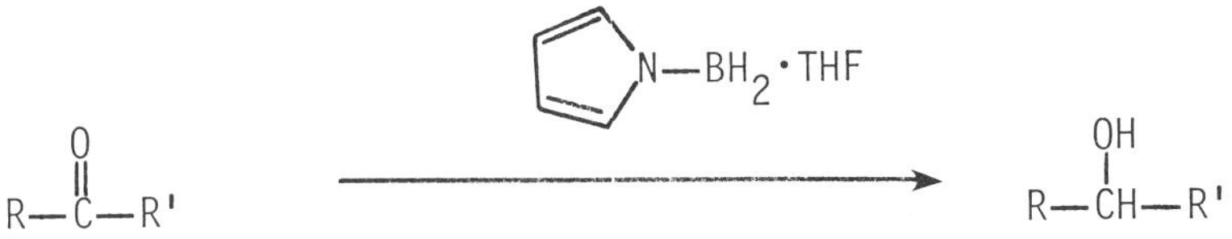

$$R—\overset{\overset{\text{O}}{\|}}{C}—R' \xrightarrow{\text{(pyrrolyl)}N—BH_2 \cdot THF} R—\overset{\overset{\text{OH}}{|}}{CH}—R'$$

∿90-100%

R, R' = H, alkyl, Ph

III.J. Other Reductions and Reviews

III.J-1   J.A. Schreifels, P.C. Maybury, and W.E. Swartz, Jr.
J. Org. Chem., 46, 1263 (1981).

Use of nickel boride (Ni_2B) as an active hydrogenation
catalyst for C≡C, C≡N, and aldehyde double bonds.
Similar to Raney nickel in most cases, but more suscep-
tible to nitrogen poisoning.

III.J-2   B.E. Maryanoff, D.F. McComsey, and S.O. Nortey,
J. Org. Chem., 46, 355 (1981).

A survey of the uses of $(CF_3\overset{\overset{\displaystyle O}{\|}}{C}-O)_2BH\cdot THF$ as a reducing

agent. Successfully reduces indoles, ketones, aldehydes,

and imines.

III.J-3   R.O. Hutchins and F. Cistone, Org. Prep. Proc. Int.,
13, 225 (1981).

Review: "Utility and Applications of Borane Dimethyl-

sulfide in Organic Synthesis"

IV

SYNTHESIS OF HETEROCYCLES

IV.A.   Aziridines

IVA-1   A. Hassner et al., Tetrahedron Lett., 22, 1863
(1981).

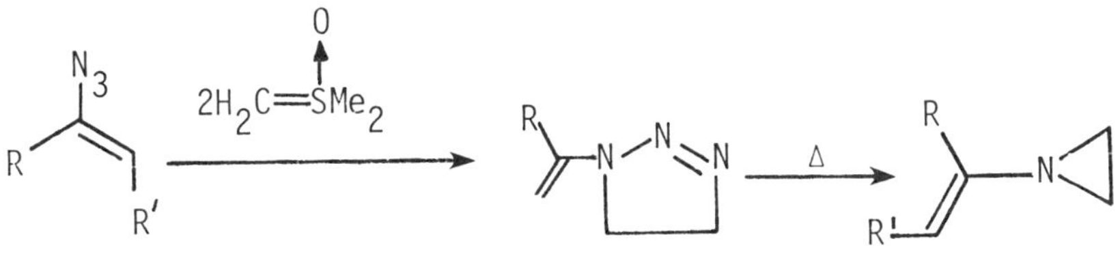

89-95%   91-94%

R, R' = H, alkyl, Ph, cyclic

IV.A-2   H. Ila et al., Synthesis, 623 (1981).

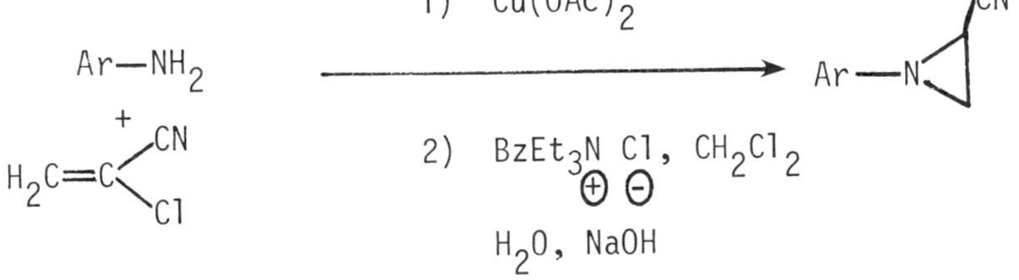

Ar = subst. Ph

IV. B. Furans, etc.

IV.B-1   T. Mandai _et al._, _Tetrahedron Lett._, 22, 2187 (1981).

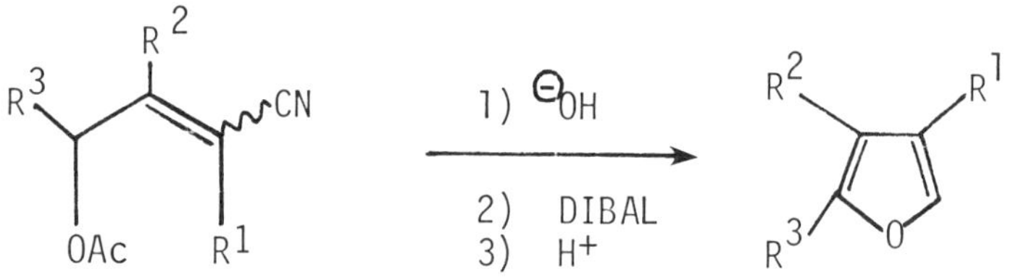

40-76%

R's = alkyl, Ph, cyclic

IV. B-2   H. Nishiyama, M. Sasaki, and K. Itoh, _Chem. Lett._, 1363 (1981).

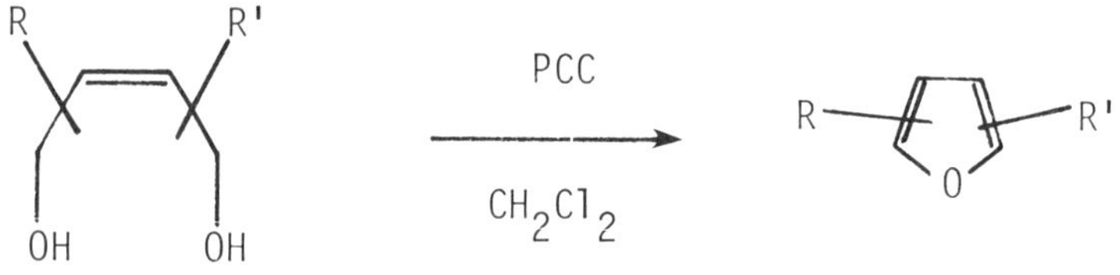

widely varying yields

R, R' = alkyl, vinyl, etc.

IV.B-3   H. König, F. Graf, and V. Weberndörfer, Liebigs Ann. Chem., 668 (1981).

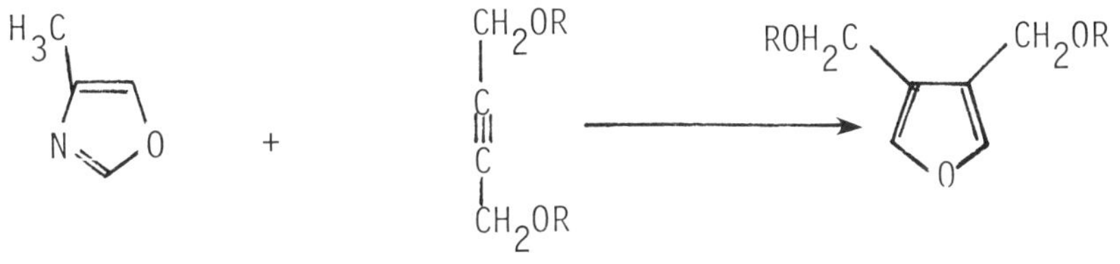

Widely varying yields.

R = alkyl, acyl, vinyl, etc.

IV. B-4   Y. Fujiwara et al., J. Org. Chem., 46, 851 (1981).

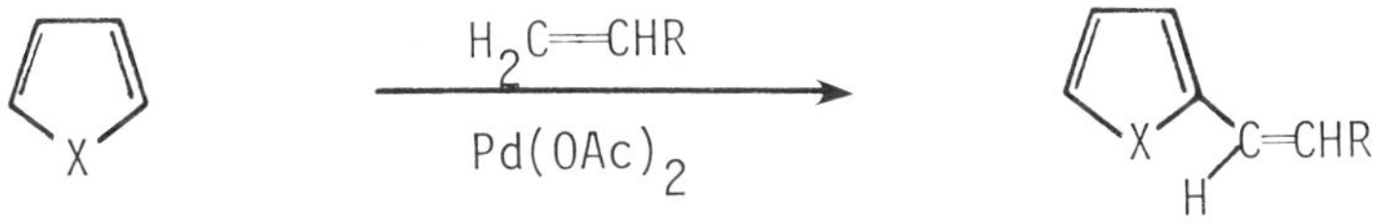

13-33%

+

11-46%

X = O, S
R = -CN, =Ph, -COOMe

IV.B-5   S.M. Nolan and T. Cohen, J. Org. Chem., 46, 2473
(1981).

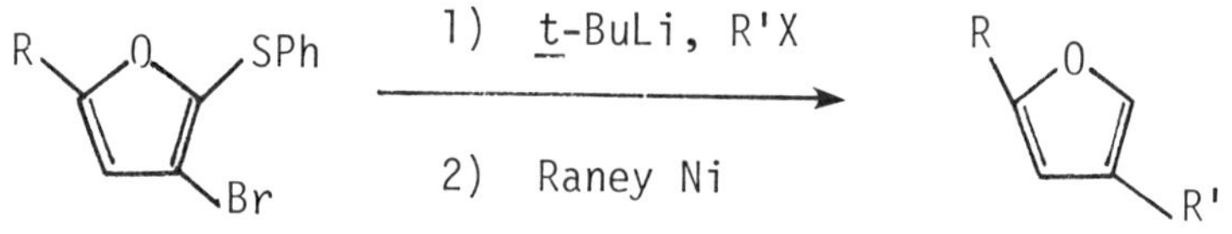

R = CH(OEt)$_2$, H, SPh
R' = Me, n-Bu, n-octyl

IV.B-6   F. Barba, M.D. Velasco, and A. Guirado, Synthesis,
625 (1981).

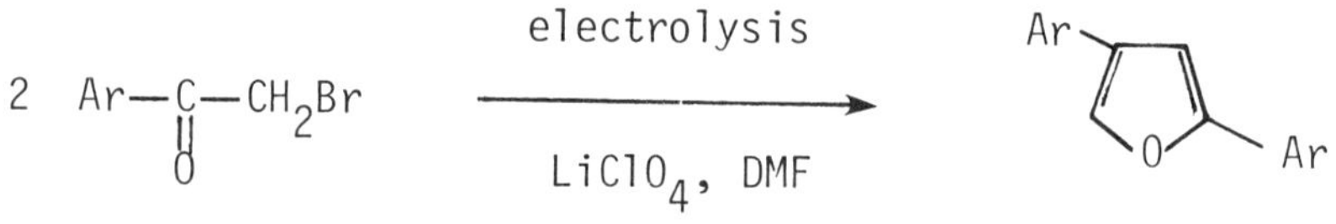

Ar = subst. Ph   64-80%

IV.B-7   T. Shono et al., Chem. Lett., 1121 (1981).

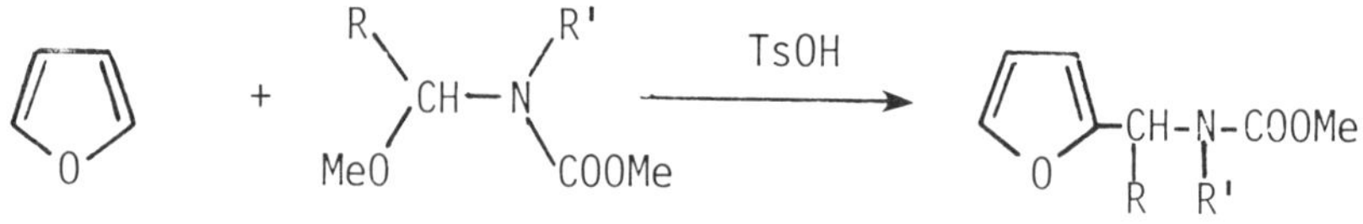

R, R' = cyclic, H, alkyl   ∿60-80%

IV.B-8   I. Akomoto, M. Sano, and A. Suzuki, <u>Bull.Chem. Soc.</u>
<u>Japan</u>, <u>54</u>, 1587 (1981).

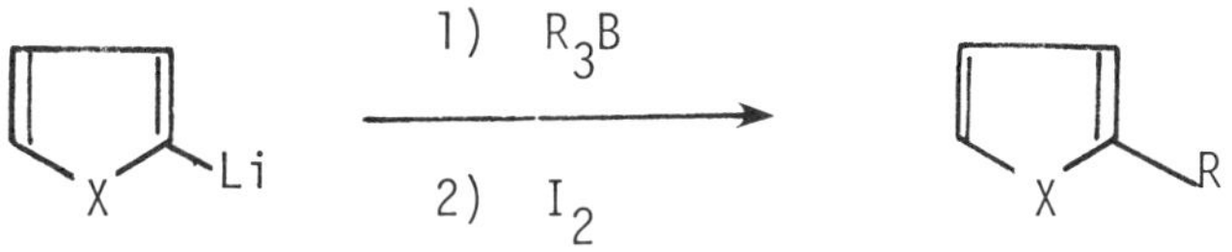

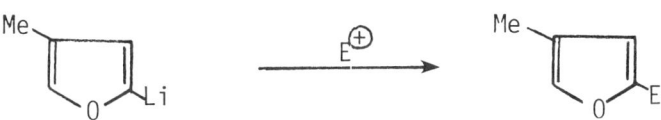

X = O,S
R = 1o, 2o alkyl, cycloalkyl

IV.B-9   D.W. Knight and D.C. Rustidge, <u>J.C.S. Perkin I</u>,
679 (1981).

E^+ = acetone, benzaldehyde, 1o halides, etc.   ~55-65%

IV.B-10   U.E. Wiersum, <u>Aldrichimica Acta</u>, <u>14</u>, 53 (1981).

Review: "Isobenzofuran and Related o-Quinonoid Systems"

IV.B-11   A. Sc ttri _et al._, _J.C.S. Perkin I_, 2398 (1981).

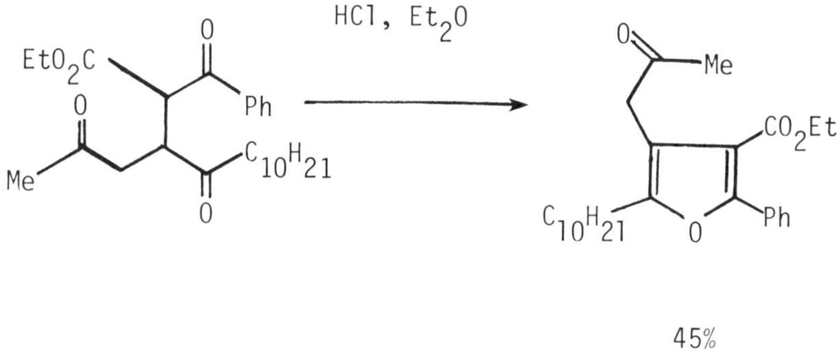

45%

Several additional examples.

IV.B-12   D.R. Williams, J.G. Phillips, and B.A. Barner, _J. Am. Chem. Soc._, _103_, 7398 (1981).

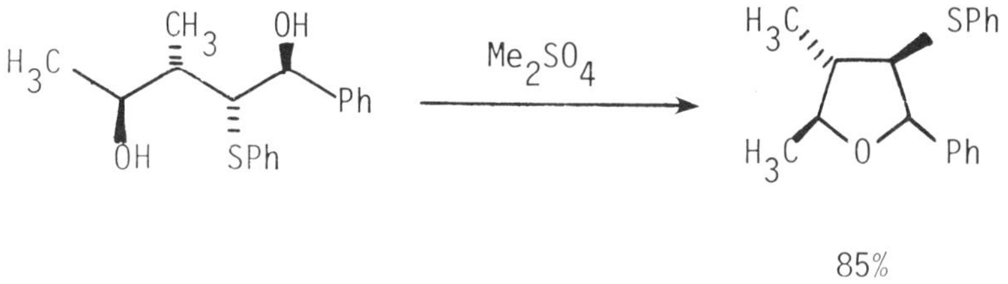

85%

Several additional examples.

IV.B-13   P. Audin et al., Bull Soc. Chim. France II, 313
(1981).

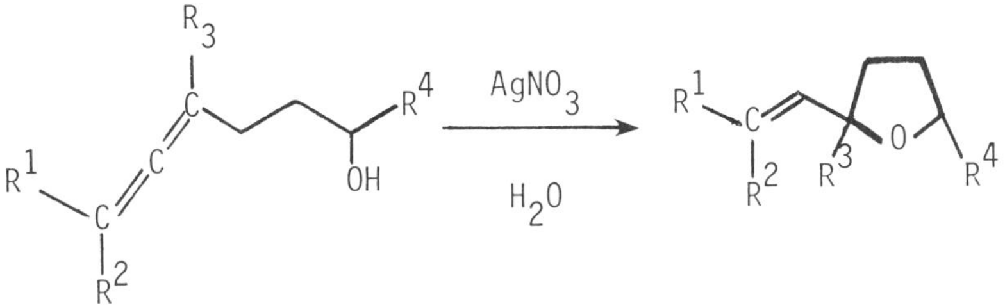

40-95%

R^1, R^2 = H, 1°, alkyl
R^3, R^4 = H, Me

IV.B-14   C.K. Bradoher and D.C. Reames, J. Org. Chem., 46,
1384 (1981).

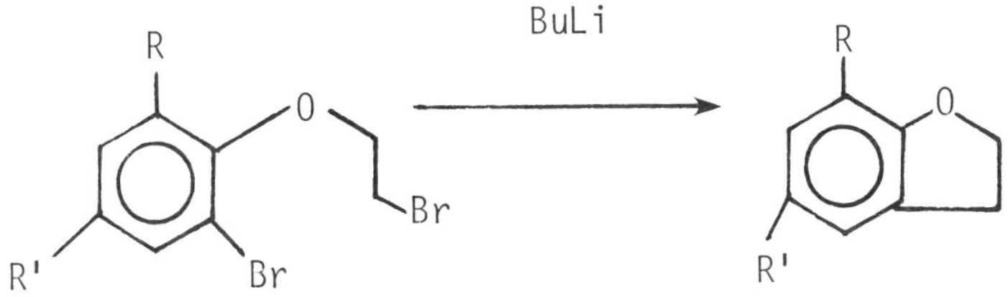

69-79%

R = H, Br
R' = H, Me, OMe, Cl, Br

IV. C. Indoles

IV.C-1 W.J. Houlihan, V.A. Parrino, and Y. Uike, J. Org. Chem., 46, 4511 (1981).

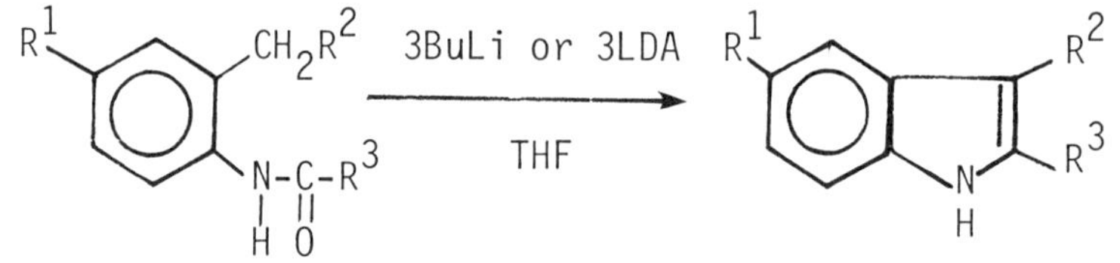

R^1 = H, OMe, Cl
R^2 = H, Me, OMe
R^3 = Ph, t-Bu, 1-adamantyl,

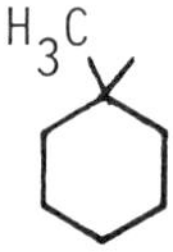

IV.C-2 P.A. Wender and A.W. White, Tetrahedron Lett., 22, 1475 (1981).

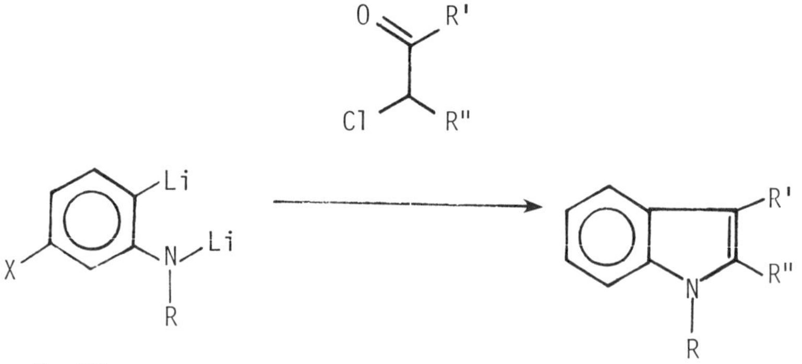

X = H, OMe
R = H, -COtBu, -COCF$_3$
R', R" = alkyl, cyclic

~50-90%

IV.C-3   Y. Watanabe et al., Chem. Lett., 603 (1981).

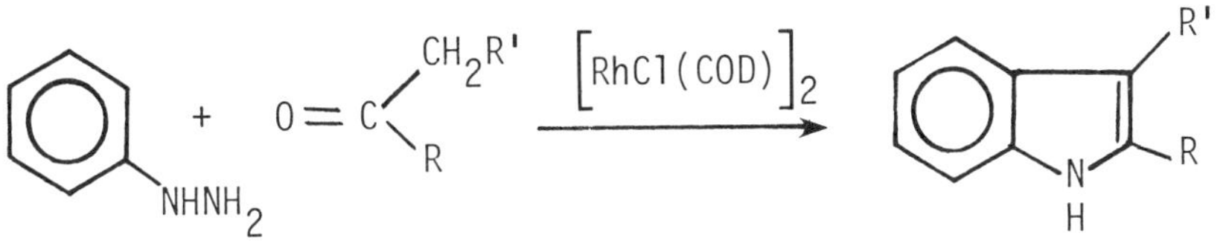

$$\text{(phenyl)-NHNH}_2 \ + \ O=C\overset{CH_2R'}{\underset{R}{\diagdown}} \ \xrightarrow{\ \left[RhCl(COD)\right]_2\ } \ \text{(indole with } R', R, N-H)$$

22-59%

R, R' = H, alkyl, cyclic

IV.C-4   M. Lecorré, A. Hercouet, and H. LeBaron, J.C.S. Chem. Comm., 14 (1981).

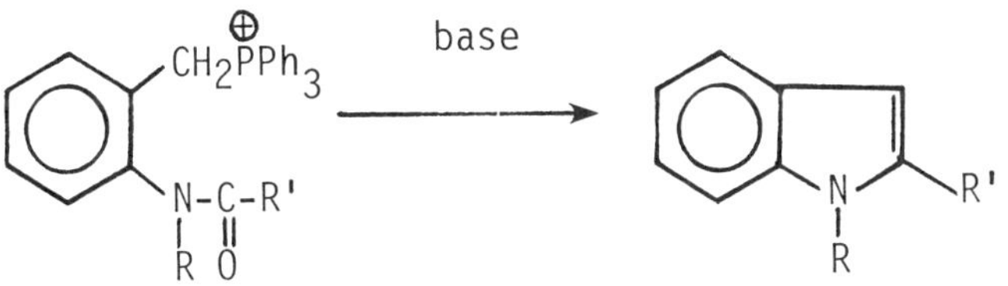

$$\text{(phenyl with }CH_2\overset{\oplus}{P}Ph_3 \text{ and } N\underset{R}{-}C\overset{\|}{\underset{O}{}}-R') \ \xrightarrow{\ base\ } \ \text{(indole with } N-R, R')$$

80-97%

R = H, Ph
R' = Me, vinyl, subst. Ph

IV.C-5   P.T. Kim <u>et al.</u>, <u>J. Het. Chem.</u>, <u>18</u>, 1373 (1981).

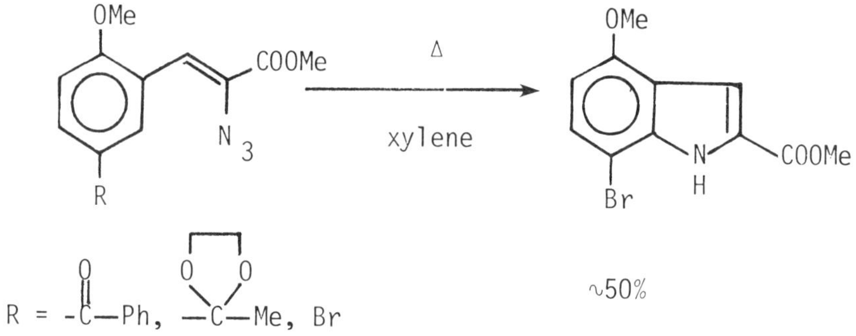

$$R = -\overset{O}{\underset{}{C}}-Ph, \quad -\overset{O}{\underset{}{\overset{|}{C}}}-Me, \quad Br$$

~50%

IV.C-6   H. Maehr and J.M. Smallheer, <u>J. Org. Chem.</u>, <u>46</u>, 1752 (1981).

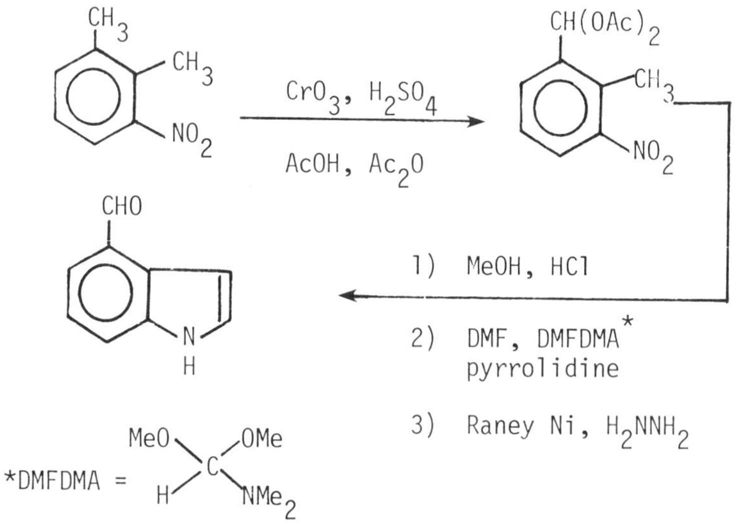

1)   MeOH, HCl

2)   DMF, DMFDMA[*]
     pyrrolidine

3)   Raney Ni, H_2NNH_2

*DMFDMA =

$$MeO\underset{H}{\overset{OMe}{\underset{}{C}}}NMe_2$$

IV.C-7   J.E. Nordlander et al., J. Org. Chem., 46, 778 (1981).

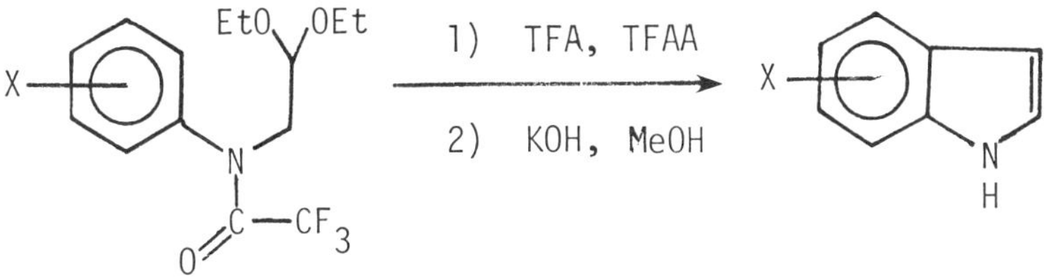

IV.C-8   Y. Kikugawa and M. Kawase, Chem. Lett., 445 (1981).

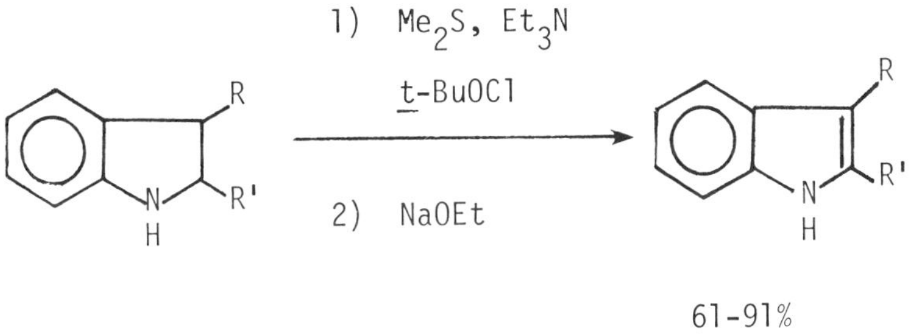

61-91%

R, R' = H, Me, cyclic,
       -(CH$_2$)$_n$CH(NHAc)COOEt, etc.

IV.C-9   E. Ucciani and A. Bonfand, J.C.S.Chem. Comm., 82 (1981).

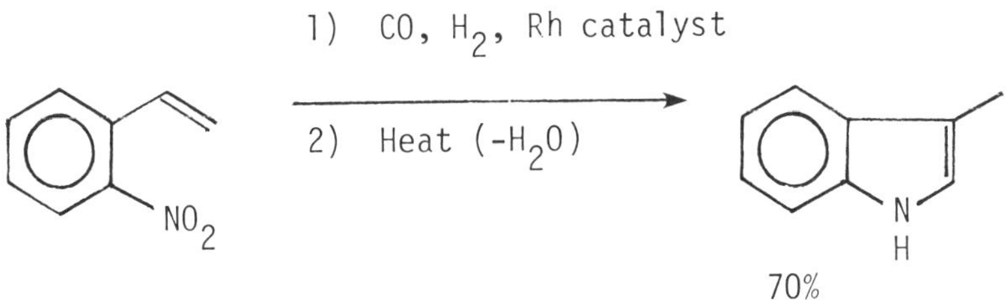

70%

IV.C-10   H. Laas, A. Nissen, and A. Nürrenbach, <u>Synthesis</u>,
958 (1981).

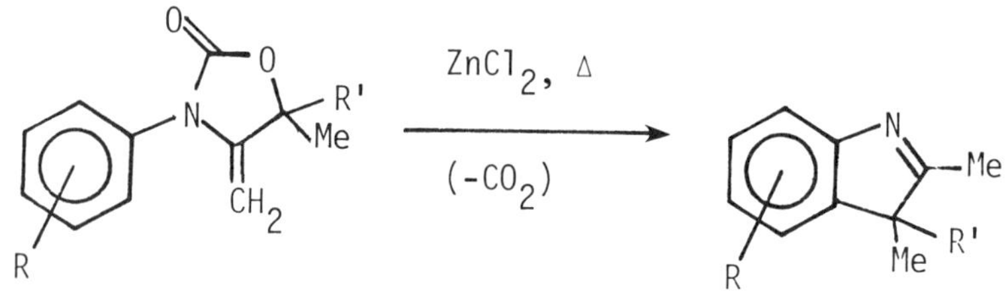

29-75%

R = H, Cl, OEt, benzo-
R' = Me, Et

IV.C-11   C.K. Bradsher and D.A. Hunt, <u>J. Org. Chem.</u>, <u>46</u>,
327 (1981).

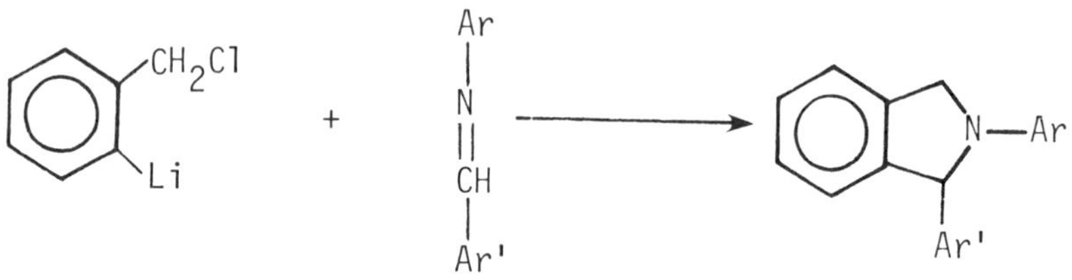

widely varying yields

Ar, Ar' = subst. Ph

IV.C-12   W.N. Speckamp et al., J. Am. Chem. Soc., 103, 4643 and 4645 (1981).

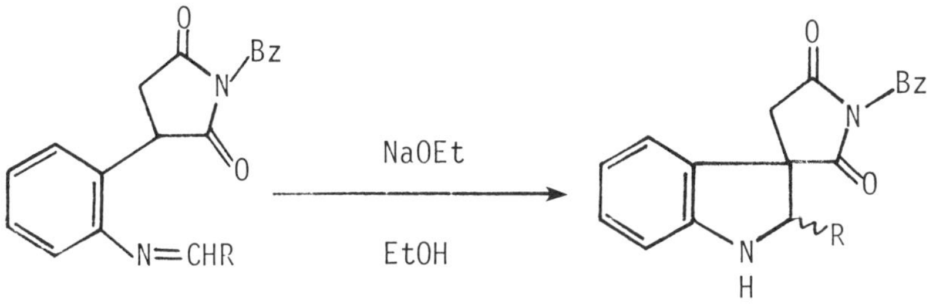

~40-90%

R = Pr, t-Bu, subst. Ph, 2-pyrrolo

IV.C-13   A. Shafiee and S. Sattari, Synthesis, 389 (1981).

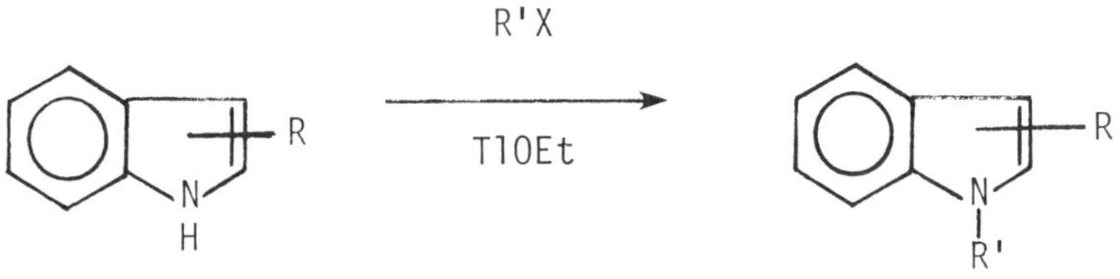

60-95%

R = H, Me, COOEt, $\overset{O}{\overset{\|}{C}}$Ph

R' = Me, Et, Ac, $\overset{O}{\overset{\|}{C}}$-Ph

IV.C-14   Y. Kikugawa and Y. Miyake, Synthesis, 461 (1981).

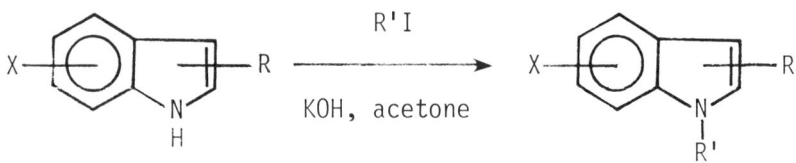

X = H, Br
R = H, Me, CH_2COOEt, $-(CH_2)_{\overline{4}}$
R' = Me, Et, Bz

IV.C-15   L.S. Hegedus, P.M. Winton, and S. Varaprath,
J. Org. Chem., 46, 2215 (1981).

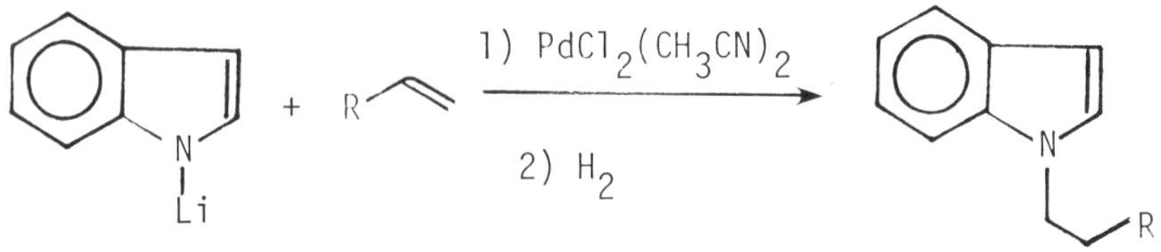

R = H, Me, Bu   ∿60-70%

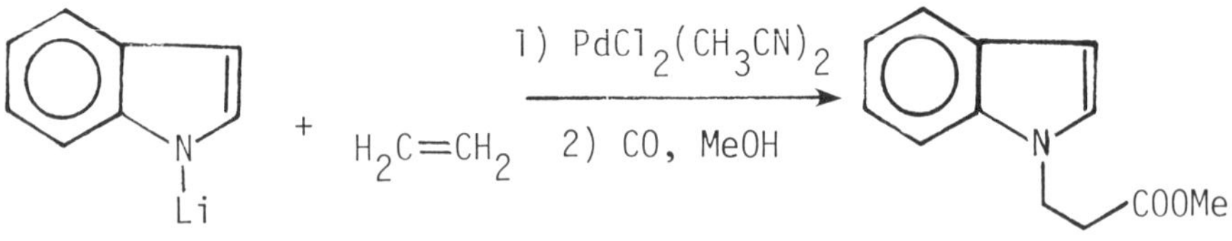

40%

IV.   D.   Lactams

IV.D-1   A. Padwa, K.F. Koehler, and A. Rodriguez, J. Am. Chem. Soc., 103, 4974 (1981).

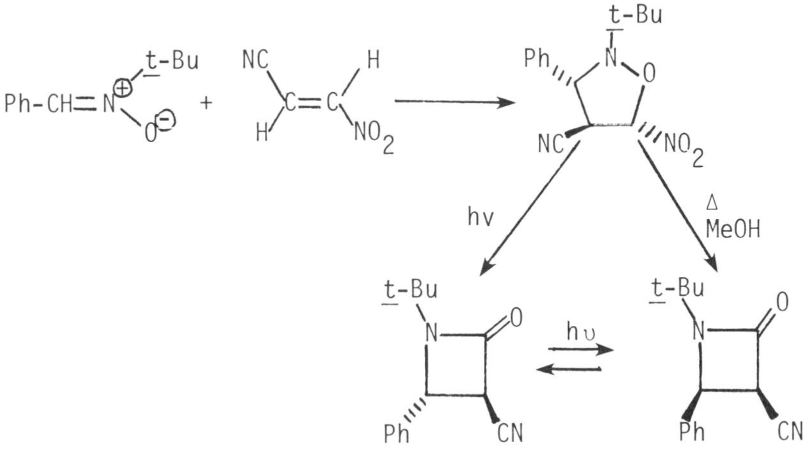

$\sim$60% overall

IV.D-2   H. Alper, C.P. Perera, and F.R. Ahmed, J. Am. Chem. Soc., 103, 1289 (1981).

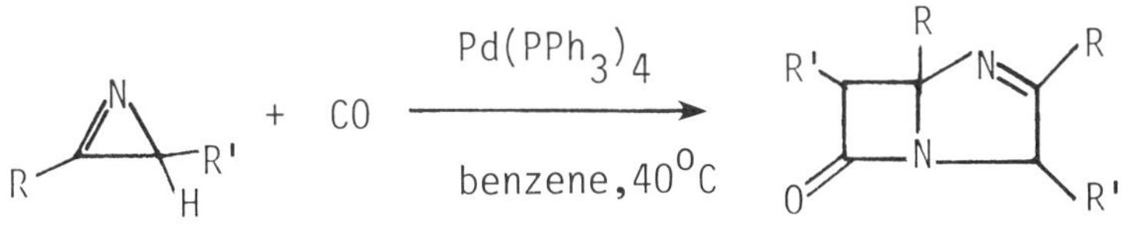

$\sim$50%

R = subst. Ph
R' = H, Me

IV.D-3   T. Yamazaki et al., Chem. Pharm. Bull., 29, 1063
(1981).

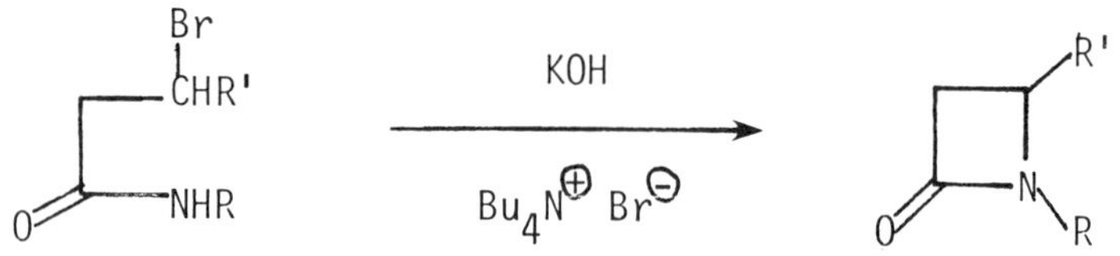

R = Subst. Ph, Bz, -CH$_2$COOEt, etc.
R' = H, COPh

IV.D-4   Y. Watanabe and T. Mukaiyama, Chem. Lett., 443
(1981).

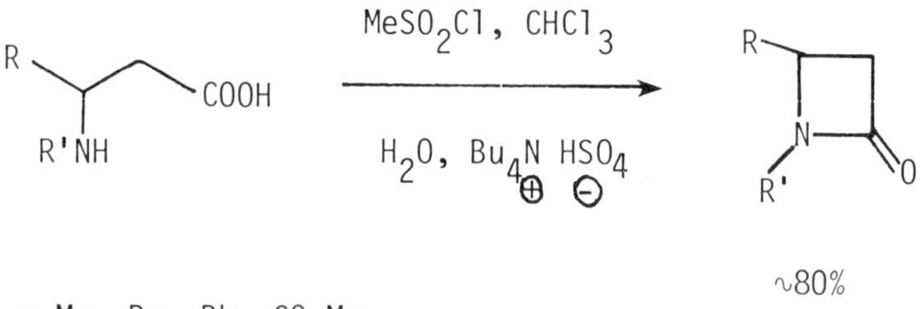

~80%

R = Me, Pr, Ph, CO$_2$Me
R' = Bz, n-Hx, c-Hx

IV.D-5   K. Ikeda, Y. Terao, and M. Sekiya, <u>Chem. Pharm.</u>
<u>Bull.</u>, <u>29</u>, 1747 (1981).

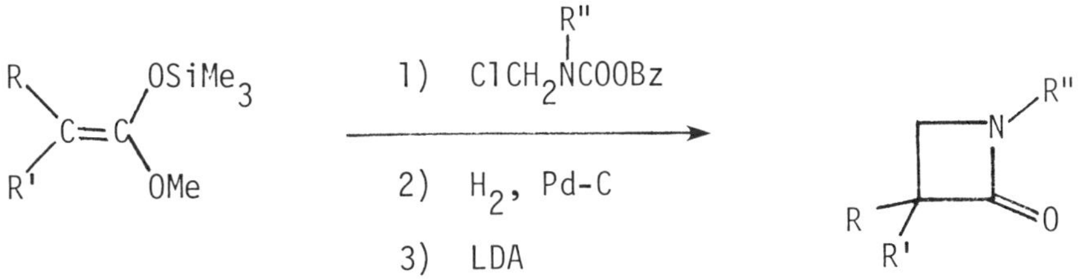

$$\text{R, R' = H, Me, OPh, } -(CH_2)_5$$

R" = Me, i-Pr

Widely varying yields.

IV.D-6   A.K. Bose <u>et al.</u>, <u>Tetrahedron</u>, <u>37</u>, 2321 (1981).

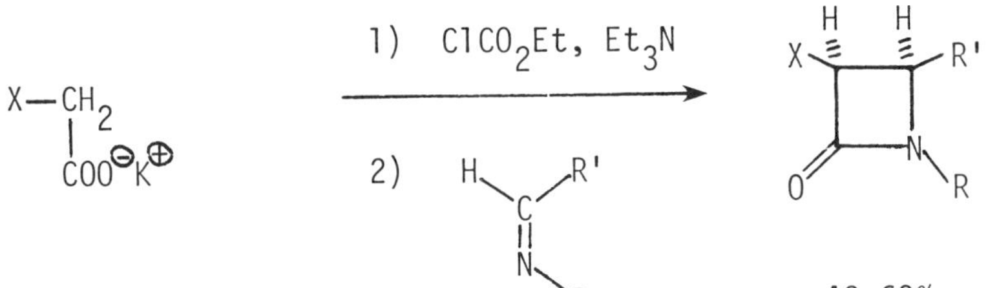

40-60%

X = protected NH_2 etc.

R, R' = alkyl, aryl, heterocyclic, etc.

IV.D-7   I. Ojima, S. Inaba, and M. Nagai, Synthesis, 545 (1981).

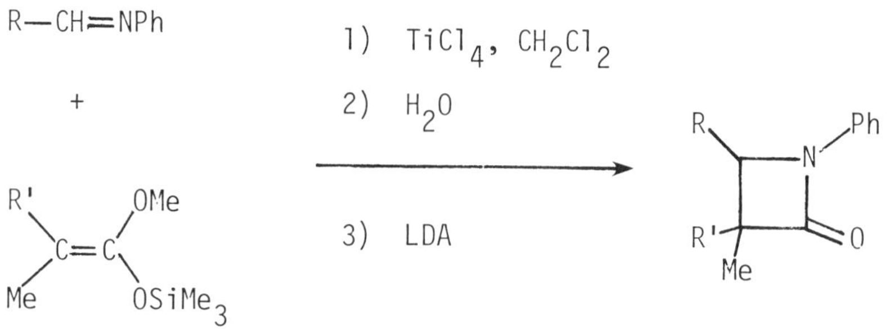

R—CH=NPh

    +

$$R'\!\!-\!\!C \overset{OMe}{\underset{OSiMe_3}{=}} C$$

Me

1) $TiCl_4$, CH_2Cl_2

2) H_2O

3) LDA

54-99%

R = Ph, 2-furyl, 2-thienyl, 2-pyridyl
R' = Me, OPh, SPh

IV.D-8   K. Piotrowska and D. Mostowicz, J.C.S. Chem. Comm., 41 (1981).

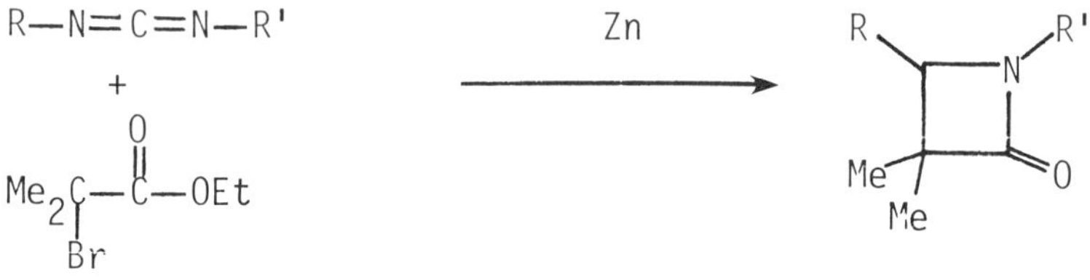

R—N=C=N—R'

    +

$$Me_2C \overset{O}{\underset{Br}{-}} C\!\!-\!\!OEt$$

Zn

51-81%

R = Ph, c-Hx

R' = 1^0, 2^0, 3^0 alkyl, p-tolyl

IV.D-9 M.S. Manhas, A.K. Bose, and M.S. Khajavi, Synthesis, 209 (1981).

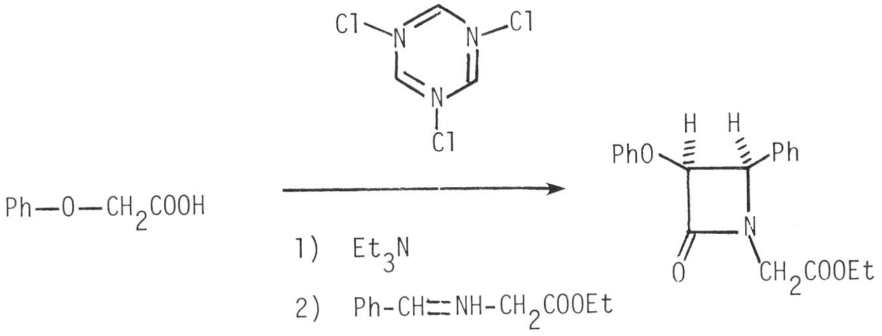
Cl
N
N
Cl
N
Cl
Ph—O—CH2COOH
1) Et3N
2) Ph-CH=NH-CH2COOEt
H
H
PhO
Ph
N
O
CH2COOEt

58%

Several additional examples.

IV.D -10 C. Belzecki and E. Rogalska, J.C.S. Chem. Comm., 57 (1981).

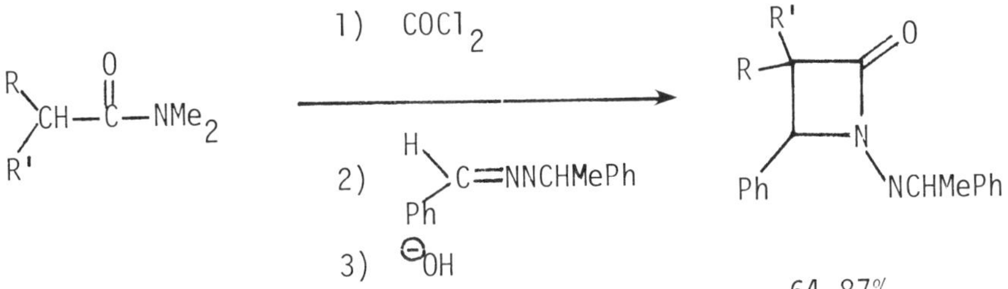
R
O
CH—C—NMe2
R'
1) COCl2
H
2) C=NNCHMePh
Ph
3) ⊖OH
R'
O
R
N
Ph
NCHMePh

R, R' = H, Me, Ph

64-87%

IV.D-11   A.K. Bose, D.P. Sahu, and M.S. Manhas, J. Org.
Chem., 46, 1229 (1981).

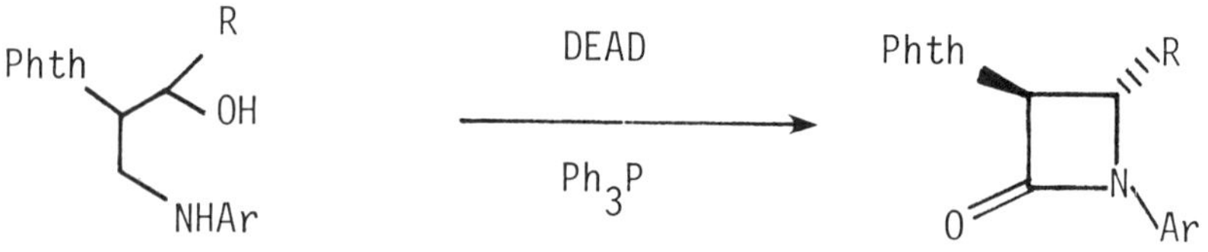

R = H, Me, Ph
Phth = phthalimido
Ar = p-tolyl

IV.D-12   R.D. Miller and P. Goelitz, J. Org. Chem., 46,
1616 (1981).

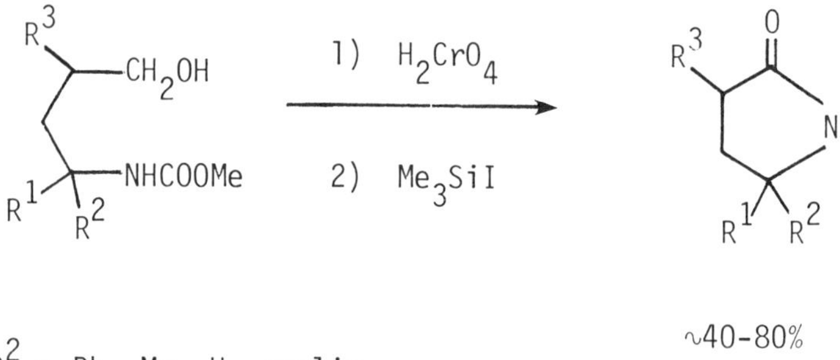

~40-80%

R^1, R^2 = Ph, Me, H, cyclic

R^3 = H, Me

IV.D-13   H. Stamm, A. Woderer, and W. Wiesert, <u>Chem. Ber.</u>,
<u>114</u>, 32 (1981).

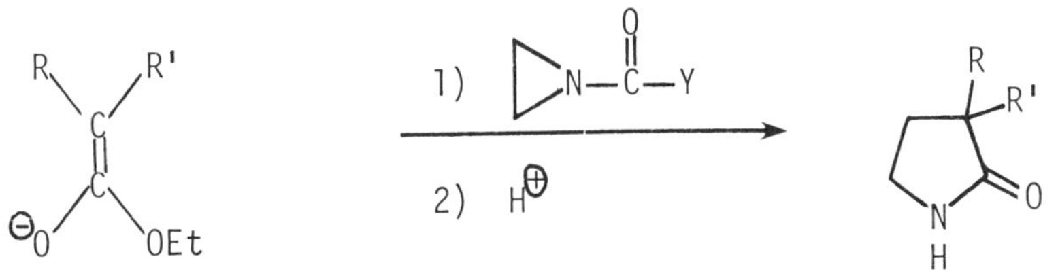

Widely varying yields.

R, R' = H, Me, Et, Ph, cyclic
Y = OEt, Ph, NR$_2$

IV.D-14   S.M. Weinreb <u>et al.</u>, <u>J. Am. Chem. Soc.</u>, <u>103</u>,
6387 (1981).

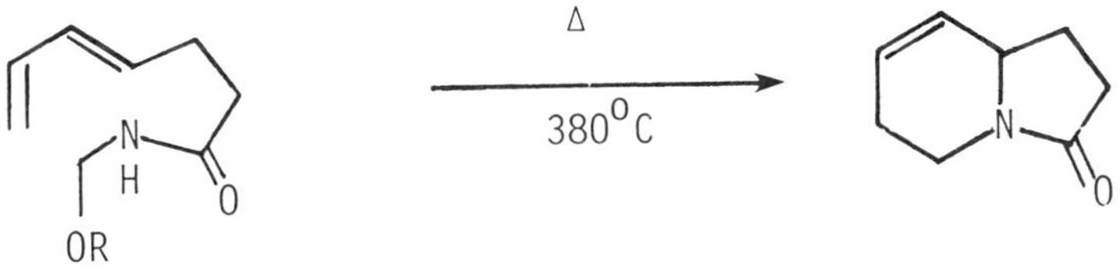

73%

Several additional examples.

IV.D-15   Y. Tamura et al., Synthesis, 534 (1981).

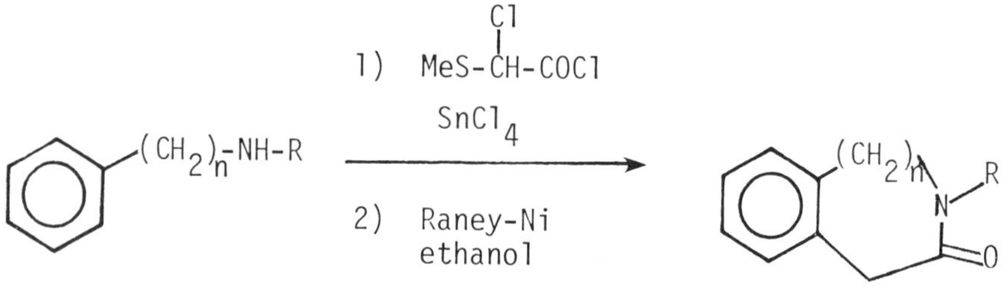

n = 0, 1
R = H, Me, Ph

~40-60% overall

IV.D-16   J.L. Soto et al., Synthesis, 529 (1981); Org. Prep.
Proc. Int, 13, 331 (1981).

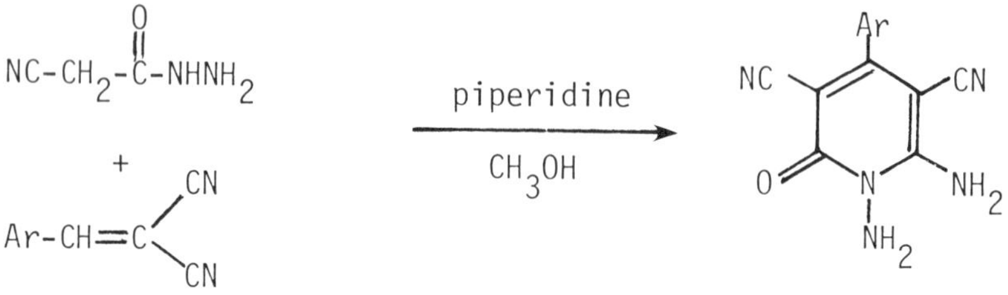

69-99%

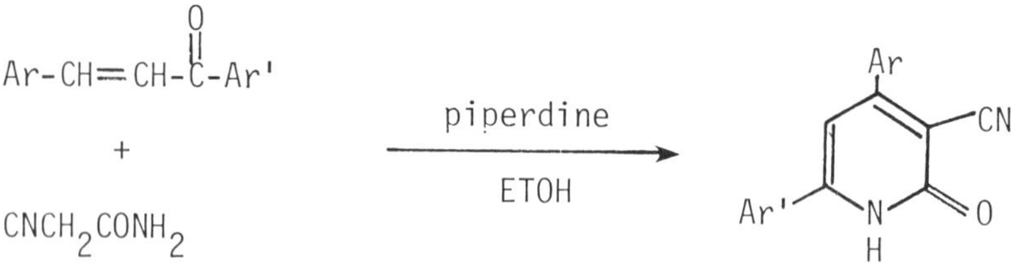

Ar, Ar' = subst. Ph

IV.D-17   J.N. Chatterjea et al., Liebigs Ann. Chem., 52
(1981).

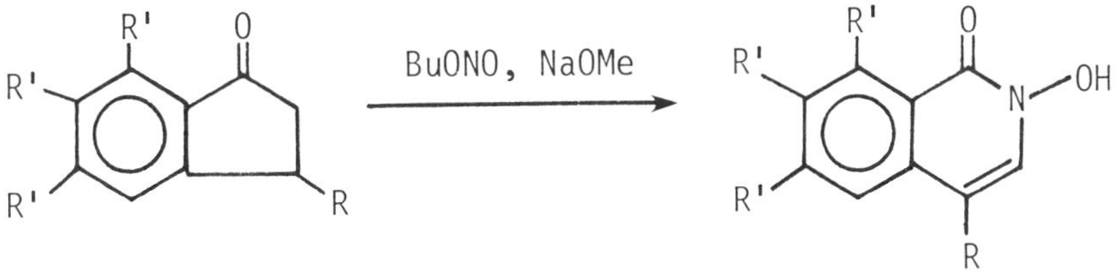

R = H, Me, Ph
R's = H, OMe

IV.D-18   E. Fujita, Pure and Appl. Chem., 53, 1141 (1981).

Review: "A New and Efficient Aminolysis and its Application to Synthesis of Macrolactam Alkaloids."

IV.D-19   G.R. Krow, Tetrahedron, 37, 1283 (1981).

Review: "Nitrogen Insertion Reactions of Bridged
Bicyclic Ketones. Regioselective Lactam Formation."

IV. E.   Lactones

IV.E-1   T. Nakano et al., Chem. Lett., 415 (1981).

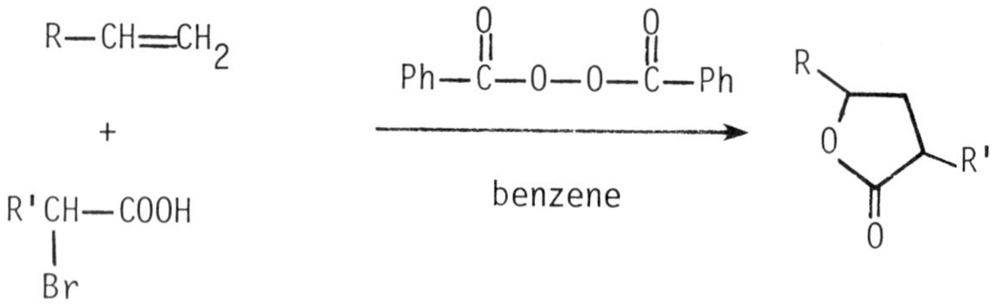

R—CH=CH$_2$

+

R'CH—COOH
 |
 Br

Ph—C(=O)—O—O—C(=O)—Ph

benzene

→

42-79%

R = alkyl
R' = H, alkyl

IV.E-2   M.P. Doyle and V. Bagheri, J. Org. Chem., 46, 4806 (1981).

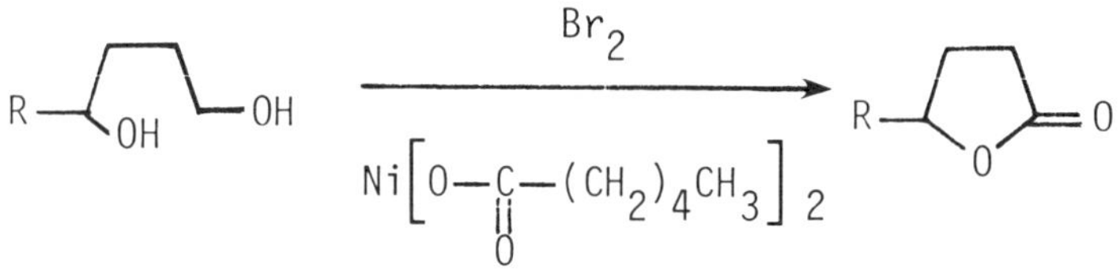

Br$_2$

Ni[O—C(=O)—(CH$_2$)$_4$CH$_3$]$_2$

R = n-alkyl, Ph   58-70%

IV.E-3   S.F. Brown, B.S. Bal, and H.W. Pinnick, <u>Tetra-hedron Lett.</u>, <u>22</u>, 4891 (1981).

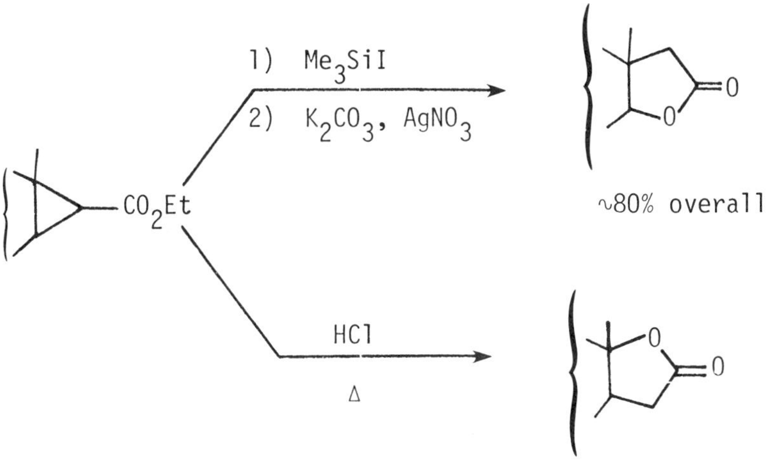

$\sim$80% overall

IV.E-4   M.M. Midland and N. H. Nguyen, <u>J. Org. Chem.</u>, <u>46</u>, 4107 (1981).

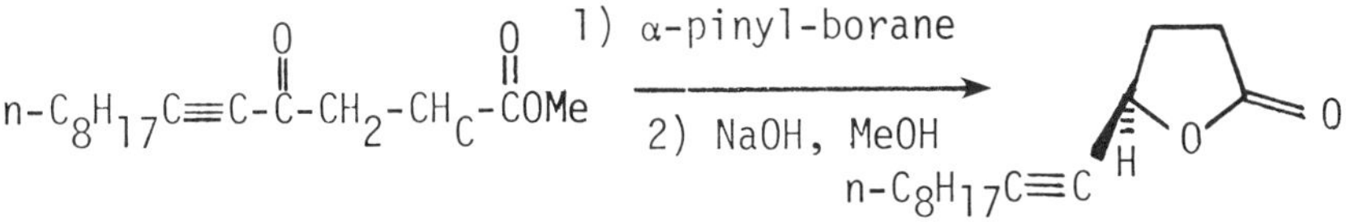

70-75%

85-90% ee

IV.E-5   T. Shono <u>et al.</u>, <u>Chem. Lett.</u>, 1217 (1981).

~50-80%

R, R' = H, alkyl, Ph, cyclic

IV.E-6   N. DeKimpe <u>et al.</u>, <u>Synth. Comm.</u>, <u>11</u>, 35 (1981).

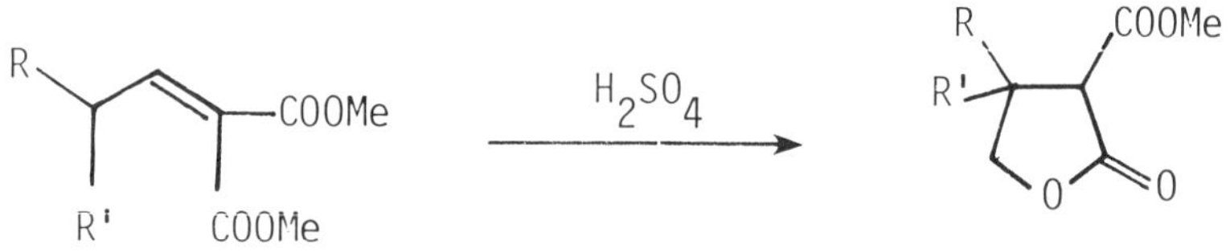

75-91%

R, R' = Me, Et, -(CH$_2$)$_{\overline{5}}$

IV.E-7   R.W. Aben, R. Hofstraat, and J.W. Scheeren, Rec. Trav. Chim. Pays-Bas, 100, 355 (1981).

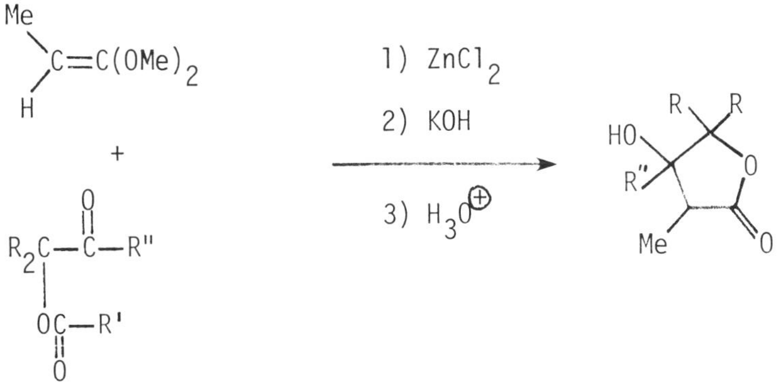

R = H, Me, $-(CH_2)_{\overline{4}}$
R' = H, Me, $CHCl_2$
R" = H, Me

IV.E-8   A.R. Chamberlain, M. Dezube, and P. Dussault, Tetrahedron Lett., 22, 4611 (1981).

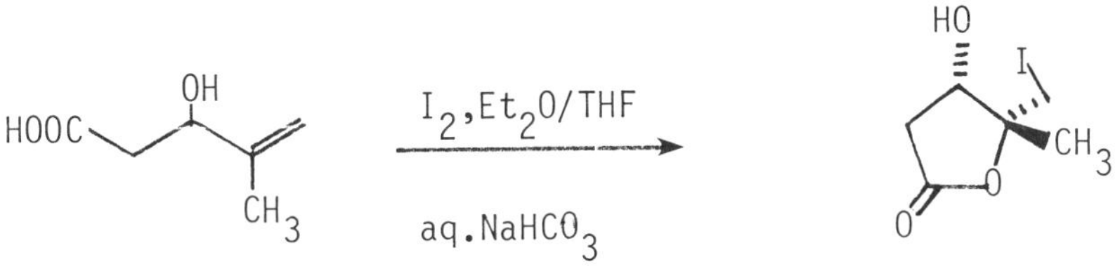

74% (95% cis)

IV.E-9   P. Canonne, M. Akssira, and G. Lemay, Tetrahedron
Lett., 22, 2611 (1981).

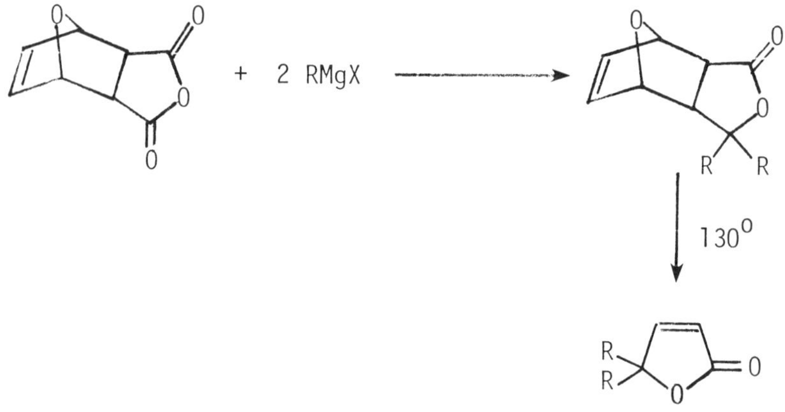

R = alkyl, Ph, Bz   95%(R=-CHMe$_2$)

IV.E-10   T. Shono, Y. Matsumura, and S. Yamane, Tetrahedron
Lett., 22, 3269 (1981).

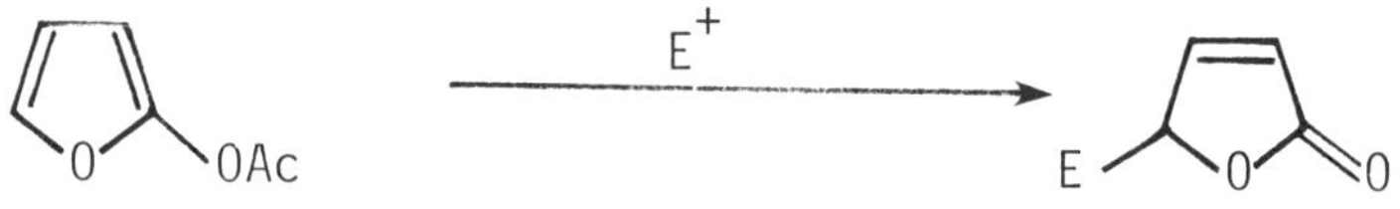

~60-90%

$E^{\oplus}$ = R-CHO, $CH_2(OCH_3)_2$, $CH(OCH_3)_3$

IV.E-11   M. Larcheveque et al., Tetrahedron Lett., 22, 1595 (1981).

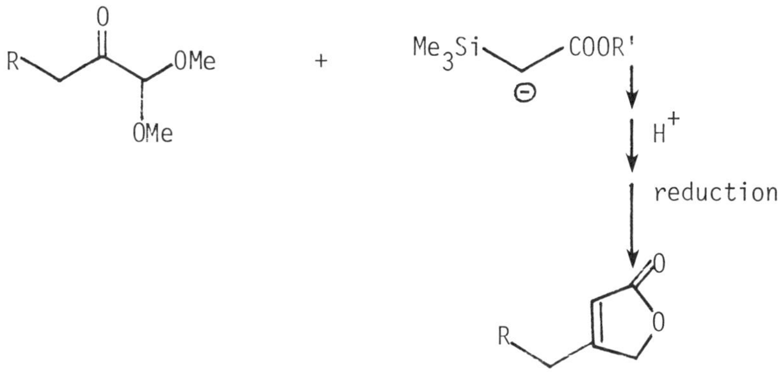

R = alkyl, allylic, etc.   ∿60-70% overall

IV.E-12   T. Mise, P. Hong, and H. Yamazaki, Chem. Lett., 993 (1981).

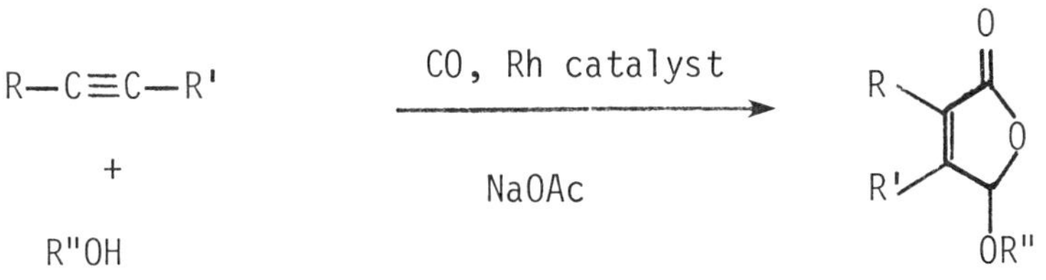

R, R' = Ph, Me   ∿50-80%
R" = alkyl

IV.E-13   A.B. Smith III et al., J. Am. Chem. Soc., 103, 1501 (1981).

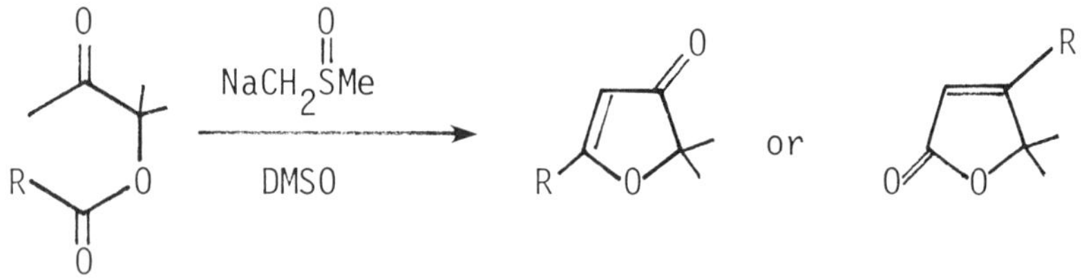

20-88%

R = alkyl, vinyl, Ph, etc.

IV.E-14   G.A. Krafft and J.A. Katzenellenbogen, J. Am. Chem. Soc., 103, 5459 (1981).

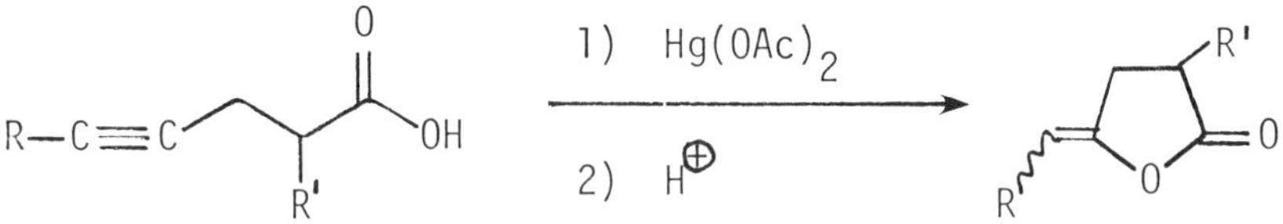

∿50-70%

R = H, Me, Br, Cl, SiMe₃
R' = H, Ph

IV.E-15 M.F. Semmelhack and S.J. Brickner, J. Org. Chem., 46, 1723 (1981).

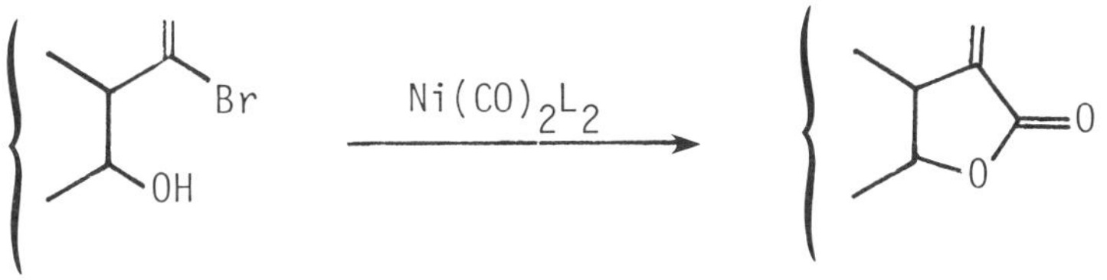

∿50-90%

IV.E-16 J.R. Norton et al., J. Am. Chem. Soc., 103, 7520 (1981).

Catalytic cyclocarbonylation of acetylenic alcohols leads to methylene lactones, for instance:

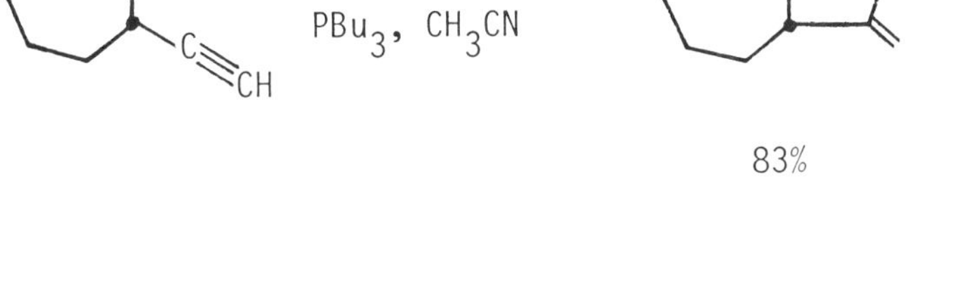

83%

IV.E-17   M.F. Semmelhack and S.J. Brickner, <u>J. Am. Chem. Soc.</u>, <u>103</u>, 3945 (1981).

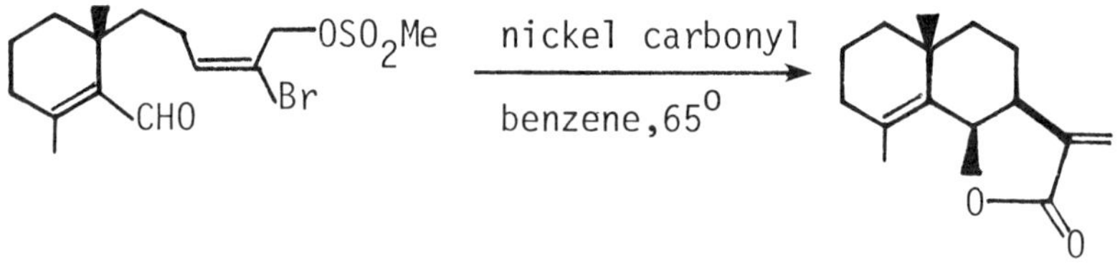

∿50%

IV.E-18   I. Kuwajima and H. Urabe, <u>Tetrahedron Lett.</u>, <u>22</u>, 5191 (1981).

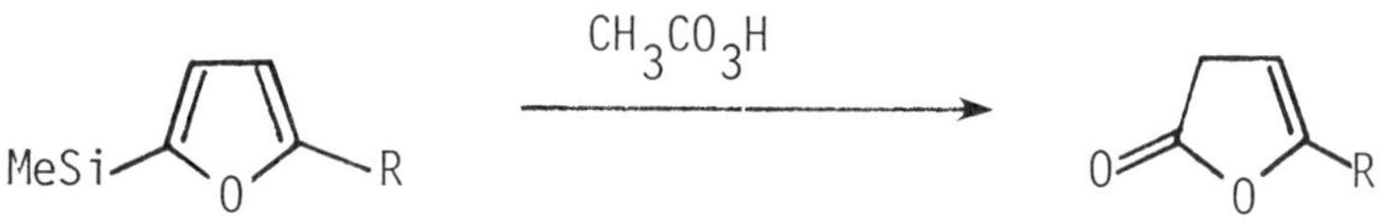

Widely varying yields.

R = alkyl, containing olefins, acetals,
    silyl ethers

IV.E-19 R.L. Danheiser *et al.*, *Tetrahedron Lett.*, **22**, 4205 (1981).

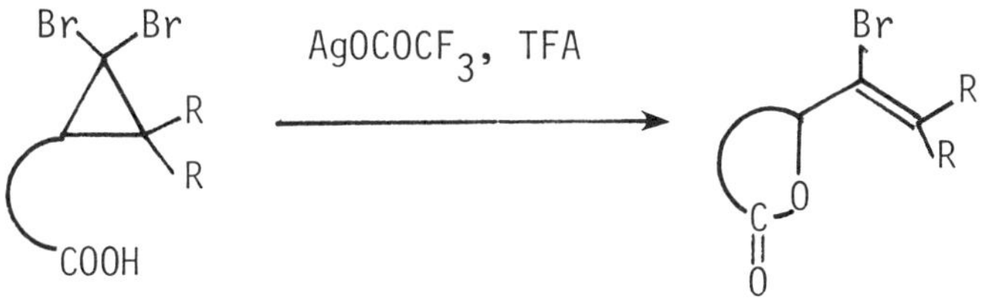
Br
Br
R
R
COOH
$AgOCOCF_3$, TFA
Br
R
R
O
C
O

R = H, Me

~70-80%

Ring size 5-6

IV.E-20 M. Chmielewski and J. Jurczak, *J. Org. Chem.*, **46**, 2230 (1981).

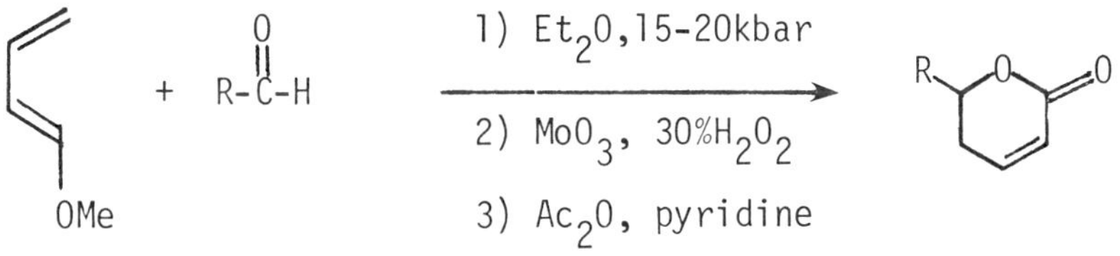
OMe
+ R-C-H
O
1) Et_2O,15-20kbar
2) MoO_3, 30%H_2O_2
3) Ac_2O, pyridine
R
O
O

~50%

R = *n*-C_5H_{11}, Ph

IV.E-21   G. Falsone and B. Spur, Liebigs Ann. Chem., 565
(1981).

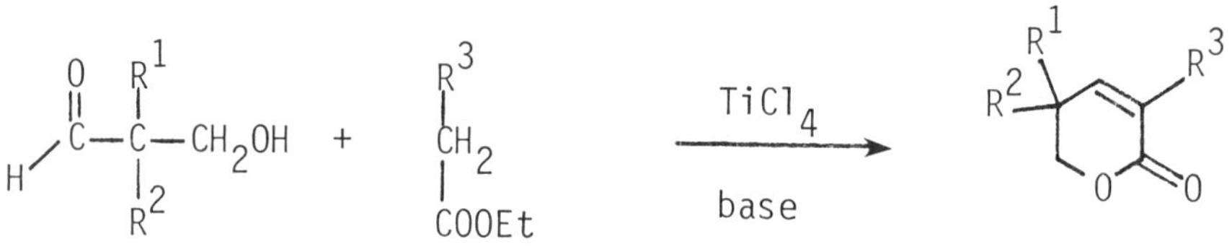

R^1, R^2 = Me, 1^0 alkyl

R^3 = alkyl, acyl, -CN

~70-80%

IV.E-22   S. -I. Murahashi et al., Tetrahedron Lett., 22,
5327 (1981).

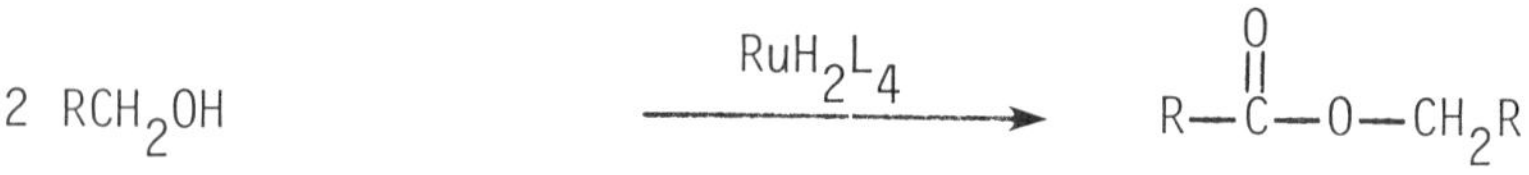

2 RCH$_2$OH $\xrightarrow{\text{RuH}_2\text{L}_4}$ R—$\overset{\overset{\text{O}}{\|}}{\text{C}}$—O—CH$_2$R

52-98%

R = alkyl, Bz, cycloalkyl, etc.

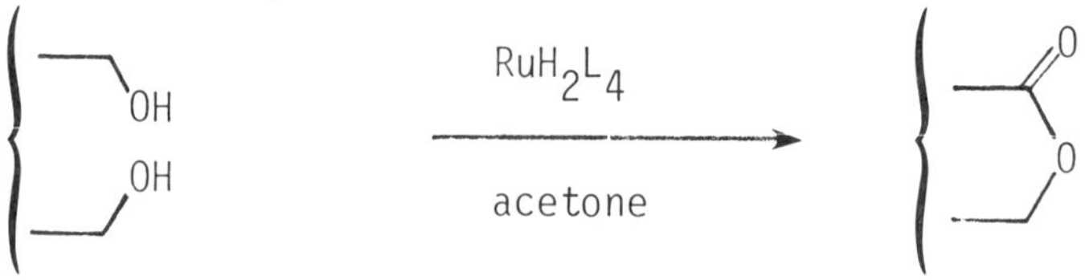

IV.E-23   U. Schmidt and M. Dietsche, <u>Angew. Chem. Int. Ed.</u>,
<u>20</u>, 771 (1981).

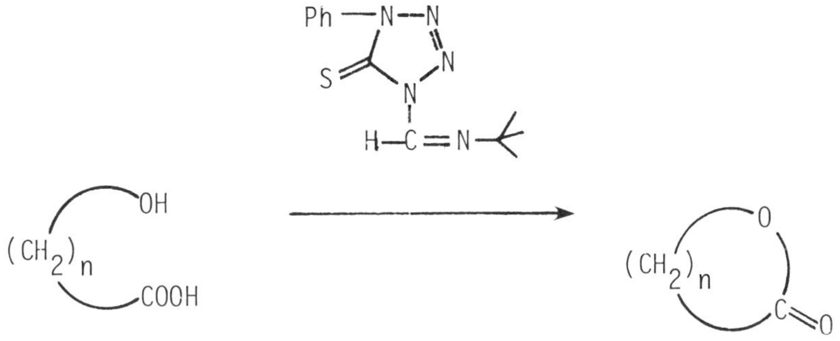

n = 15, 16, 17, 19

>90%

IV.E-24   V. Bhat and R.C. Cookson, <u>J.C.S. Chem. Comm.</u>,
1123 (1981).

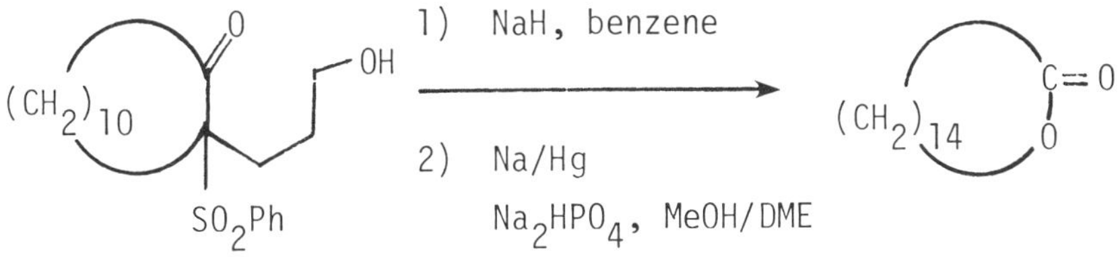

1)   NaH, benzene

2)   Na/Hg
     Na_2HPO_4, MeOH/DME

81%

IV.E-25   R. Ikan, V. Weinstein, and U. Ravid, <u>Org. Prep.</u>
<u>Proc. Int.</u>, <u>13</u>, 59 (1981).

Review: "Synthesis of Saturated γ-Lactones"

IV.E-26   P.W. Scott and I.T. Harrison, <u>J. Org. Chem.</u>, <u>46</u>,
1914 (1981).

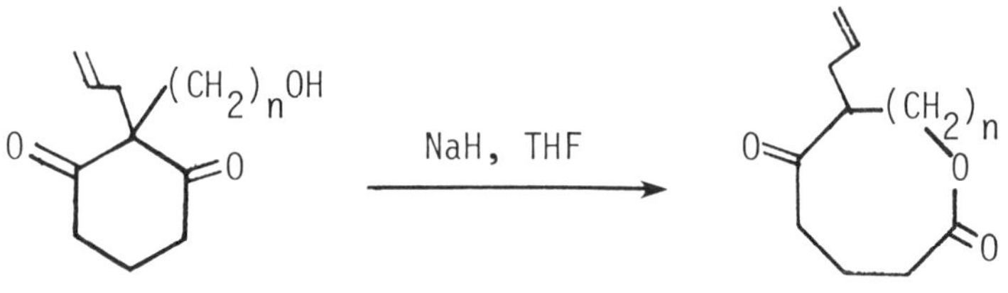

NaH, THF

11-49%

n = 4-20

IV.F.   Pyridines, Quinolines, etc.

IV.F-1   D.L. Boger and J.S. Panek, J. Org. Chem., 46, 2179 (1981).

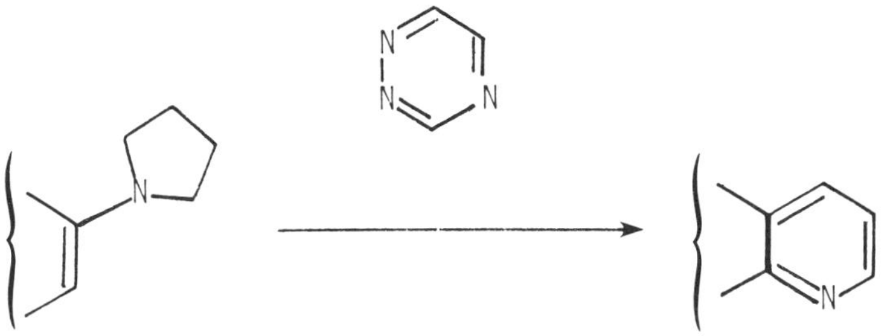

∿30-70%

IV.F-2   K. T. Potts et al., J. Am. Chem. Soc., 3584 and 3585 (1981).

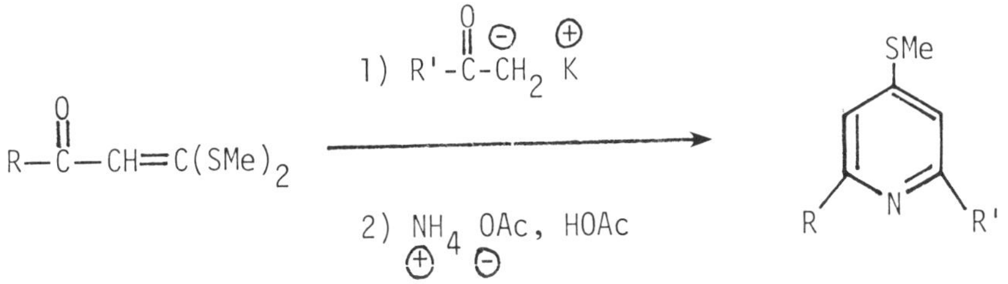

∿ 70-80%

R,R' = subst. Ph, pyridyl, thienyl, furyl

IV.F-3   D.K. Dantchev and I.C. Ivanov, Synthesis, 227 (1981).

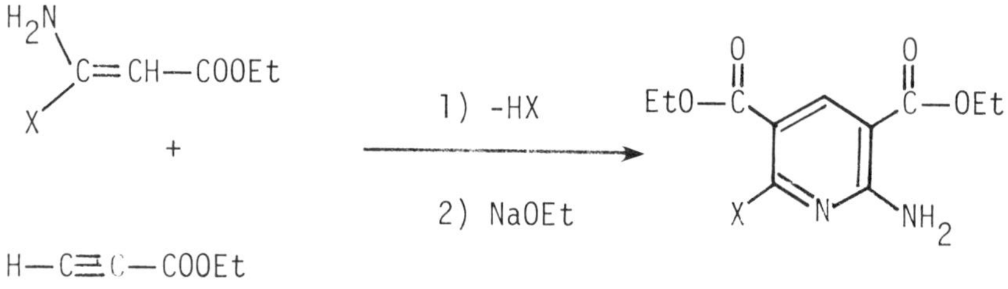

42% (X = -OEt)

X = -NH$_2$, -OEt

64% (X = -NH$_2$)

IV.F-4   S. Senda et al., J. Org. Chem., 46, 846 (1981).

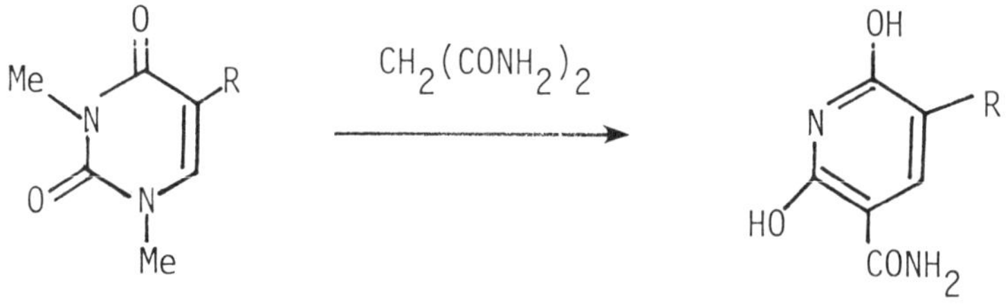

~60-90%

R = H, Me, CN, halogen

IV.F-5   S. Huybrechts and G.J. Hoornaert, <u>Synth. Comm.</u>,
<u>11</u>, 17 (1981).

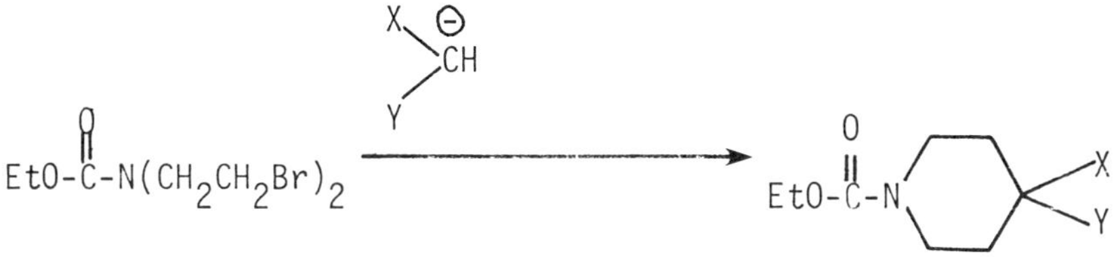

$$EtO-\overset{O}{\underset{\|}{C}}-N(CH_2CH_2Br)_2 \quad\xrightarrow{\underset{Y}{\overset{X}{\searrow}}\overset{\ominus}{\underset{}{CH}}}\quad EtO-\overset{O}{\underset{\|}{C}}-N\diagdown\diagup\diagdown\overset{X}{\underset{Y}{}}$$

53-78%

X and/or Y are electron-withdrawing,
such as $\overset{O}{\underset{\|}{\underset{\diagup C\diagdown}{}}}$ or $C\equiv N$.

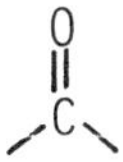

IV.F-6   A.M. van Leusen and J.W. Terpstra, <u>Tetrahedron Lett.</u>,
<u>22</u>, 5097 (1981).

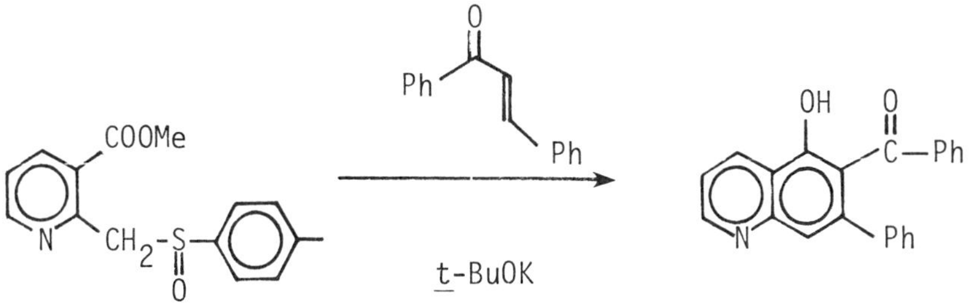

$\underline{t}$-BuOK

72%

Several additional examples

IV.F-7 Y. Watanabe, S.C. Shim, and T. Mitsudo, Bull Chem. Soc. Japan, 54, 3460 (1981).

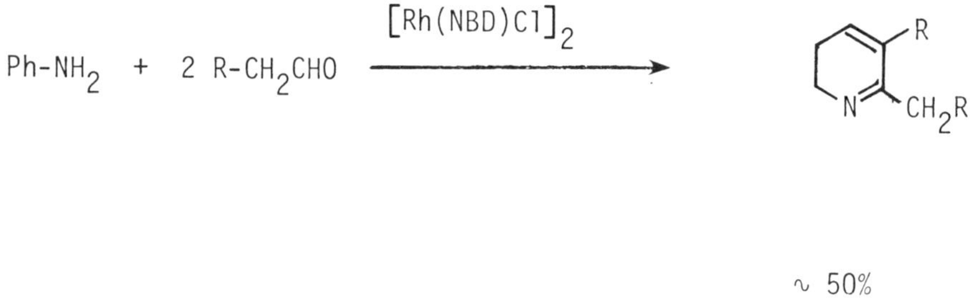
Ph-NH2
+
2 R-CH2CHO
[Rh(NBD)Cl]2
R
N
CH2R
~ 50%

R = H, Me, Et, Bu

IV.F-8 B.W. Hansen and E.B. Pedersen, Liebigs Ann. Chem., 1485 (1981).

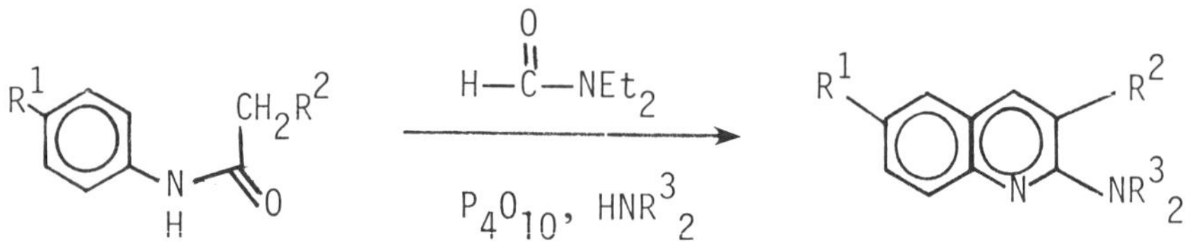
R1
CH2R2
N
H
O
H—C—NEt2
O
P4O10, HNR32
R1
R2
N
NR32

R^1 = H, Me, Cl, Br, OEt

R^2 = H, Me

R^3 = Et, Pr, Me, cyclic

widely varying yields

IV.F-9 D.L. Boger, C.E. Brotherton, and M.D. Kelley, Tetrahedron, 37, 3977 (1981).

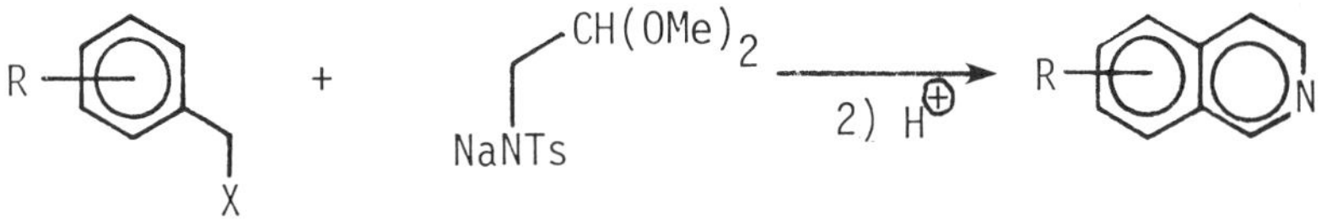
R
+
CH(OMe)2
NaNTs
X
2) H⊕
R
N

X = halogen, OMes

R = H, OMe, Br, etc.

IV.F-10 J.P. Marino and R.D. Larsen, Jr., J. Am. Chem. Soc., 103, 4642 (1981).

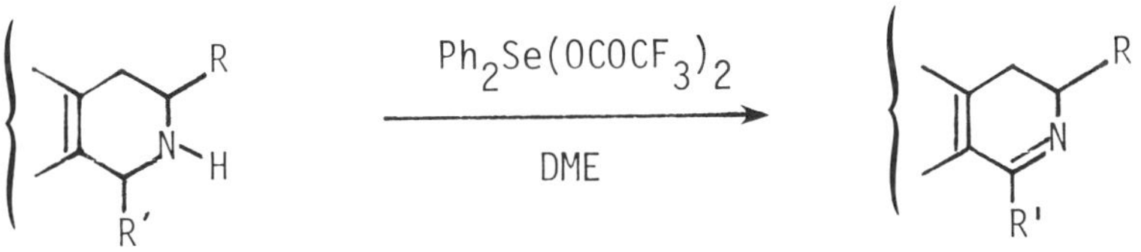
R
N
H
R'
Ph2Se(OCOCF3)2
DME
R
N
R'

~70-90%

may also be used to form

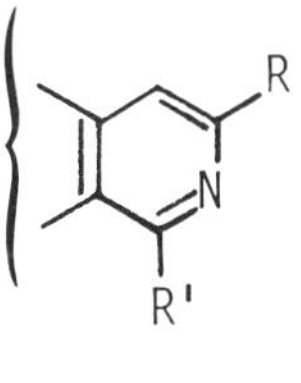
R
N
R'

IV.F-11 N.S. Narasimhan, A.C. Ranade, and B.H. Bhide, Indian J. Chem., 20B, 439 (1981).

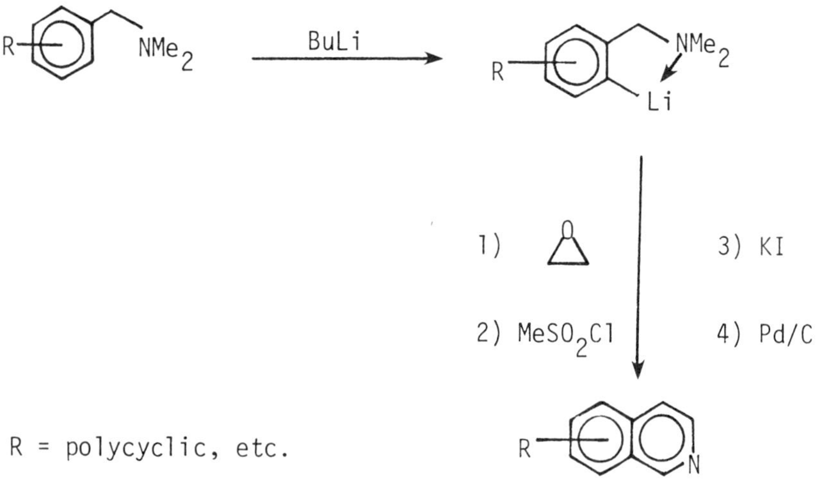

R
NMe2
BuLi
R
NMe2
Li
1)
2) MeSO2Cl
3) KI
4) Pd/C
R
N

R = polycyclic, etc.

IV.F-12 K. Akiba et al., Tetrahedron Lett., 22, 4977 (1981).

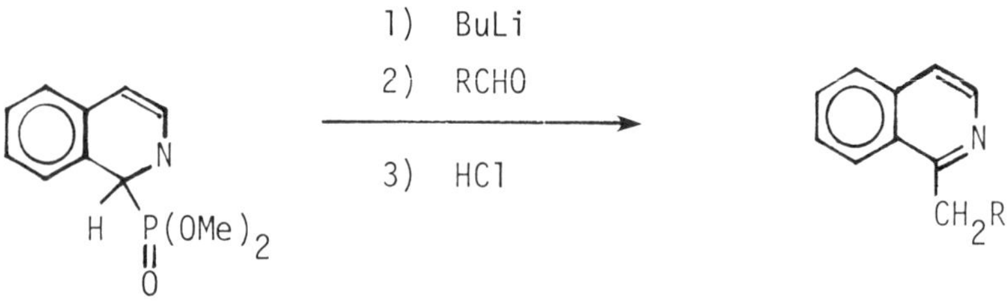

N
H
P(OMe)2
O
1) BuLi
2) RCHO
3) HCl
N
CH2R

R = i-Pr, n-Pr, t-Bu, subst. Ph

IV.F-13   L.N. Yakhontov and D.M. Krasnokutskaya, <u>Russ.</u>
<u>Chem. Rev.</u>, <u>50</u>, 565 (1981).

Review: "Advances in the Chemistry of α,α'-disubstituted

Pyridines"

IV.F-14   H. Beschke, <u>Aldrichimica Acta</u>, <u>14</u>, 13 (1981).

Review: "Chemical Reactions of Newly Available

Pyridines"

IV.F-15   A.N. Kost, S.P. Gromov, and R.S. Sagitullin,
<u>Tetrahedron</u>, <u>37</u>, 3423 (1981).

Review: "Pyridine Ring Nucleophilic Recyclizations"

IV. G. Pyrroles, etc.

IV.G-1 S. Nakanashi et al., Chem. Lett., 869 (1981).

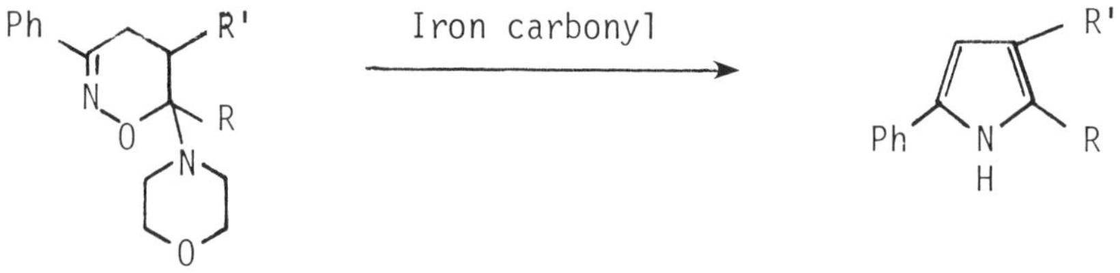

28-99%

R, R' = H, Me, Et, -(CH_2)-_n

IV.G-2 K. Utimoto, H. Miwa, and H. Nozaki, Tetrahedron Lett., 22, 4277 (1981).

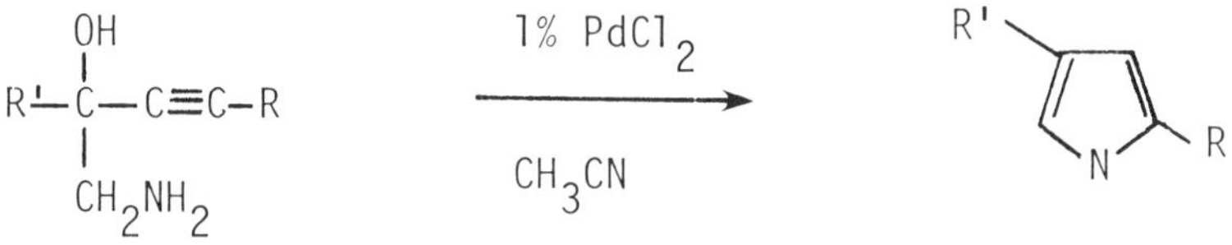

∿80-90%

R = n-hexyl, Ph
R' = H, Et, t-Bu

IV.G-3   J. Lugtenberg et al., Org. Prep. Proc. Int., 13, 97 (1981).

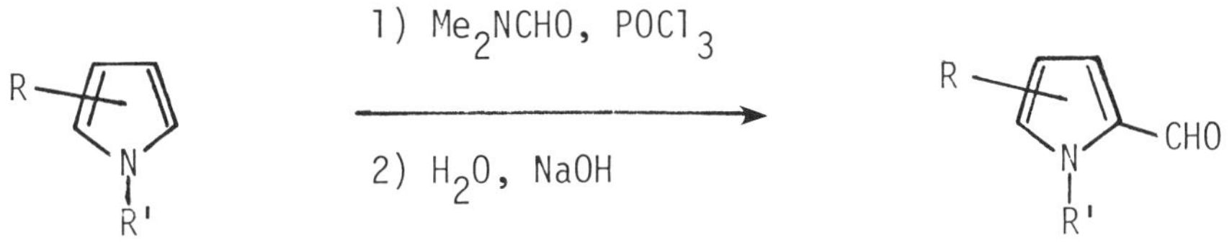

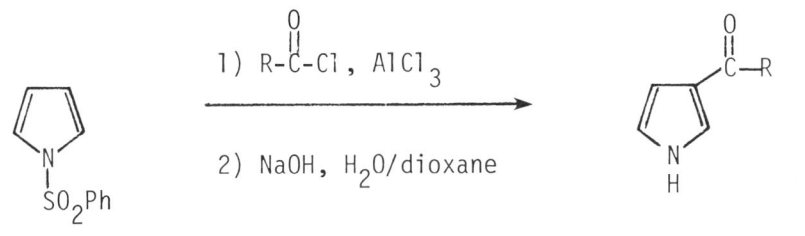

1) Me$_2$NCHO, POCl$_3$

2) H$_2$O, NaOH

R = H, Me
R' = H, Me

65-95%

IV.G-4   J. Rokach et al., Tetrahedron Lett., 22, 4901 (1981).

1) R-C-Cl, AlCl$_3$

2) NaOH, H$_2$O/dioxane

R = Me, Ph, cyclopropyl, etc.

∿80-90%

IV.G-5   K.C. Nicolaou, D.A. Claremon, and D.P. Papahatjis, Tetrahedron Lett., 22, 4647 (1981).

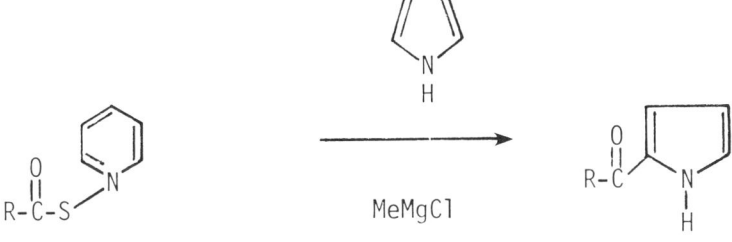

MeMgCl

R = alkyl, allyl, phenyl,
    heterocyclic, etc.

80-95%

IV.G-6   J.V. Cooney and W.E. McEwen, J. Org. Chem., 46, 2570 (1981).

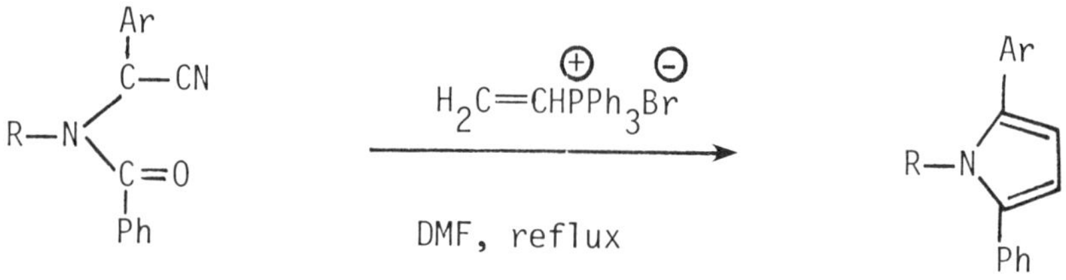

R—N structure with Ar, C—CN, C=O, Ph groups

$H_2C=CHPPh_3Br$ (with ⊕ and ⊖ charges)

DMF, reflux

→ pyrrole product with Ar, R—N, Ph

R = alkyl, subst. Ph, Bz, etc.   50-100%
Ar = subst. Ph

IV.G-7   J.O. Madsen et al., Acta Chem. Scand. B, 35, 77 (1981).

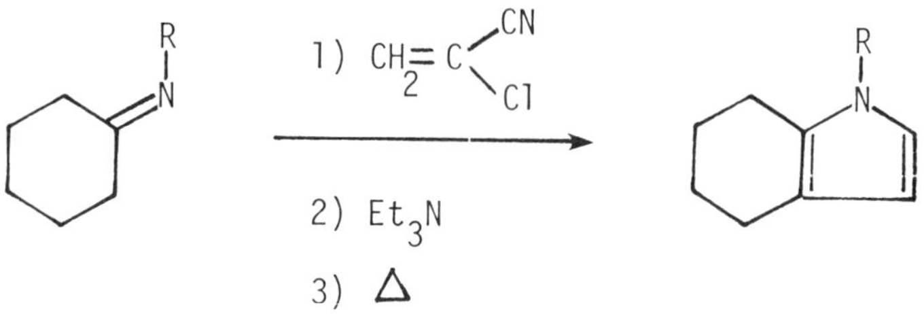

cyclohexanone imine (R—N=)

1) $CH_2=C$ with CN and Cl groups

2) Et_3N

3) △

→ fused pyrrole product with R—N

                                                          79% (R = Me)

IV.G-8   H. Quast, W.von der Saal, and J. Stawitz, <u>Angew.</u>
<u>Chem. Int. Ed.</u>, <u>20</u>, 588 (1981).

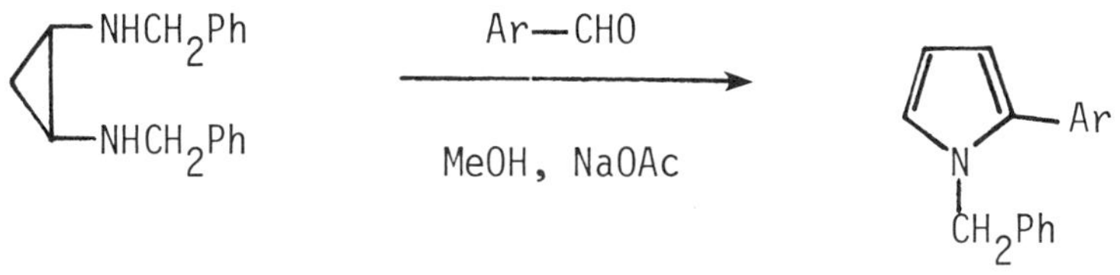

Ar = subst. Ph, 2-pyridyl   widely varying
                                                           yields

---

IV.G-9   H. Meyer, <u>Liebigs Ann. Chem.</u>, 1534 (1981).

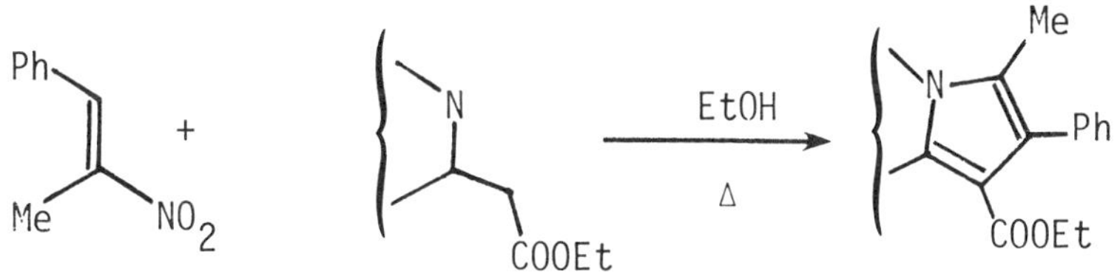

∿80-90%

IV.G-10   D.P. Schumacher and S.S. Hall, J. Org. Chem., 46, 5060 (1981).

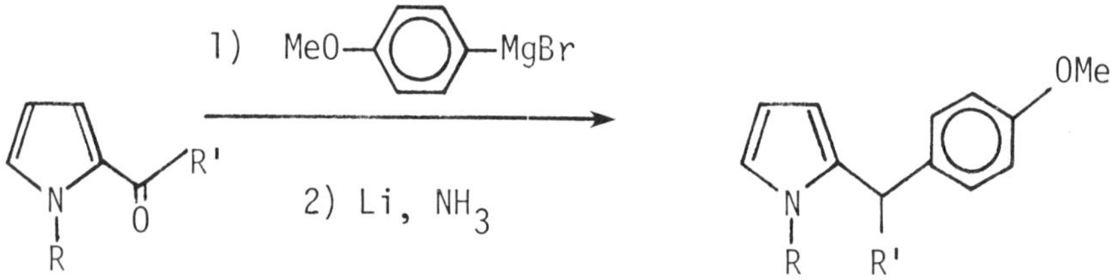

R, R' = H, Me, Ph   ∿70-90%

IV.G-11   K.W. Blake and I. Gillies, J.C.S. Perkin I, 700 (1981).

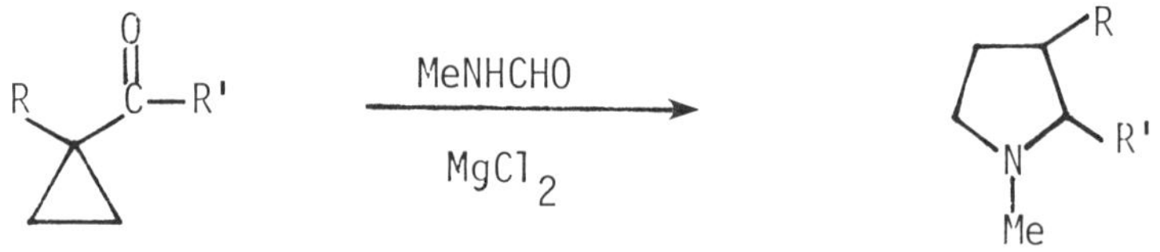

R = H, Ph
R' = H, alkyl, aryl

IV.G-12   M. Jubault et al., Tetrahedron Lett., 22, 3961
(1981).

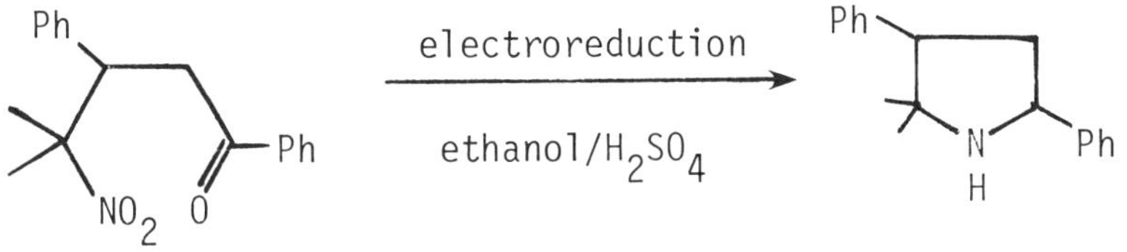

quantitative

IV.G-13   P.S. Mariano et al., J. Am. Chem. Soc., 103,
3148 (1981).

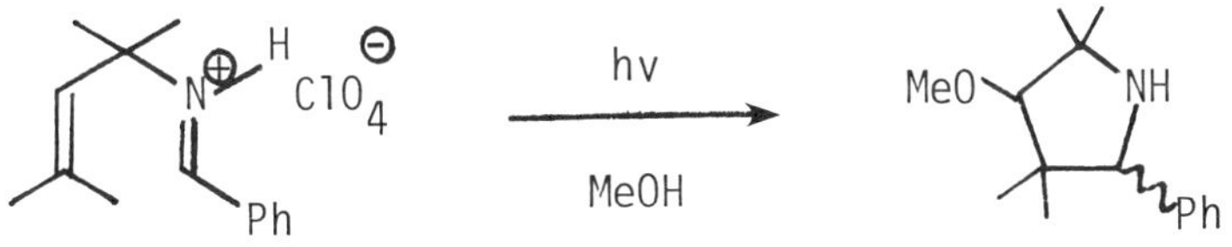

51%

IV.G-14   R.P. Evstigneeva, Pure and Appl. Chem., 53, 1129
(1981).

Reveiw: "Advances and Perspectives of Porphyrin

Synthesis"

IV. H. Other Heterocycles with One Heteroatom

(see also: II.F.1, VI.A.9)

IV.H-1   H.H. Wasserman et al., J. Org. Chem., 46, 2991 (1981).

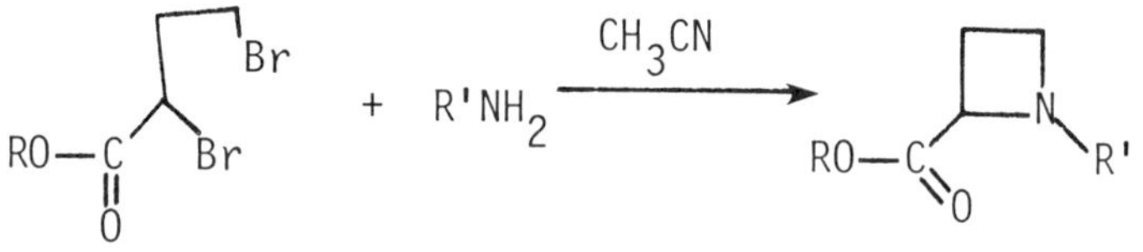

R = Me, Et, i-Pr   ∿50-80%
R' = alkyl, aryl, Bz, etc.

Several additional procedures.

IV.H-2   B.-T. Khai, C. Concilio, and G. Porzi, J. Org. Chem., 46, 1759 (1981).

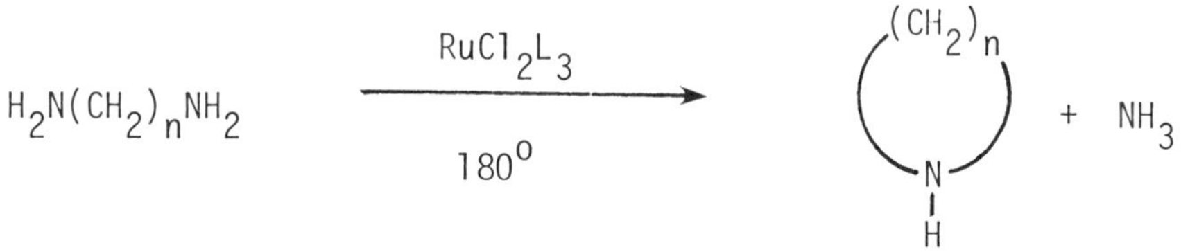

n = 4, 5, 6   78-90%

IV.H-3   R. Jeyaraman and S. Avila, Chem. Rev., 81, 149 (1981).

Review: "Chemistry of 3-Azabicyclo [3.3.1]nonanes"

IV.H-4   J. Moulines et al., Synthesis, 550 (1981).

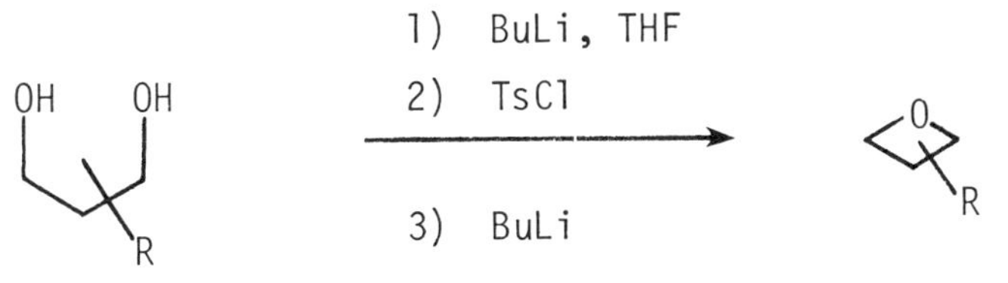

1) BuLi, THF
2) TsCl

3) BuLi

∿70-80%

R = H, Me, Ph, -(CH$_2$)$_{\overline{5}}$, etc.

IV.H-5   C.G. Bakker, J.W. Scheeren, and R.J.F. Nivard, Rec. Trav. Chim. Pays-Bas, 100, 13 (1981).

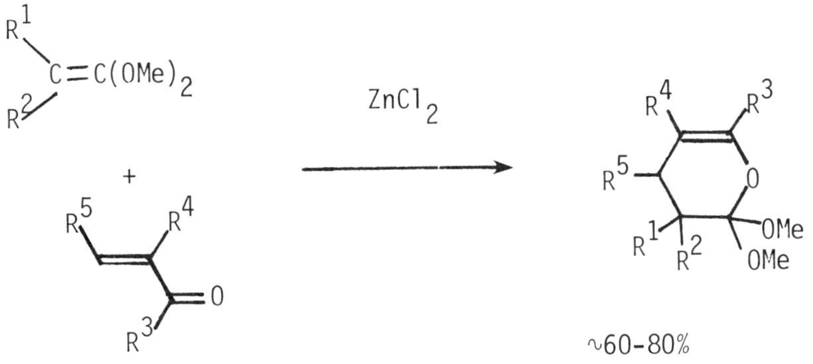

ZnCl$_2$

∿60-80%

IV.H-6   V.G. Kharchenko, I.A. Markushina, and T.I. Gubina, Doklady Chem., 255, 601 (1981).

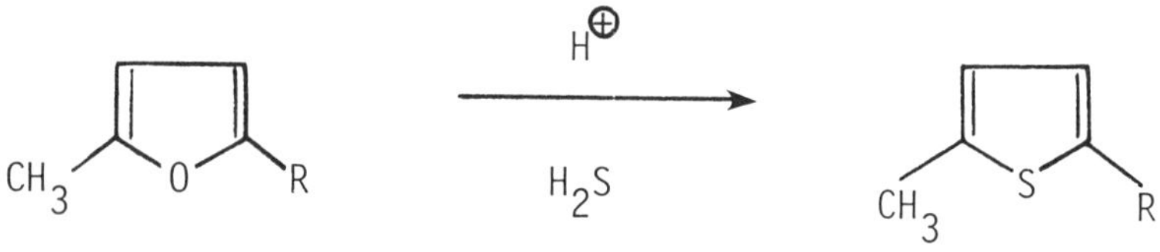

R = alkyl, containing alcohol, ketone, ester

~60-80%

IV.H-7   T. Kumamoto et al., Chem. Lett., 1341 (1981).

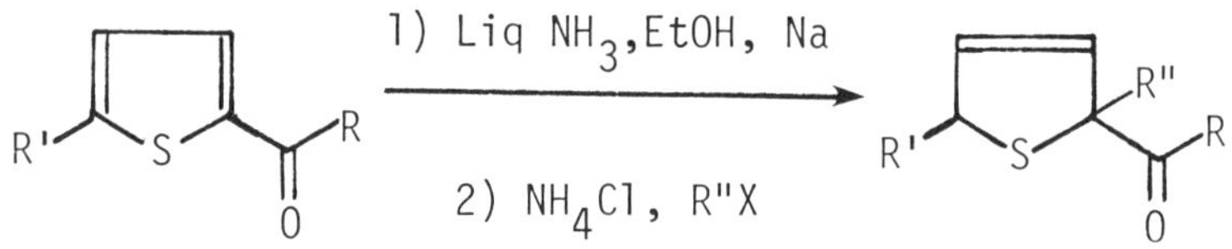

R = alkyl, cycloalkyl, OH   ~60-80%
R' = H, Me, Bu
R" = Me, Bz, allyl

IV.H-8   I. Murata and K. Nakasuji, Topics in Current Chemistry, 97, 33 (1981).

Review: "Recent Advances in Thiepin Chemistry"

IV. I. Heterocycles with Two or More Heteroatoms

1.a. 5-Membered Heterocycles with 2 N's

IV.I.1.a-1   J. Barluenga et al., J.C.S. Perkin I, 1891(1981).

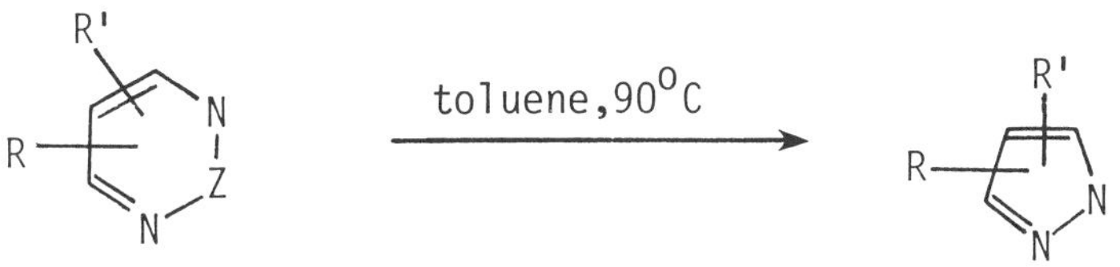

Z = S, S=O   84-94%

R, R' = Me, Ph, cyclohexyl, etc.

IV.I.1.a-2   Y. Kikugawa, Synthesis, 124 (1981).

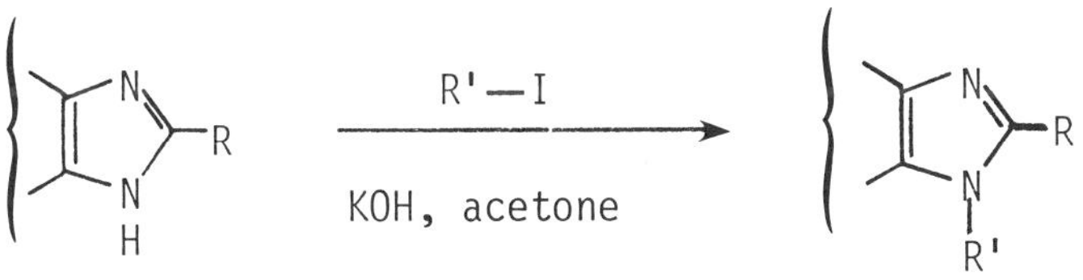

R = H, Me, Ph, NH$_2$ 70-91%

R' = Me, Et, Bz

IV.I.1.a-3   R.L. Webb and J.J. Lewis, J. Het. Chem., 18,
1301 (1981).

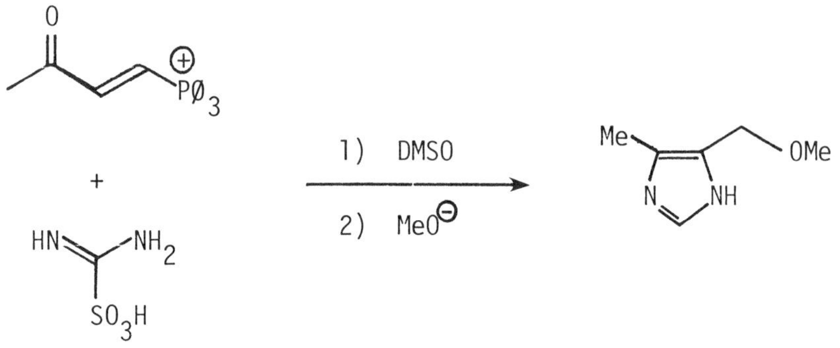

∿70% overall

IV.I.1.a-4   G. Levesque, J.-C. Gressier, and M. Proust,
Synthesis, 963 (1981).

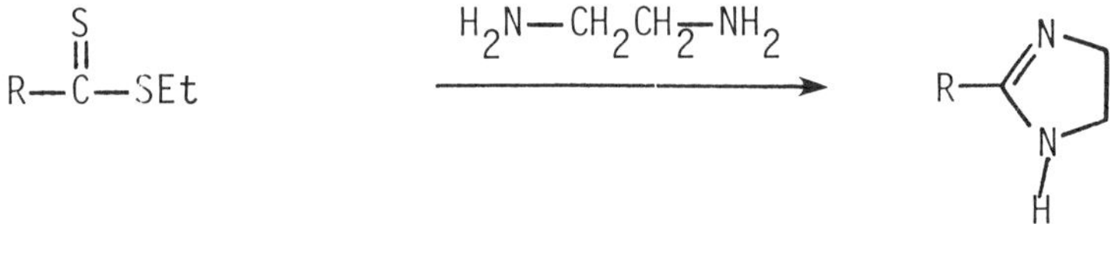

∿40-70%

R = alkyl, Ph, cyclic

IV.I.1.a-5   R. Grompper and U. Heinemann, <u>Angew. Chem. Int.</u>
<u>Ed.</u>, <u>20</u>, 296 (1981).

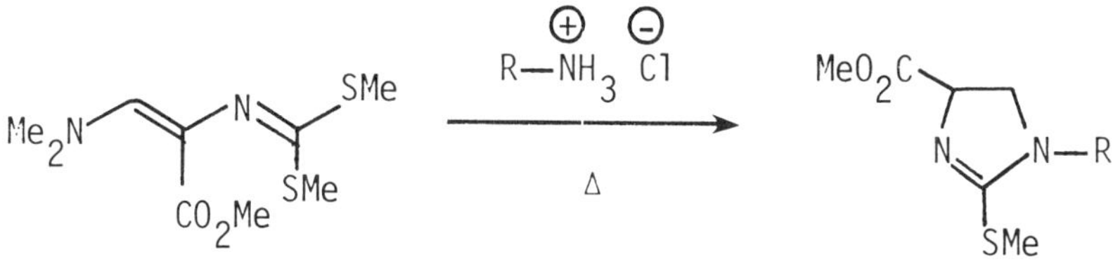

22%, 38%

R = Ph, Bz

IV.I. 1. b.   6-Membered Heterocycles with 2 N's

(see also: VI.A.15.)

IV.I.1.b-1   A.L. Weis and V. Rosenbach, <u>Tetrahedron Lett.</u>,
<u>22</u>, 1453 (1981).

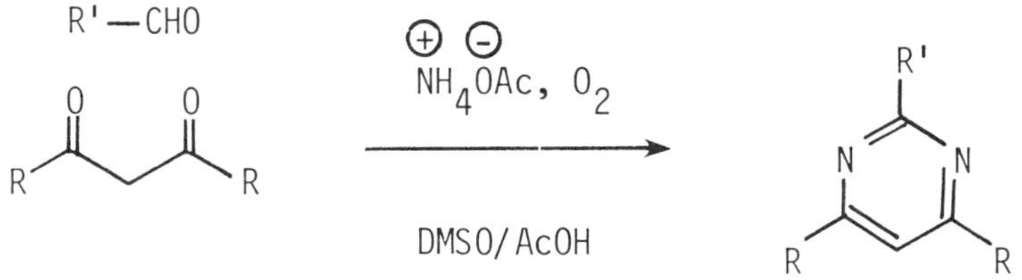

R = H, Me, Ph
R' = subst. Ph, naphthyl

12-47%

IV.I.1.b-2   B. Corain, M. Basato, and H. -F. Klein, <u>Angew.</u>
<u>Chem. Int. Ed.</u>, <u>20</u>, 972 (1981).

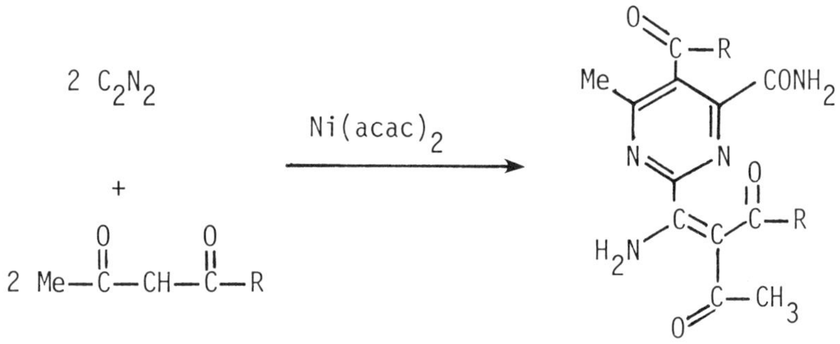

$2\ C_2N_2$

+

$$2\ Me-\overset{O}{\overset{\|}{C}}-CH-\overset{O}{\overset{\|}{C}}-R$$

$\xrightarrow{\text{Ni(acac)}_2}$

R = OEt, Ph   Widely varying yields.

IV.I.1.c.   Other Heterocycles with 2 N's

IV.I.1.c-1   P.N. Anderson, C.B. Argo, and J.T. Sharp,
<u>J.C.S. Perkin I</u>, 2761 (1981).

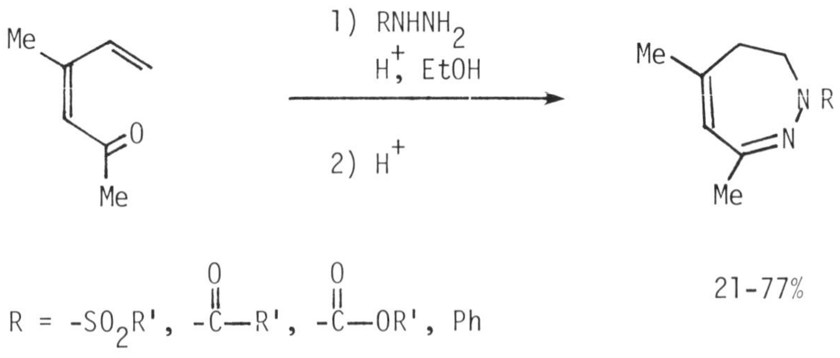

1) $RNHNH_2$
   H^+, EtOH

2) H^+

21-77%

$$R = -SO_2R',\ -\overset{O}{\overset{\|}{C}}-R',\ -\overset{O}{\overset{\|}{C}}-OR',\ Ph$$

IV.I.1.c-2   T.Tsuchiya et al., Chem. Pharm. Bull., 29, 1539 (1981).

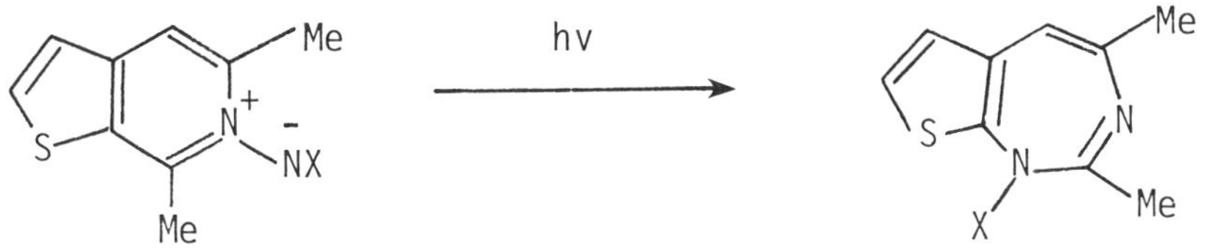

Widely varying yields.

IV.I.1.c-3   C.G. Overberger and T.F. Merkel, J. Org. Chem., 46, 442 (1981)

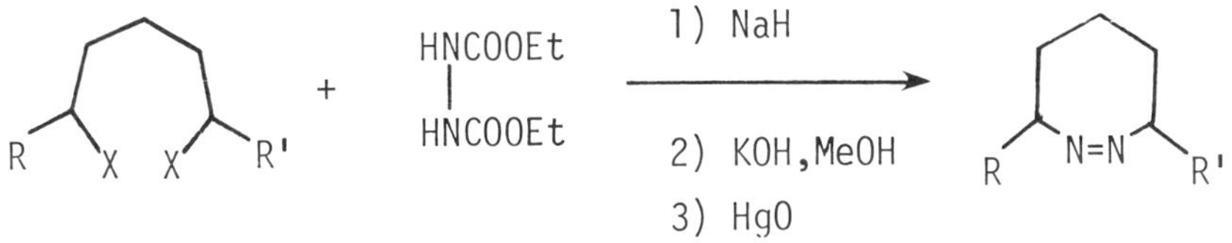

1) NaH

2) KOH,MeOH

3) HgO

∿40-60%

R, R' = H, Me, Ph

IV.I.1.c-4   V. Snieckus and J. Streith, Accounts Chem. Res., 14, 348 (1981).

Review: "1,2-Diazepines: A New Vista in Heterocyclic Chemistry"

IV. I. 2. Heterocycles with 1 N and 1 O

IV.I.2-1 Y. Tamura *et al.*, *Chem. Pharm. Bull.*, **29**, 3226 (1981).

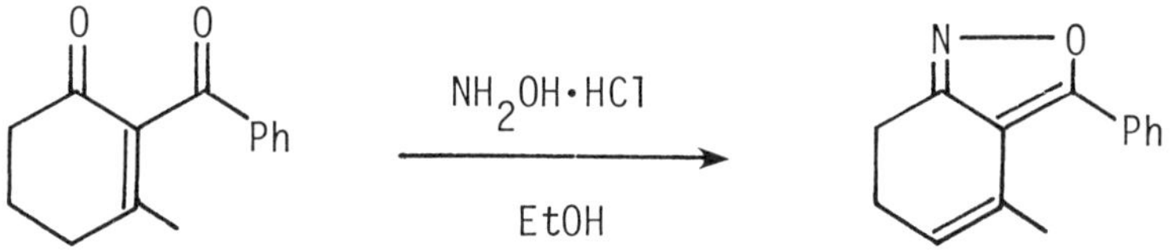
O
O
Ph
$NH_2OH \cdot HCl$
EtOH
N
O
Ph

72%

IV.I.2-2 A.M. van Leusen *et al.*, *J. Org. Chem.*, **46**, 2069 (1981).

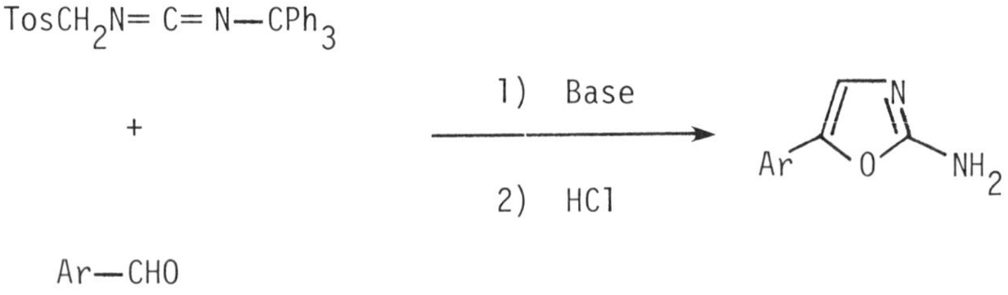
$TosCH_2N{=}C{=}N{-}CPh_3$
+
Ar—CHO
1) Base
2) HCl
N
Ar
O
NH_2

~60-80%

Ar = subst. Ph

IV.I.2-3   G.S. Reddy and M.V. Bhatt, Indian J. Chem., 20B, 322 (1981).

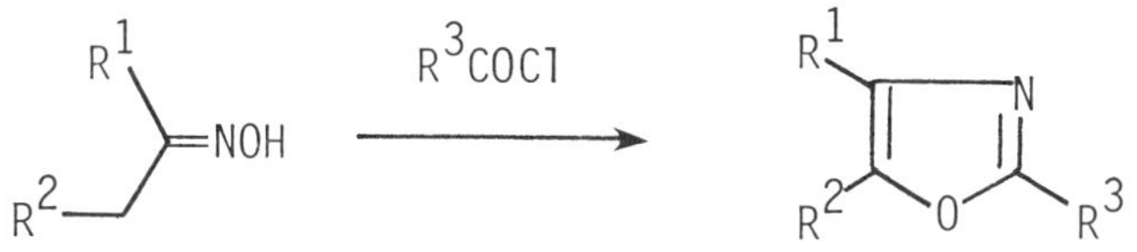

R^1, R^2 = Me, Ph, $\underline{i}$-Pr, $-(CH_2)_{\overline{4}}$

R^3 = Me, Et, Ph

68-85%

IV.I.2-4   H. Ogura, S. Mineo, and K. Nakagawa, Chem. Pharm. Bull., 29, 1518 (1981).

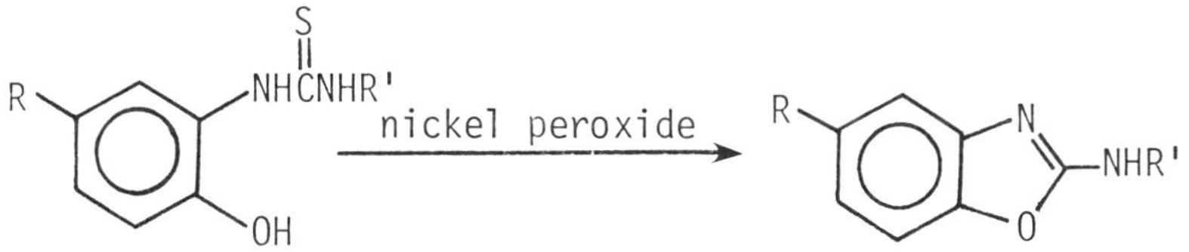

R = H, Me, Cl

R' = Me, Ph

59-85%

IV.I.2-5 H. Vorbrüggen and K. Krolikiewicz, Tetrahedron Lett., 22, 4471 (1981).

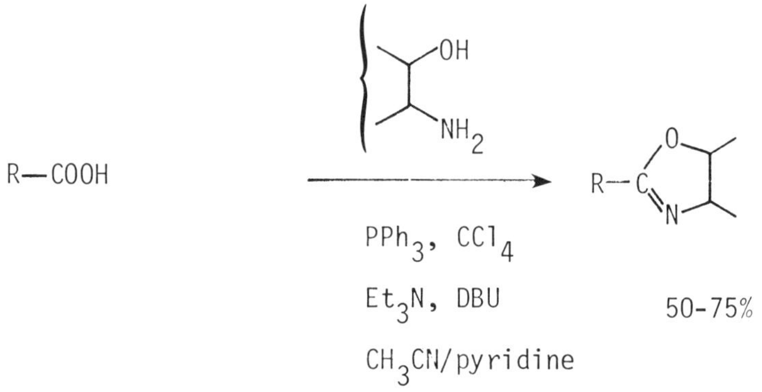
OH
NH2
R—COOH
PPh3, CCl4
Et3N, DBU
CH3CN/pyridine
R—C
O
N
50-75%

R = Ph, Bz, alkyl

IV.I.2-6 T. Hirao *et al.*, Angew. Chem. Int. Ed., 20, 126 (1981).

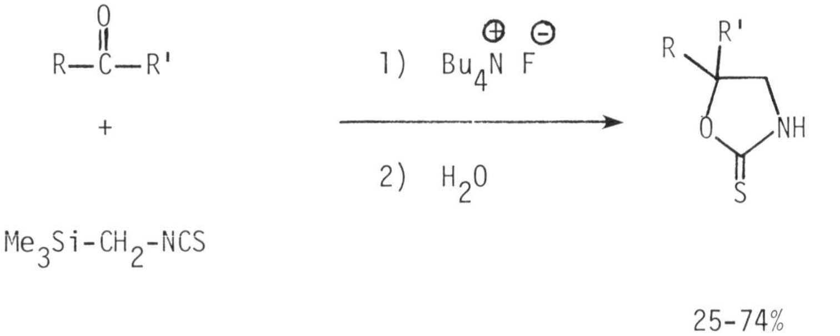
O
R—C—R'
+
Me3Si-CH2-NCS
1) Bu4N F
2) H2O
R
R'
O
NH
S
25-74%

R, R' = H, Me, *i*-Pr, Ph

IV.I.2-7 L.N. Pridgen and L.B. Killmer, J. Org. Chem., 46, 5402 (1981).

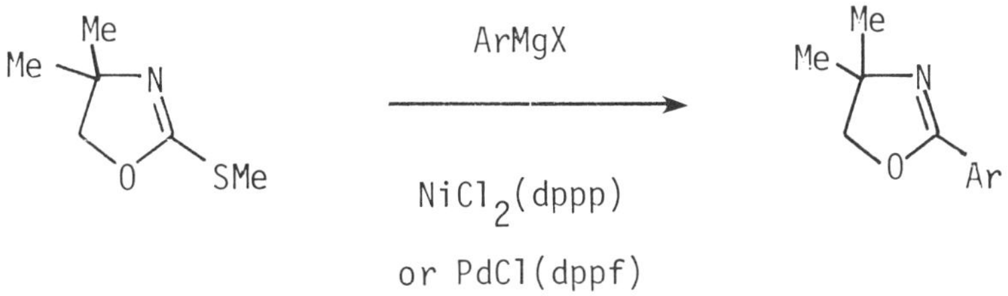
Me
Me
N
O
SMe
ArMgX
$NiCl_2$(dppp)
or PdCl(dppf)
Me
Me
N
O
Ar

Ar = Subst. Ph ∿70-90%

IV.I.2-8/IV.I.3-1 R. Grompper and U. Heinemann, Angew. Chem. Int. Ed., 20, 296 (1981).

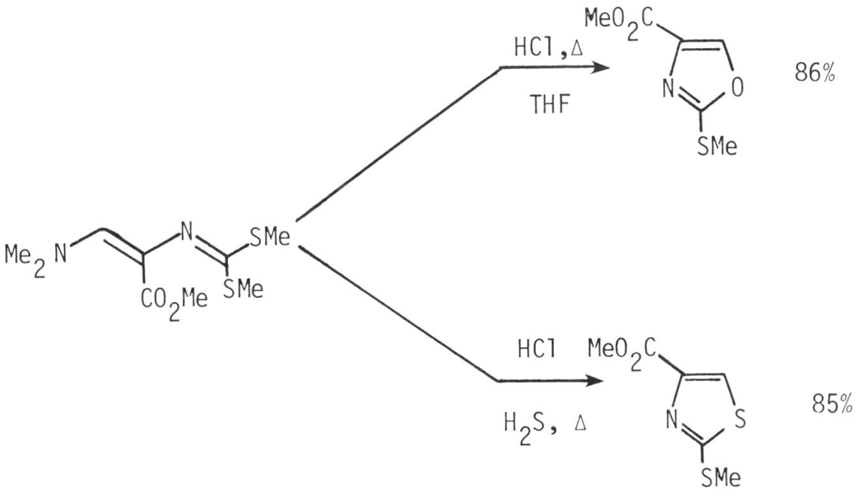
Me_2N
N
SMe
CO_2Me
SMe
HCl,Δ
THF
MeO_2C
N
O
SMe
86%
HCl
H_2S, Δ
MeO_2C
N
S
SMe
85%

IV.I.2-9/IV.I.3-2   M.Z.A. Badr et al., Bull. Chem. Soc.
Japan, 54, 1844 (1981)

$$R-\underset{\underset{XH}{|}}{CH}-\underset{\underset{NH_2}{|}}{CH}-COOEt \xrightarrow[\Delta]{Ar-CHO} \begin{array}{c} R \diagdown \diagup COOEt \\ X \diagdown NH \\ | \\ Ar \end{array}$$

R = H, Me, Ph   80-98%
X = 0, S
Ar = subst. Ph

IV. I. 3.   Heterocycles with 1 N and 1 S

IV.I.3-3   J. Gasteiger and C. Herzig, Tetrahedron, 37,
2607 (1981).

$$\begin{array}{c} Cl \\ \diagup \\ 0 \end{array} \xrightarrow[\text{(R = Me, Ph)}]{\underset{\underset{\displaystyle R-\overset{S}{\overset{||}{C}}-NH_2}{}}{}} \begin{array}{c} N \\ \diagup \diagdown R \\ S \end{array}$$

~50-90%

IV.I.3-4   V.I. Stenberg et al., Chem. Rev., 81, 175 (1981).

Review: "Chemistry and Biological Activity of
Thiazolidinones"

IV. I. 5. Heterocycles with 3 N's

IV.I.5-1 L.G. Tikhonova et al., J. Org. Chem. (USSR), 17, 1244 (1981).

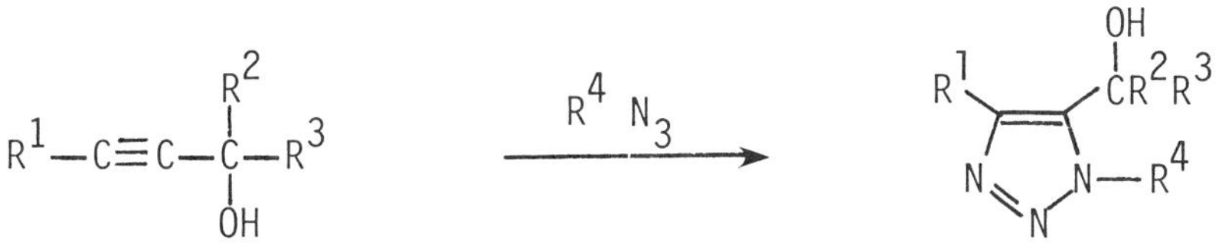

$$R^1-C\equiv C-\underset{\underset{OH}{|}}{\overset{\overset{R^2}{|}}{C}}-R^3 \quad \xrightarrow{R^4 \ N_3} \quad$$

~50-80%

R's = H, alkyl, Bz, etc.

IV.I.5-2 D. Martin, F. Tittelbach, and K. Nadolski, J. Prakt. Chem., 323, 694 (1981).

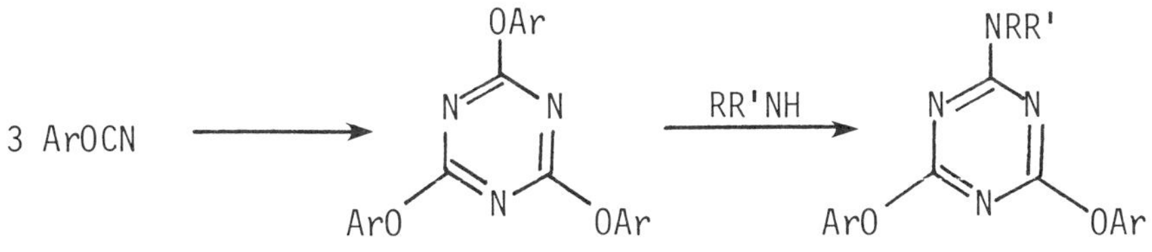

3 ArOCN $\longrightarrow$

$\xrightarrow{RR'NH}$

Ar = subst. Ph

R, R' = H, alkyl, Bz, cyclic

IV.I.5-3   R.L.N. Harris, Synthesis, 907 (1981); Aust. J. Chem., 34, 623 (1981).

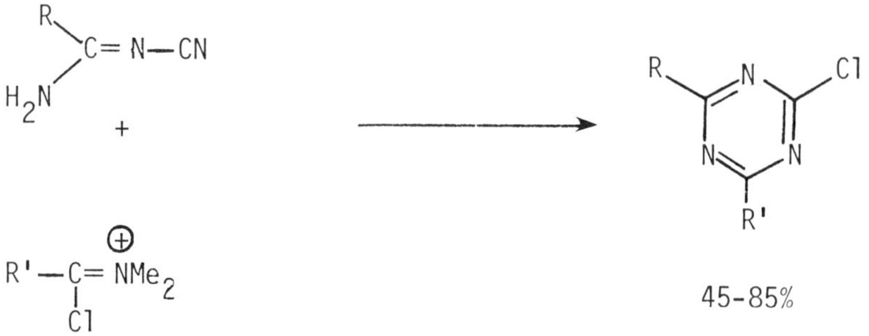

R = OPh, OEt, SMe, Cl

R' = H, Ph, Me, 2-furyl, etc.

IV. I. 6.   Other Heterocycles with Two or More Heteroatoms

IV.I.6-1   J. Plenkiewicz and T. Zdrojewski, Bull. Soc. Chim. Belges, 90, 193 (1981).

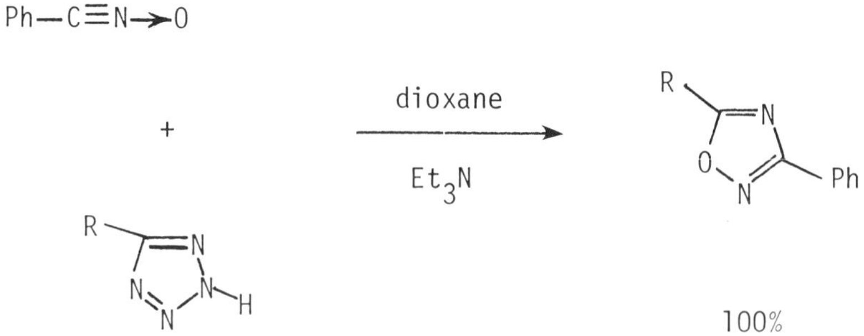

R = i-Pr, Ph

IV.I.6-2   H. Schubert, Z. Chem., 30 (1981).

Widely varying yields.

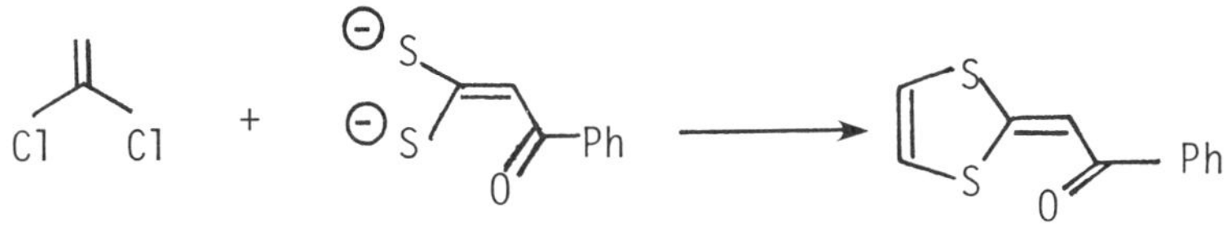

IV. J. General Heterocyclic Reviews

IV.J-1   K.C. Nicolaou, Tetrahedron, 37, 4097 (1981).

Review: "Organoselenium-Induced Cyclizations in

Organic Synthesis."

Ring closures are included which lead from unsaturated sub-
strates to lactones, cyclic ethers, cyclic thioethers,
N-heterocycles, and carbocycles.

IV.J-2   R. Gallo, H.J.M. Dou, and P. Hassanaly, Bull. Soc.
Chim. Belges, 90, 849 (1981).

Review: "Phase Transfer Catalysis in Heterocyclic

Chemistry"

V

PROTECTING GROUPS

V. A. Hydroxyl Protecting Groups

(see also: VI.A.10., VI.A.11.)

V.A-1   G.A. Olah et al., Synthesis, 471 (1981).

$$R{-}OH \xrightarrow[\text{Nafion---H, } \triangledown]{Me{-}OCH_2O{-}Me} R{-}O{-}CH_2OMe$$

57-96%

V.A-2   F.M. Menger and C.H. Chu, J. Org. Chem., 46, 5044 (1981).

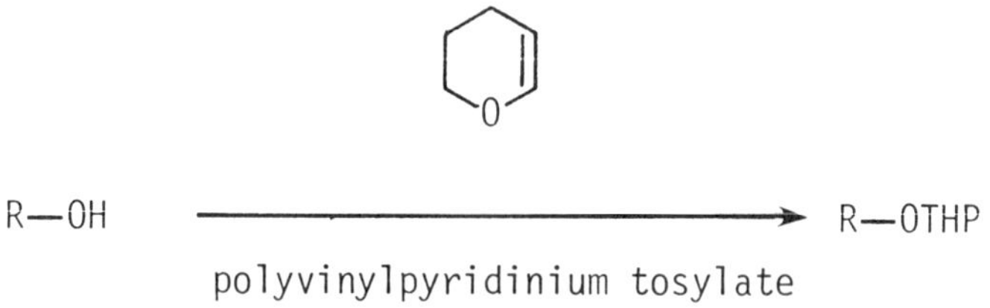

$$R{-}OH \xrightarrow{\text{polyvinylpyridinium tosylate}} R{-}OTHP$$

R = 1^0, 2^0, 3^0 alkyl, benzylic, etc.

∿80-90%

V.A-3   T. Hiyama et al., Synthesis, 899 (1981).

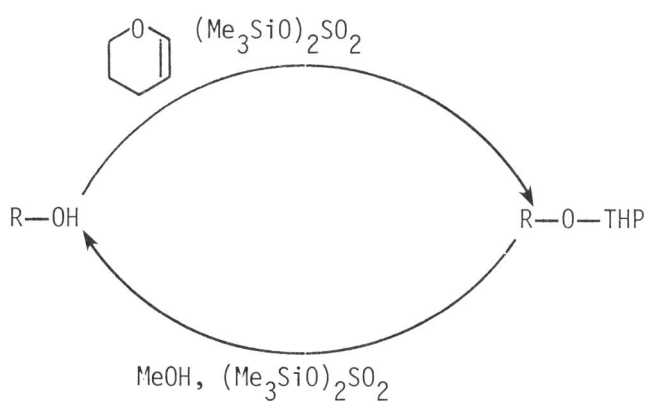

$$R—OH \quad \xrightarrow{(Me_3SiO)_2SO_2} \quad R—O—THP$$

MeOH, $(Me_3SiO)_2SO_2$

V.A-4   A. Hosomi and H. Sakurai, Chem. Lett., 85 (1981).

$$R—OH \xrightarrow[\quad I_2 \quad]{\quad Me_3SiCH_2CH=CH_2 \quad} R—O—SiMe_3$$

>90%

R = alkyl, aryl

V.A-5   T. Morita, Y. Okamoto, and H. Sakurai, Synthesis, 745 (1981).

$$R—OH \xrightarrow[\quad CH_2Cl_2 \quad]{\quad F_3C—SO_3H, \ H_2C=CH—CH_2SiMe_3 \quad} R—O—SiMe_3$$

92-96%

R = alkyl, Bz, etc.

V.A-6   T. Veysoglu and L.A. Mitscher, <u>Tetrahedron Lett.</u>, <u>22</u>, 1299 and 1303 (1981).

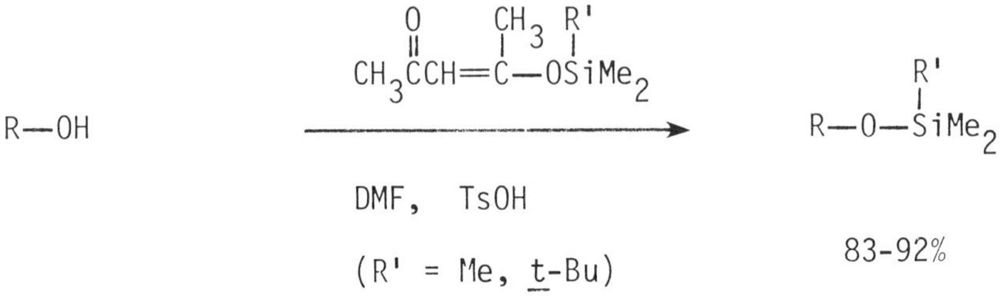

R—OH
$$
\xrightarrow[\text{(R' = Me, } \underline{t}\text{-Bu)}]{\underset{\text{DMF,   TsOH}}{\overset{\overset{\displaystyle O}{\overset{\displaystyle \|}{CH_3\overset{}{C}CH=\overset{\overset{\displaystyle CH_3}{|}}{C}}}-\overset{\overset{\displaystyle R'}{|}}{}OSiMe_2}}
$$
R—O—SiMe$_2$ with R' substituent

83-92%

R = alkyl, cycloalkyl, etc.

V.A-7   J.B. Chattopadhyaya <u>et al.</u>, <u>Tetrahedron Lett.</u>, <u>22</u>, 969 (1981).

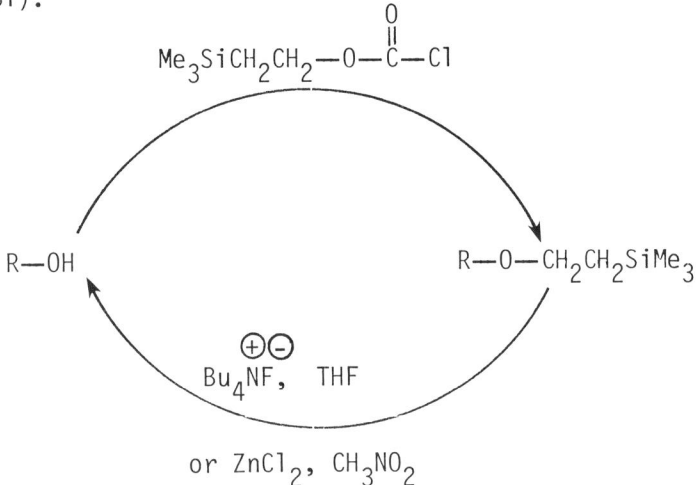

V.A-8   B. Loubinoux, G. Coudert, and G. Guillaumet, Tetrahedron Lett., 22, 1973 (1981).

Use of guiacylmethyl (GuM) ethers as a hydroxyl-protection method:

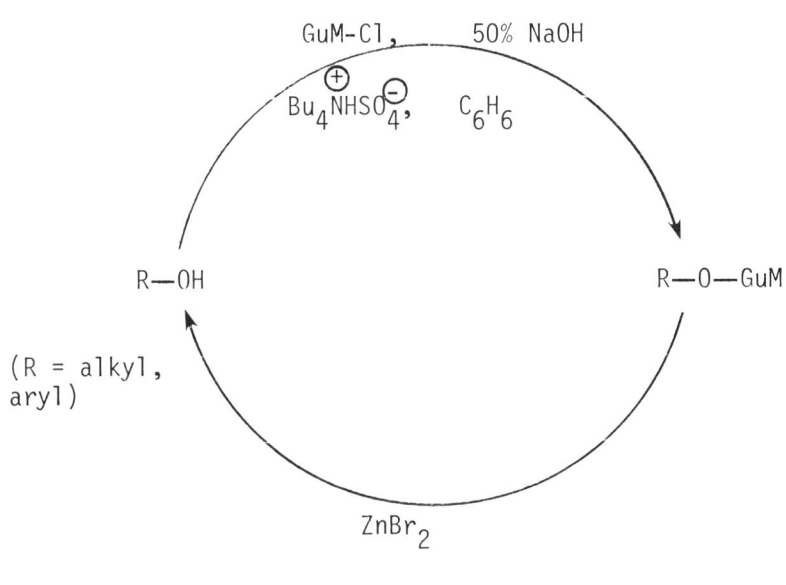

$$\text{GuM-Cl,} \quad 50\% \text{ NaOH}$$
$$\text{Bu}_4\overset{\oplus}{N}\text{HSO}_4^{\ominus}, \quad C_6H_6$$

R—OH

R—O—GuM

(R = alkyl, aryl)

ZnBr$_2$

V.A-9   G.A. Olah et al., Angew. Chem. Int. Ed., 20, 690 (1981).

$$R—O—R' \xrightarrow[\text{NaI}]{\text{MeSiCl}_3} R—OH \quad (+ R'I)$$

R = 1°, 2° alkyl

R' = Me, Bz, CPh$_3$,   THP

~80-100%

V.A-10   H. Niwa, T. Hida, and K. Yamada, Tetrahedron Lett.,
22, 4239 (1981).

$$R—OMe \quad \xrightarrow[\text{CH}_2\text{Cl}_2, \text{ 15-crown-5}]{\text{BBr}_3, \text{ NaI}} \quad R—OH$$

~70-100%

R = 1°, 2°, 3°, cyclic, etc.

V.A-11   T. Iversen and D.R. Bundle, J.C.S. Chem. Comm.,
1240 (1981).

$$R—OH \quad \xrightarrow[\text{TfOH, hexane/CH}_2\text{Cl}_2]{\overset{\overset{\text{NH}}{\|}}{\text{PhCH}_2\text{O}—\text{C}—\text{CCl}_3}} \quad R—O—CH_2Ph$$

82-98%

Acetals survive the reaction conditions.

V.A-12   S. Hanessian, T.J. Liak, and B. Vanasse, Synthesis,
396 (1981).

Benzyl ethers may be cleaved by catalytic transfer hydrogenation:

$$R—OBz \quad \xrightarrow[\text{20\%Pd(OH)}_2/\text{C}]{\text{, EtOH}} \quad R—OH$$

Many examples using protected sugars.   ~98%

V.A-13   P.M. Dewick, <u>Synth. Comm.</u>, <u>11</u>, 853 (1981).

Conversion of phenols into O-benzyl derivatives is accomplished by benzyl tosylate. This reagent shows little tendency towards C-benzylation.

V.A-14   F. Guibe and Y.S. M'Leux, <u>Tetrahedron Lett.</u>, <u>22</u>, 3591 (1981).

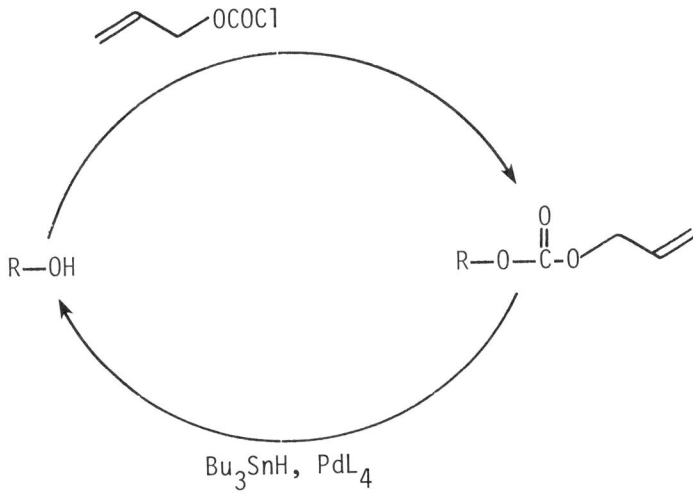

V.A-15   S.S. Jones, C.B. Reese, and S. Sibanda, Tetrahedron
Lett., 22, 1933 (1981).

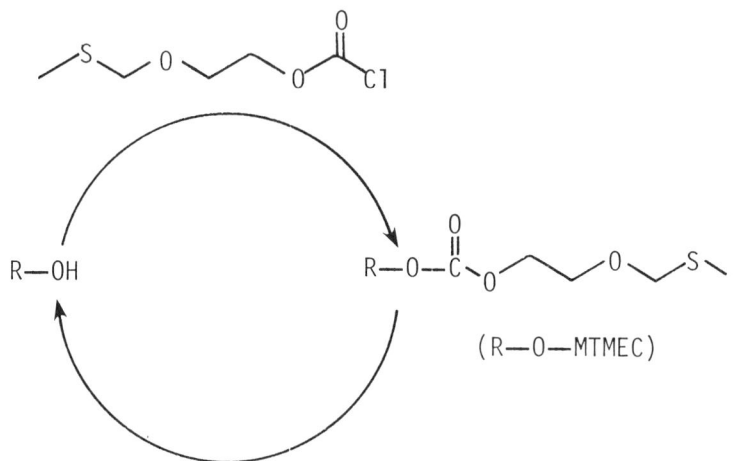

(R—O—MTMEC)

1) Hg(ClO$_4$)$_2$, pyridine or collidine

2) NH$_4$OH
   $\oplus$ $\ominus$

Used in oligonucleotide synthesis.

V.A-16   N. Balgobin, S. Josephson, and J.B. Chattopadhyaya,
Tetrahedron Lett., 22, 3667 (1981).

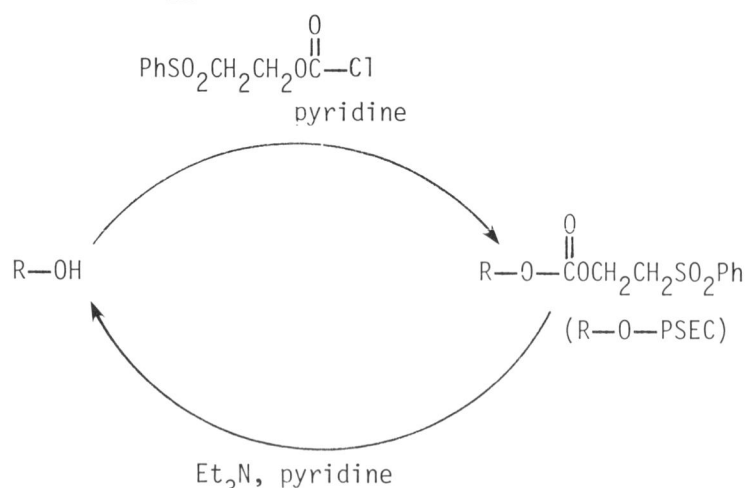

(R—O—PSEC)

V.A-17   H.J. Koerners, J. Verhoeven, and J.H. vanBoom, Rec. Trav. Chim. Pays-Bas, 100, 65 (1981).

Use of the levulinoyl group as an OH-protecting group in the synthesis of oligosaccharides. Removed by hydrazine.

V.A-18   A.G. Gonzalez et al., Tetrahedron Lett., 22, 335 (1981).

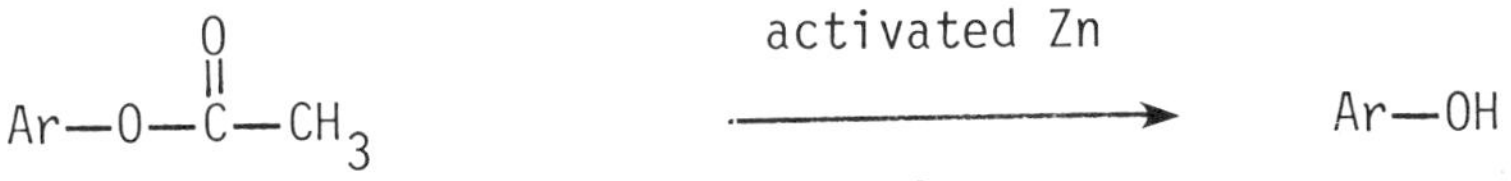

$$Ar-O-\overset{\overset{\textstyle O}{\|}}{C}-CH_3 \quad\xrightarrow[\text{MeOH}]{\text{activated Zn}}\quad Ar-OH$$

91-100%

Ar = coumarin, estrone, etc.

V.A-19   B.M. Trost and C.G. Caldwell, Tetrahedron Lett., 22, 4999 (1981).

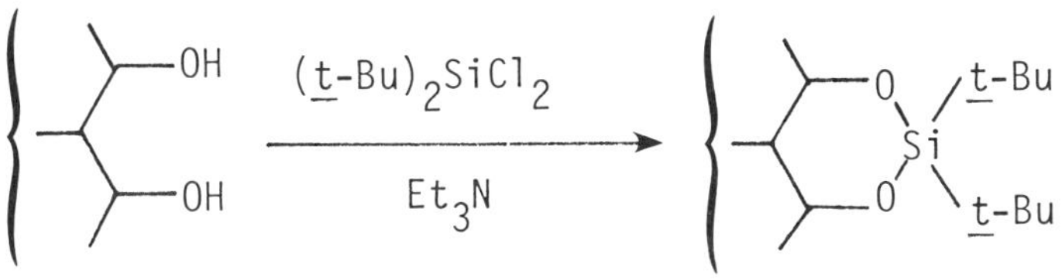

Removed using pyridinium hydrofluoride. May be removed in the presence of t-butyldimethylsilyl ethers.

~50-80%

V.A-20 B.H. Lipshutz and M.C. Morey, J. Org. Chem., 46, 2419 (1981).

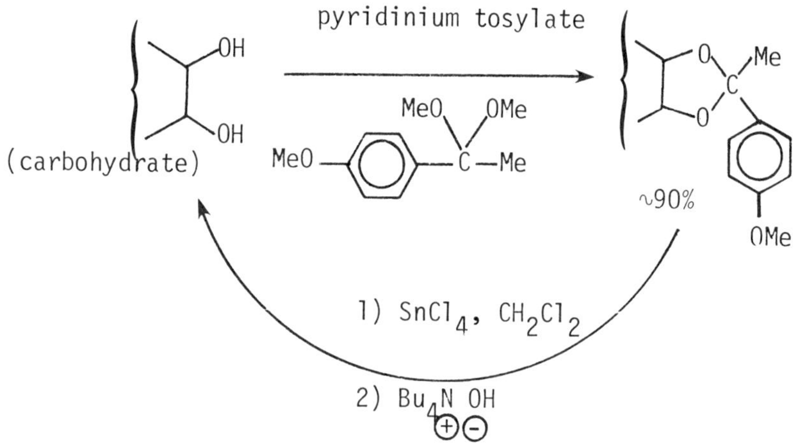
pyridinium tosylate
OH
OH
(carbohydrate)
MeO
MeO
OMe
C
Me
O
O
C
Me
OMe
~90%
1) $SnCl_4$, CH_2Cl_2
2) $Bu_4\overset{\oplus}{N}\ \overset{\ominus}{O}H$

V.A-21 C. van der Stouwe and H.J. Schräfer, Chem. Ber., 114, 946 (1981).

Use of a variety of electroactive protecting groups to protect and deprotect diols and polyhydroxy compounds in a specific and selective manner. Protection is accomplished using standard techniques ($BrCH_2C_6H_4CN$ etc.), and selective deprotection uses controlled potential electrolysis.

V.A-22/V.B-1   R.W. Johnson, E.R. Grover, and L.J. MacPherson,
Tetrahedron Lett., 22, 3719 (1981).

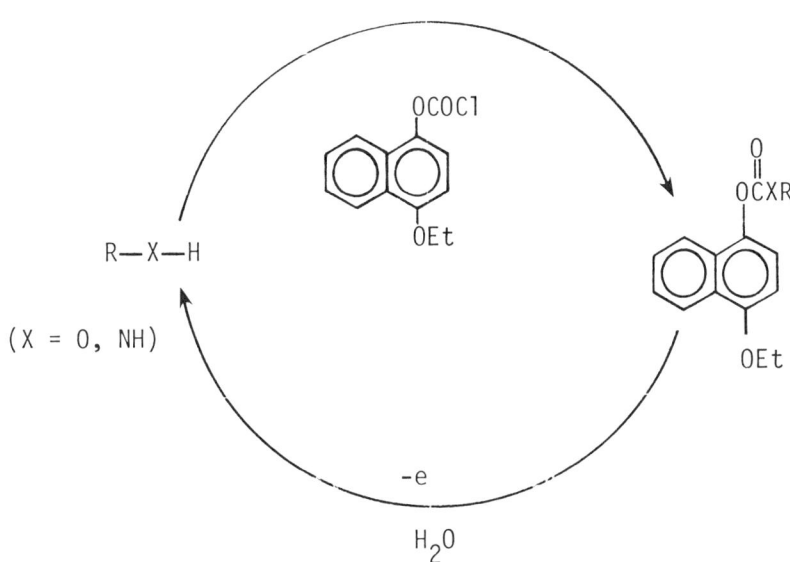

An electrochemically labile protecting group for alcohols
and amines

V. B.   Amine Protecting Groups

   (see also: VI.A.4.)

V.B-2   B.C. Laguzza and B. Ganem, Tetrahedron Lett., 22,
1483 (1981).

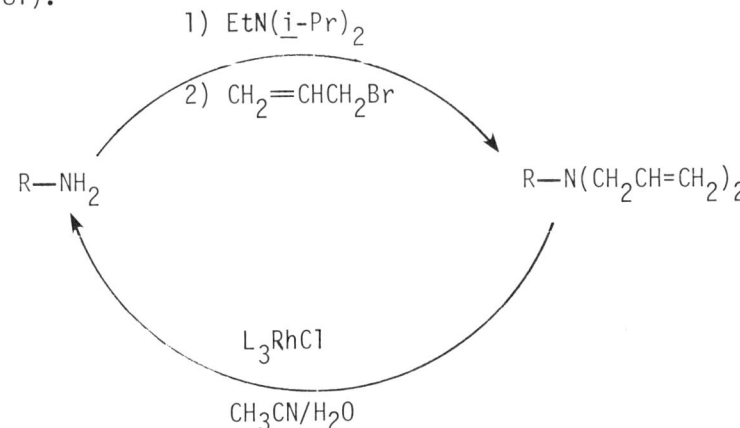

V.B-3   D.S. Kemp and G. Hanson, J.Org. Chem., 46, 4971
(1981).

Use of the Tcroc amine-protecting group in peptide synthesis. Stable to ethyl alaninate and many acidic reagents.
Cleaved by propylamine (neat) or hydrazine in ethanol.

$$Tcroc = (R-NH)-\overset{O}{\overset{\|}{C}}-O$$

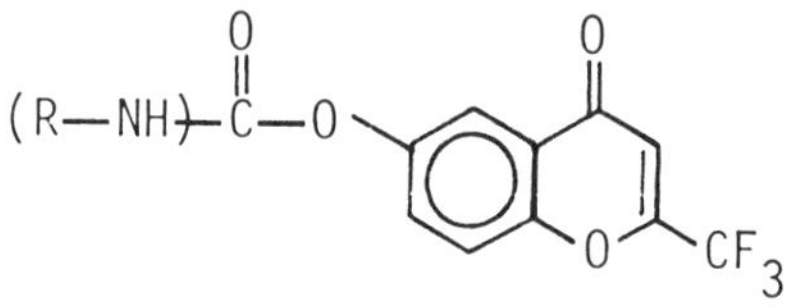

V.B-4   G.T. Young et al., J.C.S. Perkin I, 522 (1981).

Use of the DPPM group as a protecting group for histi-
dine in peptide synthesis. Stable to acid, but cleaved by
Zn/HOAc or electrolytic reduction.

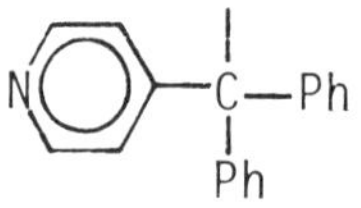

V.B-5   M. Fujino, M. Wakimasu, and C. Kitada, Chem. Pharm. Bull., 29, 2825 (1981).

A study of various multisubstituted benzenesulfonyl protecting groups for the guanidino function of arginine. Removed by TFA-thioanisole.

V.B-6   F. Brtnik, T. Barth, and K. Jost, Coll. Czech. Chem. Comm., 46, 1983 (1981).

Use of the ε-phenylacetyl derivative of lysine in peptide synthesis. Deprotected by a penicillin amidohydrolase enzyme.

V.B-7   H. Köster et al., Tetrahedron, 37, 363 (1981).

A study of N-acyl protecting groups for deoxynucleosides, including substituted phenylacetyl, phenoxyacetyl, and enzoyl protecting groups.

V.B-8   K. Itakura et al., Tetrahedron Lett., 22, 3761 (1981).

The N-acyl protecting group in nucleoside derivatives can be selectively removed by treatment with $ZnBr_2$ in the presence of alcohols to give O-protected nucleosides.

V.B-9   N.G. Kundu, Synth. Comm., 11, 787 (1981).

N-benzyl and N-benzyloxymethyl derivatives of uracil may be deblocked by Me_3SiI. Yields are 40-84%.

V. C. Sulfhydryl Protecting Groups

(see also: VI. A. 19.)

V.C-1   J.J. Pastuszak and A. Chimiak, J. Org. Chem., 46, 1868 (1981).

Use of the t-butyl group for thiol protection of cysteine. Protected using HCl/t-butanol. Deprotected using NpsCl, (2-nitrophenyl) sulfenyl chloride, followed by reduction.

V.C-2   G.T. Young et al., J.C.S. Perkin I, 522 (1981).

   Use of the DPPM group as a protecting group for cysteine
in peptide synthesis. Stable to acid, but cleaved by Zn/HOAc
$Hg(OAc)_2$, I_2, or electrolytic reduction.

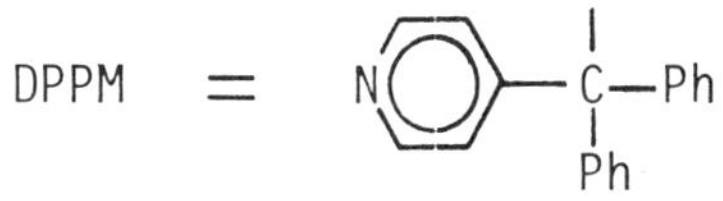

$$DPPM \quad = \quad$$

V. D. Carboxyl Protecting Groups

   (see also: VI.A.4., VI.A.10.)

V.D-1   T. Uchimaru, K. Narasaka, and T. Mukaiyama, Chem.
Lett., 1551 (1981).

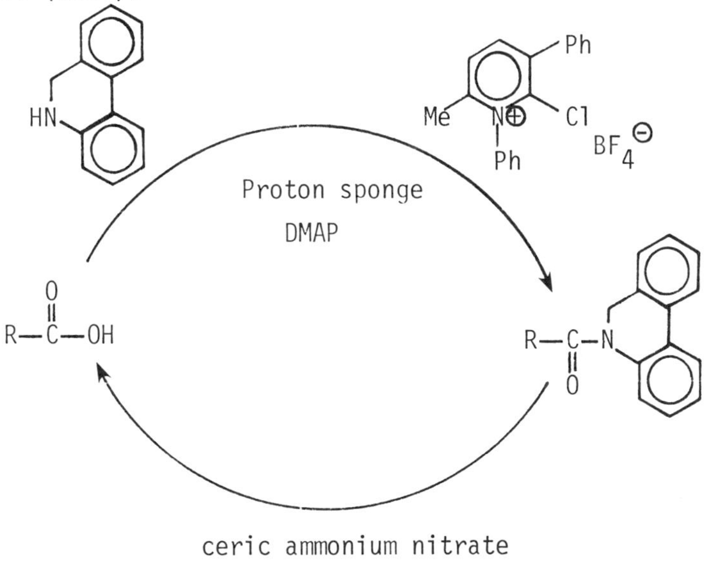

ceric ammonium nitrate

V.D-2   G.L. Larson, M. Ortiz, and M.R. deRoca, Synth. Comm., 11, 583 (1981).

Carboxylic acids may be protected as their TMS esters during hydroboration. For example:

$$CH_2{=}CHCH_2CH_2 \overset{\overset{\displaystyle O}{\|}}{-C}-OSiMe_3 \quad \xrightarrow[\text{3) } H_3O^{\oplus}]{\begin{array}{l}\text{1) } BH_3/THF \\ \text{2) } H_2O_2,\ NaOH \end{array}} \quad HO{-}(CH_2)_4{-}COOH$$

87%

V.D-3   E. Fujita et al., J. Org. Chem., 46, 1991 (1981).

$$R{-}\overset{\overset{\displaystyle O}{\|}}{C}-OR' \quad \xrightarrow[\text{2) } H_3O^{\oplus}]{\text{1) } AlBr_3,\ EtSH} \quad R{-}COOH + Et{-}S{-}R'$$

∼80-90%

R = alkyl, aryl, cycloalkyl, etc.
R' = Me, Bz

V.D-4   G. Losse and H. -U. Stiehl, Z. Chem., 188 (1981).

Use of $K_3[Co(CN)_5]$ as a hydrogenation catalyst for removal of benzyl esters from amino acids and peptides. Yields are 83-94%.

V. E.   Protecting Groups for Ketones and Aldehydes

(see also: VI.A.18.)

V.E-1   D.L. Coumins and J.D. Brown, <u>Tetrahedron Lett.</u>, <u>22</u>, 4213 (1981).

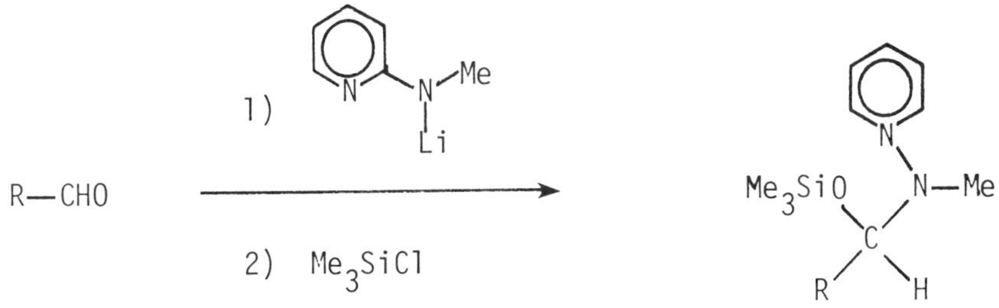

This protected aldehyde is stable to organolithium and organomagnesium reagents, and deprotection occurs during the normal workup.

V.E-2   J.K. Michie and J.A. Miller, <u>Synthesis</u>, 824 (1981).

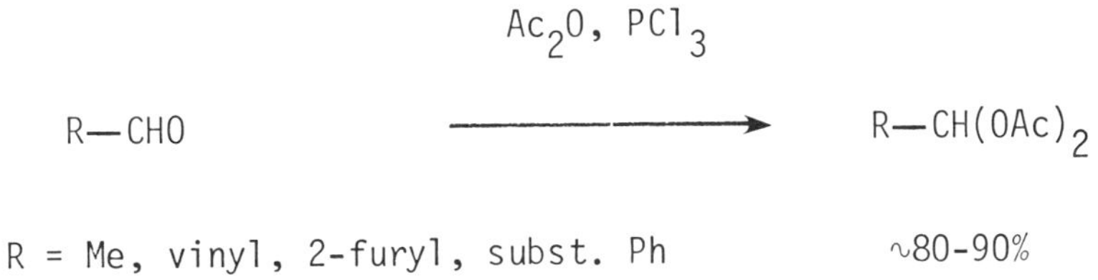

R = Me, vinyl, 2-furyl, subst. Ph   ∿80-90%

V.E-3   J. Yoshida, J. Hashimoto, and N. Kawabata, <u>Bull.</u>
<u>Chem. Soc. Japan</u>, <u>54</u>, 309 (1981).

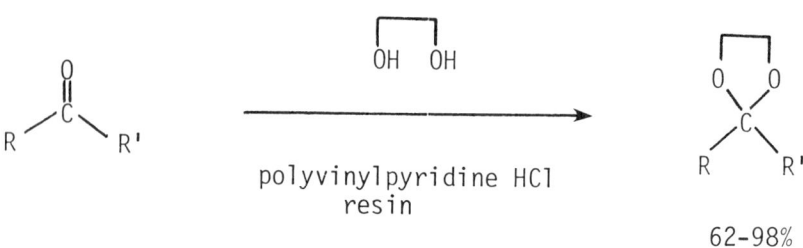

polyvinylpyridine HCl
resin

62-98%

R, R' = H, alkyl

V.E-4   G.A. Olah <u>et al</u>., <u>Synthesis</u>, 282 (1981).

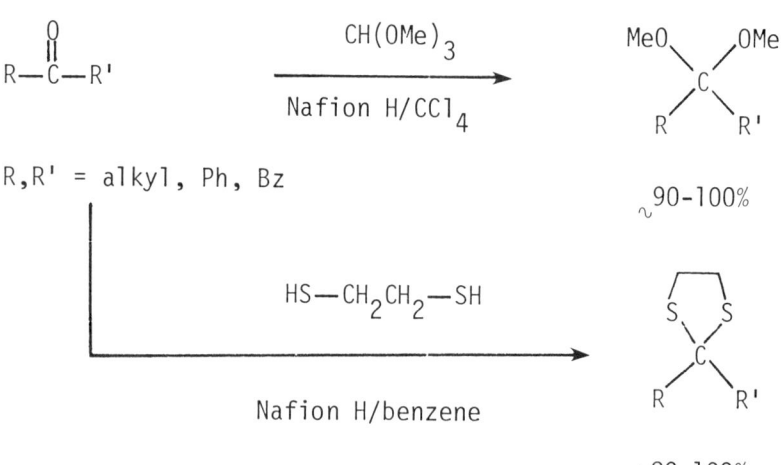

R,R' = alkyl, Ph, Bz

~90-100%

~80-100%

V.E-5   I. Degani, R. Fochi, and V. Regondi, <u>Synthesis</u>, 51
(1981).

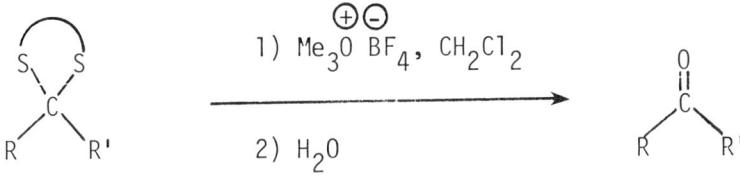

HgO, 35% aq. HBF$_4$

R—C—R' (structure with dithiane) → $R-\overset{O}{\overset{\|}{C}}-R'$

$\sim$80-100%

R, R' = H, alkyl, aryl, etc.
X = -(CH$_2$)$_2$-, -(CH$_2$)$_3$-, benzo.

V.E-6   I. Stahl, <u>Synthesis</u>, 135 (1981).

1) Me$_3$O$^{\oplus}$ BF$_4$$^{\ominus}$, CH$_2Cl_2$

2) H$_2$O

→

$\sim$90%

R, R' = Me, Et, -CH$_2$COOEt, subst. Ph

V.E-7   G.A. Olah <u>et al.</u>, <u>Synthesis</u>, 146 (1981).

R—C—R' (dithiolane)   Cl—SO$_2$—F → $R-\overset{O}{\overset{\|}{C}}-R'$

ether/H$_2$O

56-86%

R, R' = H, Me, cyclic, subst. Ph, -COOEt

V.E-8   F.A.J. Meskens, Synthesis, 501 (1981).

Review: "Methods for the Preparation of Acetals from

Alcohols or Oxiranes and Carbonyl Compounds"

V. F. Phosphate Protecting Groups

V.F-1   N. Balgobin, S. Josephson, and J.B. Chattopadhyaya,
Tetrahedron Lett., 22, 1915 (1981).

Use of the 2-phenylsulfonylethyl (PSE) phosphate-

protecting group in the synthesis of dodecathymidylic acid.

$$R-O-\overset{\overset{O}{\|}}{\underset{\underset{OAr}{|}}{P}}-O-CH_2CH_2-\overset{\overset{O}{\|}}{\underset{\underset{O}{\|}}{S}}-Ph$$

PSE derivative   Removed with

                                                  Et$_3$N in pyridine.

V.F-2   S. Honda and T. Hata, <u>Tetrahedron Lett.</u>, <u>22</u>, 2093
(1981).

Use of the 2-diphenylmethylsilylethyl (TPS) phosphate-
protecting group in oligonucleotide synthesis.

$$RO-\overset{\overset{O}{\|}}{\underset{\underset{Me}{\overset{|}{OCH_2CH_2-SiPh_2}}}{P}}-OR' \quad \xrightarrow{\overset{\oplus\ominus}{Et_4\ N\ F}} \quad RO-\overset{\overset{O}{\|}}{\underset{\underset{}{\overset{|}{O^{\ominus}}}}{P}}-OR'$$

V.F-3   H. Takaku, M. Koto, and S. Ishikawa, <u>J. Org. Chem.</u>,
<u>46</u>, 4062 (1981).

Use of the $-S-\overset{\overset{S}{\|}}{C}-NMe_2$ group as a phosphate-protecting
group in the phosphotriester approach to peptide synthesis.
Protection is effected using $(Me_2NCSS)_2 -Ph_3P$ on the
QS derivative. The group is removed by BF_3.

V. G. Pi-Bond Protecting Groups

V.G-1   D.J. Ager, I. Fleming, and S.K. Patel, <u>J.C.S. Perkin</u>
<u>I</u>, 2520 (1981).

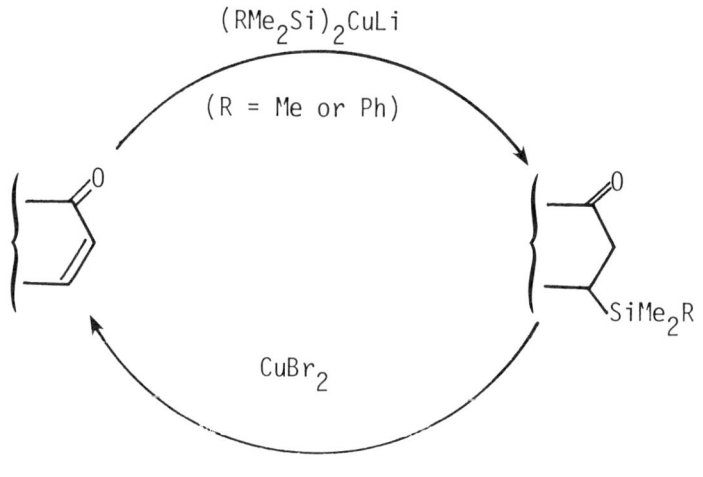

V.G-2   U. Husstedt and H.J. Schäfer, <u>Tetrahedron Lett.</u>, <u>22</u>,
623 (1981).

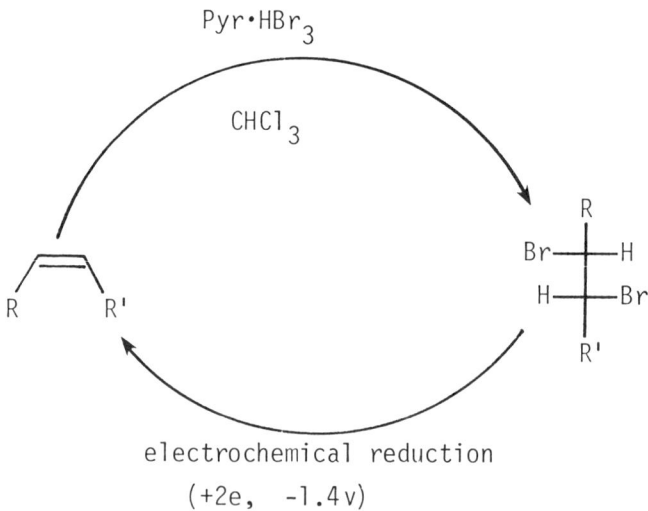

V.G-3   G.R. Knox and I.G. Thom, <u>J.C.S. Chem. Comm.</u>, 373
(1981).

Use of $(diene)Fe(CO)_3$ complexes in the synthesis of
insect pheromones. The $Fe(CO)_3$ complex locks in the <u>E</u>
configuration in 1,3-dienes, while other synthetic
transformations are effected. Cleaved by trimethylamine
N-oxide.

V. H. Miscellaneous Protecting Groups

V.H-1   J.M.J. Fréchet, <u>Tetrahedron</u>, <u>37</u>, 663 (1981).

Review: "Synthesis and Applications of Organic Polymers
as Supports and Protecting Groups"

V.H-2   P.M. Maitlis, <u>Chem. Soc. Rev.</u>, <u>10</u>, 1 (1981).

Review: "η^5-Cyclopentadienyl and η^6-Arene as Protecting
Ligands towards Platinum Metal Complexes"

VI

USEFUL SYNTHETIC PREPARATIONS

VI. A. Functional Group Preparations

1. Acids, Acid Halides, etc.

(see also: II.A.2.)

VI.A.1-1   M.W. Anderson, R.C.F. Jones, and J. Saunders, Tetrahedron Lett., 22, 261 (1981).

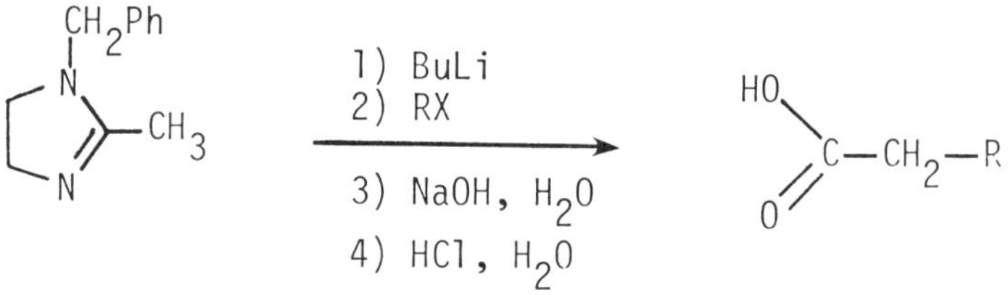

R = 1°, 2° alkyl, allylic, Bz

∿80-90%

(alkylation step)

VI.A.1-2   D. Liotta et al., J. Org. Chem., 46, 2605 (1981).

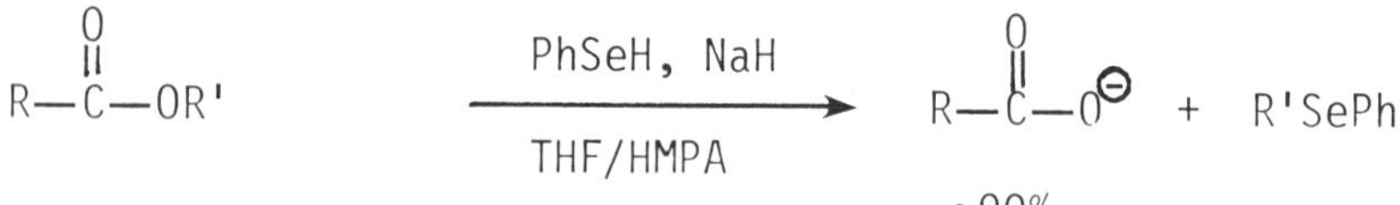

$$R-\overset{O}{\underset{||}{C}}-OR' \xrightarrow[\text{THF/HMPA}]{\text{PhSeH, NaH}} R-\overset{O}{\underset{||}{C}}-O^{\ominus} + R'SePh$$

∿90%

R = 1°, 2°, 3° alkyl, Ph

R' = Me, Et, i-Pr, Bz, cyclic

VI.A.1-3   S.P. McManus, J. Org. Chem., 46, 3097 (1981).

Use of polymer-bound exazolines for synthesis of carboxylic
acids, e.g.:

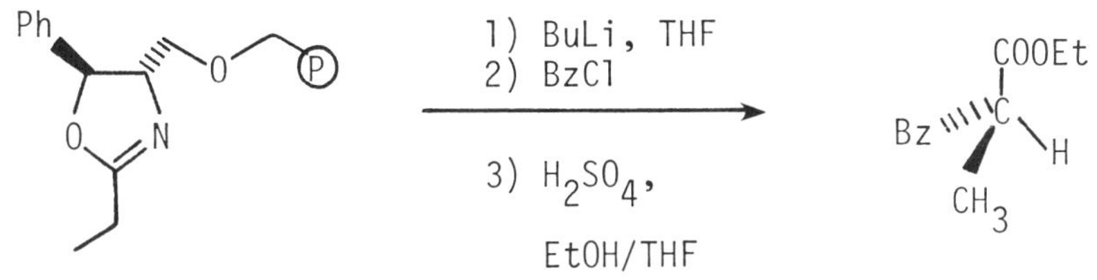

1) BuLi, THF
2) BzCl

3) H₂SO₄,
   EtOH/THF

45%
56% ee

VI.A.1-4   A.H. Schmidt, M. Russ, and D. Grosse, Synthesis,
216 (1981).

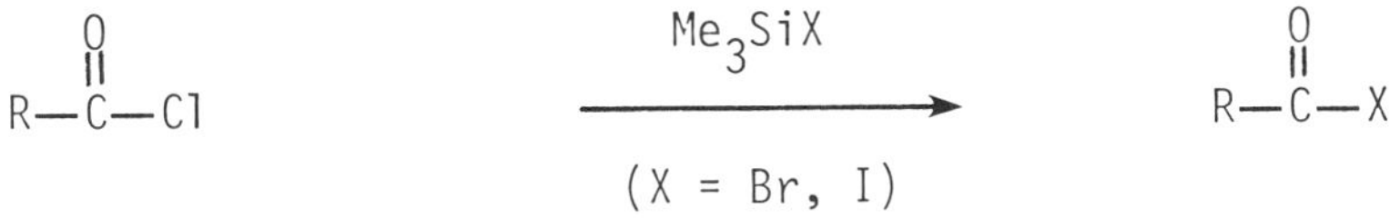

$$R-\overset{O}{\underset{||}{C}}-Cl \xrightarrow{Me_3SiX} R-\overset{O}{\underset{||}{C}}-X$$

(X = Br, I)

R = alkyl, Ph

∿80-90%

VI.A.1-5   H.M.R. Hoffmann and K. Haase, Synthesis, 715(1981).

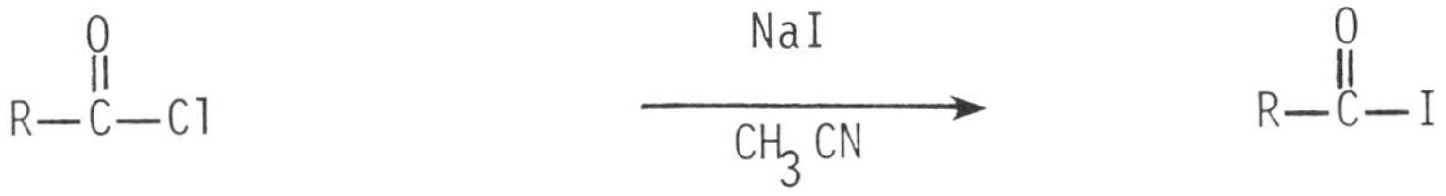

$$R-\overset{O}{\underset{||}{C}}-Cl \xrightarrow[CH_3CN]{NaI} R-\overset{O}{\underset{||}{C}}-I$$

R = alkyl, Ph, benzyl, vinyl, etc.

∿80-95%

VI.A.1-6   R. Mestres and C. Palomo, Synthesis, 218 (1981).

$$PhO-\overset{\overset{O}{\|}}{\underset{PhNH}{P}}-Cl$$

2 R—COOH $\xrightarrow{\hspace{4cm}}$ $R-\overset{\overset{O}{\|}}{C}-O-\overset{\overset{O}{\|}}{C}-R$

∿90-98%

R = alkyl, subst. Ph, styryl, etc.

VI.A.1-7/VI.A.2-1   S.L. Regen and A.K. Mehrotra, Synth.
Comm., 11, 413 (1981).

$R-\overset{\overset{O}{\|}}{C}-OR'$ $\xrightarrow[\text{2) H}^{\oplus}]{\text{1) KOH, Al}_2\text{O}_3}$ R—COOH + R'OH

Both acid and alcohol
may be isolated,
∿80-100%

R = alkyl, aryl
R' = alkyl

VI. A. 2. Alcohols and Phenols

(see also: II.B.1, III.A., III.F.1)

VI.A.2-2   A. Pelter and J.M.Rao, J.C.S. Chem. Comm., 1149
(1981).

R_3B $\xrightarrow[\text{2) Hg}^{II}\text{ of MeO}_2\text{SF}]{\text{1) Li-C(SPh)}_3}$ R_3C—OH

R = alkyl, cycloalkyl   71-91%

VI.A.2-3   G. Cardillo, et al., Synthesis, 793 (1981).

$$R\!-\!X \xrightarrow[\text{benzene or THF}]{\text{polymer-bound carbonate}} R\!-\!OH$$

$\sim$90%

R = 1° alkyl, allyl, Bz

X = Br, Cl

VI.A.2-4   A.R. Katritzky, A. Saba, and R.C. Patel, J.C.S.
Perkin I, 1492 (1981).

$R\!-\!CH_2NH_2 \xrightarrow{\hspace{3cm}} R\!-\!CH_2OH$

1) [pyrylium $BF_4^{\ominus}$ reagent]

2) [sodium 2-(hydroxymethyl)benzoate]   phase-transfer

VI.A.2-5   R.O. Hutchins, I.M. Taffer, and W. Burgoyne,
J. Org. Chem., 46, 5214 (1981).

$$R\!-\!CH\!=\!CH_2 \xrightarrow{BH_3CN^{\ominus}/BF_3 \cdot Et_2O} R\!-\!CH_2\!-\!CH_2OH$$

R = alkyl, Ph, etc. $\sim$80%

VI.A.2-6   H.C. Brown and P.K. Jadhav, J. Org. Chem., 46,
5048 (1981).

Use of isopinocamphenylborane for asymmetric hydroboration

of alkenes. Yields are ∿70-80%, and optical yields are

above 70%.

VI.A.2-7   H.C. Brown et al., J. Org. Chem., 46, 531 and 930
(1981).

Studies of the regiochemistry of oxymercuration-

demercuration of alkenes substituted with methoxy, hydroxy,

acetoxy, chloro, epoxy, and thiomethyl groups.

H.C. Brown et al., J. Org. Chem., 46, 3810 (1981).

More solvomercuration studies, comparing mercuric acetate,

trifluoroacetate, nitrate, and methanesulfonate.

VI.A.2-8   P.K. Jadhav and H.C. Brown, J. Org. Chem., 46,
2988 (1981).

Dilongifolylborane, Lgf$_2$BH, effects asymmetric hydro-

boration of cis, trisubstituted acyclic, and trisubstituted

cyclic prochiral olefins to provide chiral alcohols in the

range of 60-78% ee.

VI. A. 3. Alkyl and Aryl Halides

(see also: II.B.2.)

VI.A.3-1   G.W. Kabalka et al., J. Org. Chem., 46, 2582 and
3113 (1981); Synth. Comm., 11, 521 (1981).

halide source = $I^{\ominus}$, Chloramine-T, Br_2 or BrCl,
                 ICl, NaOAc

VI.A.3-2   K. Smith et al., J. Chem. Research(S), 376 (1981).

$$R_3B \ + \ Cl_2NTs \ \longrightarrow \ R-Cl$$

R = alkyl   64-100%

VI.A.3-3   S. Oae and H. Togo, Synthesis, 371 (1981).

$$R-SO_3H \ \xrightarrow{\ I_2,\ PPh_3\ } \ R-I$$

R = 1^0 alkyl

(several other sulfur acids work)

VI.A.3-4   G.A. Olah, S.C. Narang, and L.D. Field, J. Org. Chem., 46, 3727 (1981).

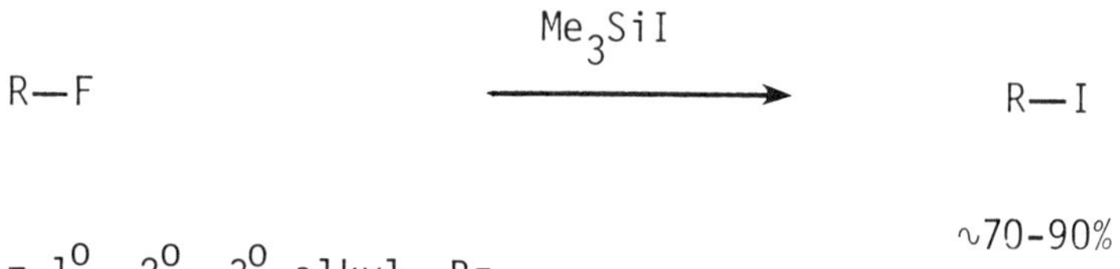

$$R\text{—}F \xrightarrow{\quad Me_3SiI \quad} R\text{—}I$$

$\sim$70-90%

R = 1°, 2°, 3° alkyl, Bz

VI.A.3-5   A. Citterio and A. Arnoldi, Synth. Comm., 11, 639 (1981).

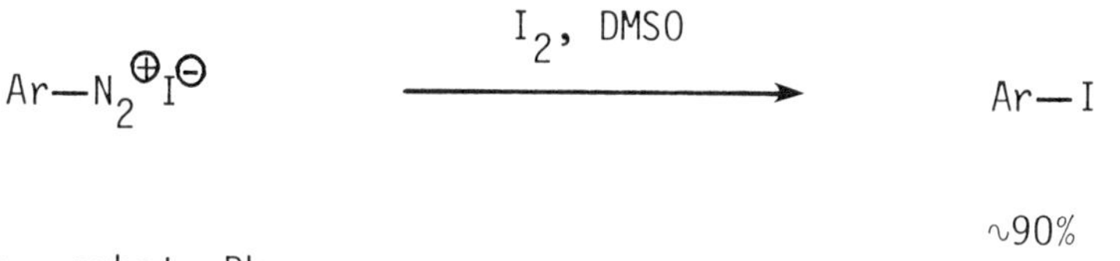

$$Ar\text{—}N_2^{\oplus}I^{\ominus} \xrightarrow{\quad I_2, DMSO \quad} Ar\text{—}I$$

$\sim$90%

Ar = subst. Ph

VI.A.3-6   T.B. Patrick, J.J. Scheibel, and G.L. Cantrell, J. Org. Chem., 46, 3917 (1981).

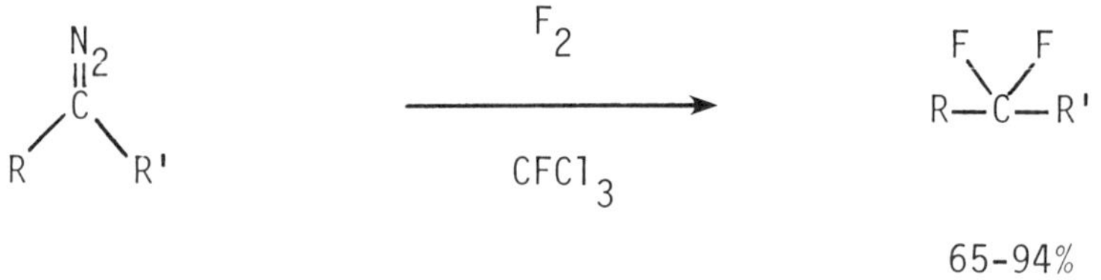

$$\underset{R}{\overset{N_2}{\underset{\displaystyle R'}{\parallel}}} C \xrightarrow[CFCl_3]{\quad F_2 \quad} R\text{—}\underset{F}{\overset{F}{C}}\text{—}R'$$

65-94%

R, R' = alkyl, Ph, acyl

VI.A.3-7   F.A. Bloshchitsa et al., J. Org. Chem. (USSR), 17, 1260 (1981).

$$CH_3-\overset{O}{\overset{\|}{C}}-(CH_2)_n-\overset{O}{\overset{\|}{C}}-OEt \xrightarrow{SF_4,\ HF} CH_3-CF_2-(CH_2)_n-\overset{O}{\overset{\|}{C}}-OEt$$

n = 0, 1, 2   76-85%

VI.A.3-8   R.M. Magid, B.G. Talley, and S.K. Souther, J. Org. Chem., 46, 824 (1981).

~80-90%

R = α, β-methyl   Also works with saturated alcohols.

VI.A.3-9   J.N. Denis and A. Krief, Tetrahedron Lett., 22, 1429 (1981).

1) Me$_3$SiX
2) CrO$_3$, H$_2$SO$_4$

(X = Br, I)

$$R-\underset{X}{\overset{}{CH}}-\overset{O}{\overset{\|}{C}}-R'$$

R = n-alkyl,
R' = H, actyl   } —(CH$_2$)$_n$—
                        (n = 4, 6, 10)

77-93%

VI.A.3-10   R.D. Miller and D.R. McKean, J. Org. Chem., 46, 2412 (1981).

$$R-\overset{\overset{\displaystyle O}{\|}}{C}\overset{R''}{\underset{R'}{\diagdown}} \xrightarrow{\text{Me}_3\text{SiI}} R-\overset{\overset{\displaystyle O}{\|}}{C}-CH_2-\underset{I}{\overset{}{C}}H-R'$$

R   = alkyl, Ph
R' = alkyl
R" = H, alkyl, cyclic with R'

~90%

VI.A.3-11   M. Kolb and J. Barth, Synth. Comm., 11, 763(1981).

$$(CH_2)_n \underset{C=O}{\overset{O}{\diagup\diagdown}} \xrightarrow[\text{R'OH}]{\text{Me}_3\text{SiI}} I-(CH_2)_n-COOR'$$

>90%

n   = 2 - 5
R' = alkyl, Bz, etc.

VI.A.3-12   H.R. Kricheldorf, G. Mörber, and W. Regel, Synthesis, 383 (1981).

$$\underset{O}{\diagup\diagdown}CH_2OAc \xrightarrow{\text{Me}_3\text{SiBr}} Br-(CH_2)_3-\underset{\overset{|}{Br}}{\overset{CH_2OAc}{\overset{|}{C}H}}$$

87%

VI.A.3-13   C.G. Krespan and D.C. England, <u>J. Am. Chem. Soc.</u>, <u>103</u>, 5598 (1981).

$$(CF_3)_2C{=}O \ + \ F_2C{=}CFCF_2OSO_2F \xrightarrow[0^\circ]{KF} (CF_3)_2CFOCF_2CF{=}CF_2$$

VI.A.3-14   S.G. Hegde and J. Wolinsky, <u>Tetrahedron Lett.</u>, <u>22</u>, 5019 (1981).

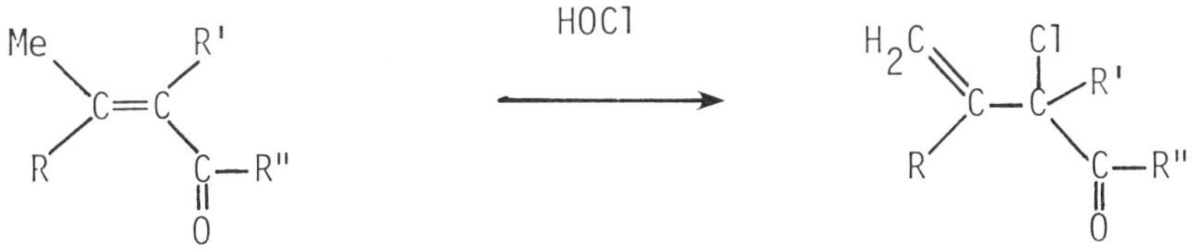

$\sim 70\%$

R, R' = alkyl, cyclic, etc.

VI.A.3-15   J.W. Gillard and M. Israel, <u>Tetrahedron Lett.</u>, <u>22</u>, 513 (1981).

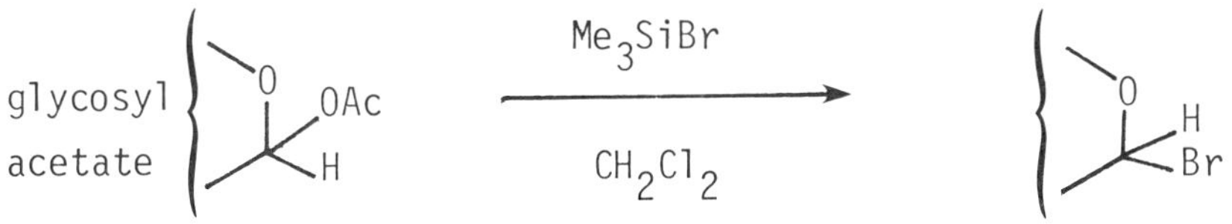

$\sim 100\%$

VI.A.3-16   M.R.C. Gerstenberger and A. Haas, Angew. Chem.
Int. Ed., 20, 647 (1981).

Review: "Methods of Fluorination in Organic Chemistry"

VI. A. 4. Amides

(see also: IV.D., VI.A.17.)

VI.A.4-1   K. Ito et al., Synthesis, 287 (1981).

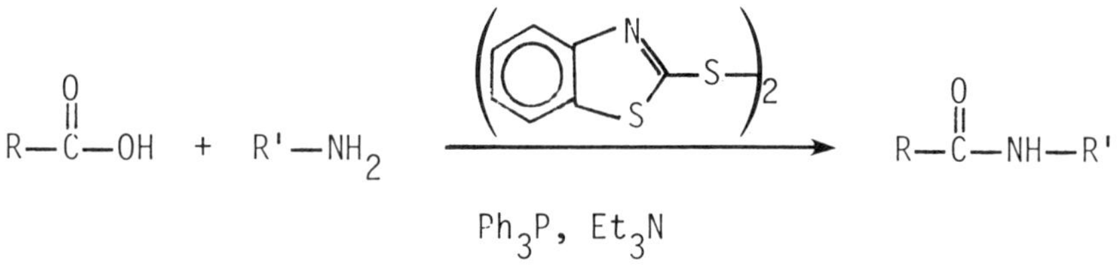

$R = 1^{o}, 2^{o}$ alkyl, Ph, styryl

$R' = $ Ph, morpholino, cyclohexyl, etc.

VI.A.4-2   M. Ueda, K. Seki, and Y. Imai, Synthesis, 991
(1981).

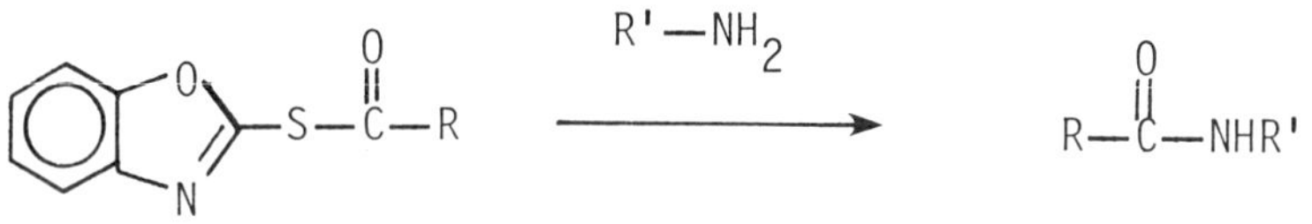

R = Me, Ph   81-98%
R' = Ph, Bz

VI.A.4-3   Y. Watanabe and T. Mukaiyama, Chem. Lett., 285 (1981).

$$R^2-\overset{O}{\underset{\|}{C}}-OR^1 \xrightarrow[\substack{R^3R^4NH, \; PhP\left(-O-\underset{NO_2}{\underset{\|}{\bigcirc}}\right)_2 \\ \underset{O}{\|}}]{\overset{\oplus \quad \ominus}{KOH, \; Bu_4N \; HSO_4}} R^2-\overset{O}{\underset{\|}{C}}-NR^3R^4$$

$\sim$80-90%

R^1= H, alkyl, etc.

R^2= alkyl, aryl, benzyl, vinyl, etc.

R^3, R^4 = H, alkyl, cyclic, Bz, etc.

VI.A.4-4   S. Uemura et al., J. Org. Chem., 46, 4727 (1981).

1) PhSeCl, CH_3CN
   CF_3SO_3H

2) 30% H_2O_2

NHCOCH$_3$

$\sim$40-80% overall

VI.A.4-5   J. Barluenga et al., J.C.S. Chem. Comm., 670(1981).

R—CH=CH$_2$
+
MeCONH$_2$

1) Hg(NO$_3$)$_2$

2) NaBH$_4$

$$Me-\overset{O}{\underset{\|}{C}}-NH-\underset{R}{\underset{|}{CH}}-CH_3$$

R = Ph, Bz, n-alkyl, cycloalkene

VI.A.4-6   G.C. Crockett _et al._, Synth. Comm., 11, 447 (1981).

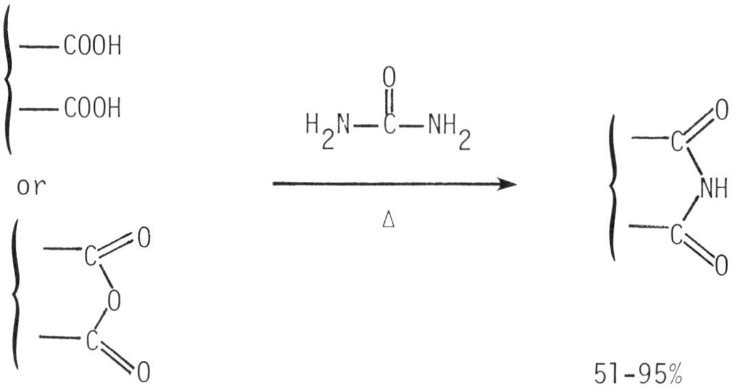

51-95%

VI.A.4-7   I. Akimoto and A. Suzuki, Synth. Comm., 11, 475
(1981).

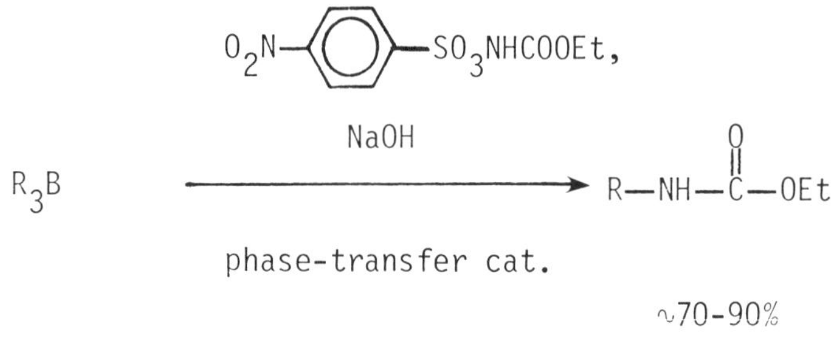

O_2N—⟨benzene⟩—$SO_3NHCOOEt$,

NaOH

R_3B → $R—NH—\overset{O}{\overset{\|}{C}}—OEt$

phase-transfer cat.

~70-90%

R = _n_-alkyl

VI.A.4-8   J. March and J.S. Engenito, Jr., <u>J. Org. Chem.</u>, <u>46</u>, 4304 (1981).

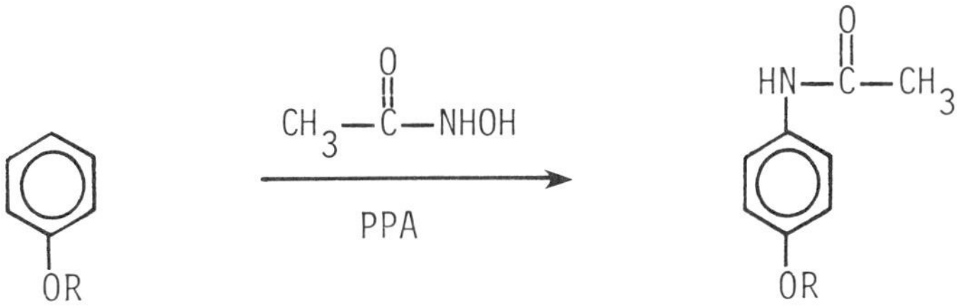

R = Me, Et   ~50%

VI.A.4-9   T. Gajda and A. Zwierzak, <u>Synthesis</u>, 1005 (1981).

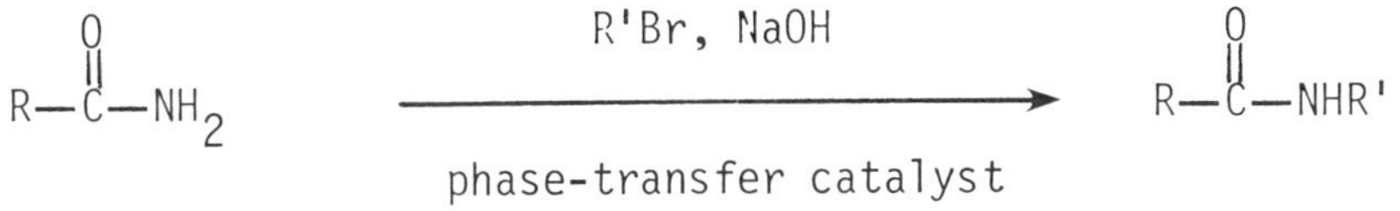

R = Ph, Et

R' = 1° alkyl

VI.A.4-10   J. Yamawaki, T. Ando, and T. Hanafusa, <u>Chem. Lett.</u>, 1143 (1981).

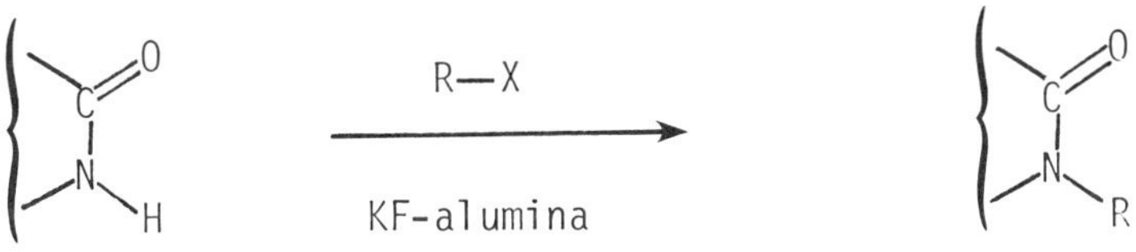

R—X = MeI,   BzCl   ~80-90%

Uracil and xanthine derivatives may be methylated similarly.

VI. A. 5. Amines

(see also: III.D.)

VI.A.5-1   R.O. Hutchins and M. Moskowitz, J. Org. Chem.,
46, 3571 (1981).

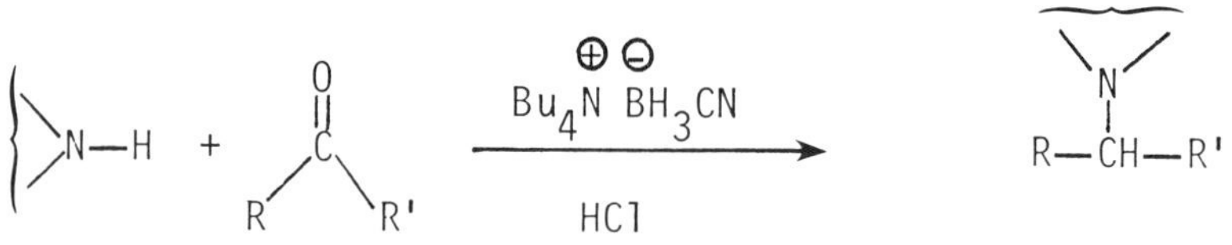

1^0, 2^0, benzyl, cyclic, etc.   ~60-80%
R, R' = H, alkyl, aryl, cyclic

VI.A.5-2   G.W. Kabalka et al., J. Org. Chem., 46, 4296
(1981).

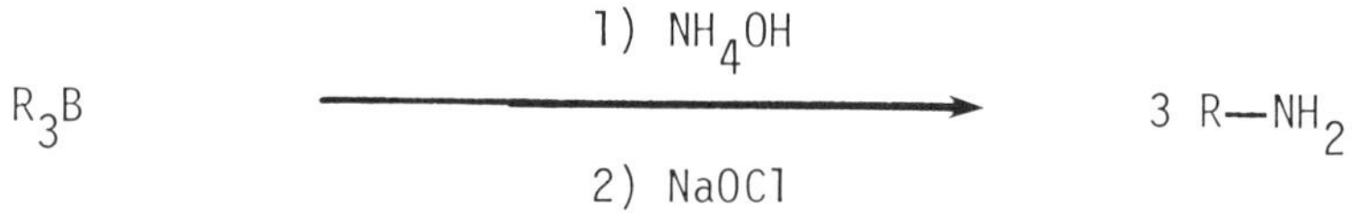

$$R_3B \xrightarrow[\text{2) NaOCl}]{\text{1) NH}_4\text{OH}} 3 \; R\text{—NH}_2$$

49-92%

R = 1^0, 2^0 alkyl, cyclic; may contain ester, sulfide

VI.A.5-3   A.I. Meyers, J.P. Lawson, and D.R. Carver, $\underline{J.\ Org.}$ $\underline{Chem.}$, $\underline{46}$, 3119 (1981).

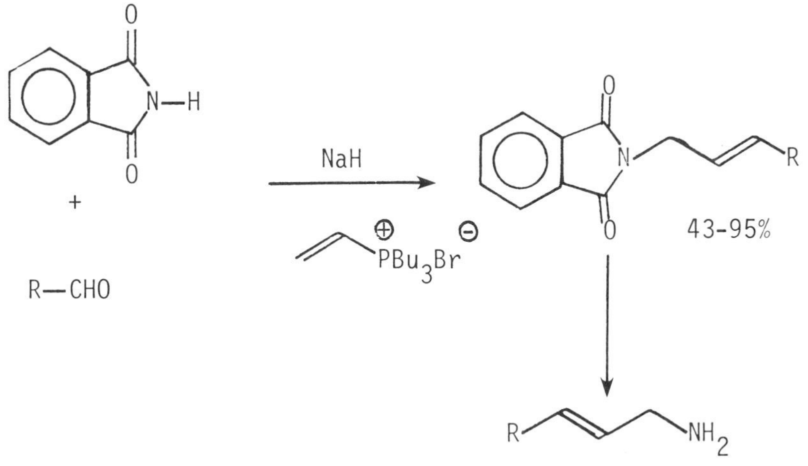

R = alkyl, aryl, vinyl, furyl, etc.

VI.A.5-4   R. Grigg $\underline{et\ al.}$, $\underline{J.C.S.\ Chem.\ Comm.}$, 611 (1981).

$$R-NH_2 \xrightarrow[\text{RhHL}_4, \ 5\%]{\text{MeOH}} R-NHMe$$

R = Bu, Ph, $\underline{c}$-Hx   39-98%

$$R_2NH \xrightarrow[\text{RhHL}_4, \ 5\%]{\text{R'OH}} R_2N-R'$$

R = alkyl, Ph, cyclic   52-99%
R' = Me, Et, Bz

VI.A.5-5   Y. Watanabe, Y. Tsuji, and Y. Ohsugi, Tetrahedron
Lett., 22, 2667 (1981).

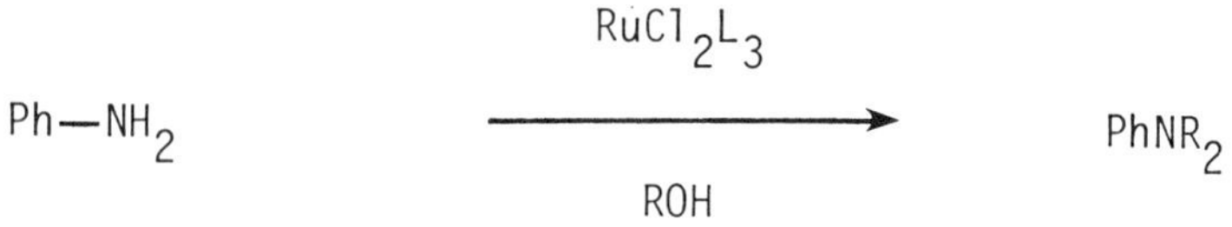

$$Ph-NH_2 \xrightarrow[ROH]{RuCl_2L_3} PhNR_2$$

74-88%

R = Et, Pr, Bu

VI.A.5-6   J.J. Bozell and L.S. Hegedus, J. Org. Chem., 46,
2561 (1981).

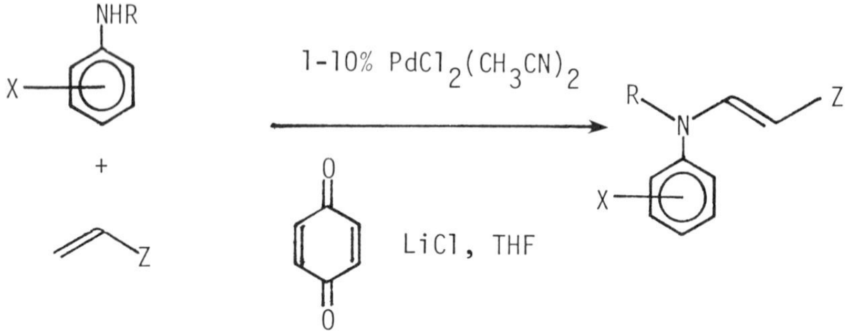

Widely varying yields.

R = H, Me
Z = COOMe, CN
X = H, o-Br, o-NO$_2$, p-OMe

VI.A.5-7   S.R. Wann, P.T. Thorsen, and M.M. Kreevoy, <u>J. Org. Chem.</u>, <u>46</u>, 2579 (1981).

$$R-\overset{O}{\underset{\|}{C}}-N\overset{R'}{\underset{R''}{}} \quad \xrightarrow[\text{acidic DMSO}]{\text{NaBH}_4} \quad R-CH_2N\overset{R'}{\underset{R''}{}}$$

R = alkyl, Ph, Bz
R', R" = H, alkyl, Ph

VI.A.5-8   H.C. Brown, S. Narasimhan, and Y.M. Choi, <u>Synthesis</u>, 439 and 996 (1981).

$$R-\overset{O}{\underset{\|}{C}}-NR_2' \quad \xrightarrow[\substack{H_3B \cdot SMe_2 \\ TMEDA}]{BF_3 \cdot Et_2O} \quad R-CH_2-NR_2'$$

72-89%

R = alkyl, aryl, cycloalkyl
R' = H, Me, <u>i</u>-Pr

VI.A.5-9   J.E. Wrobel and B. Ganem, <u>Tetrahedron Lett.</u>, <u>22</u>, 3447 (1981).

$$\xrightarrow{\text{LiEt}_3\text{BH}}$$

84%

(97% <u>cis</u>)

VI.A.5-10 R. Grigg, T.R.B. Mitchell, and N. Tongpenyai, Synthesis, 442 (1981).

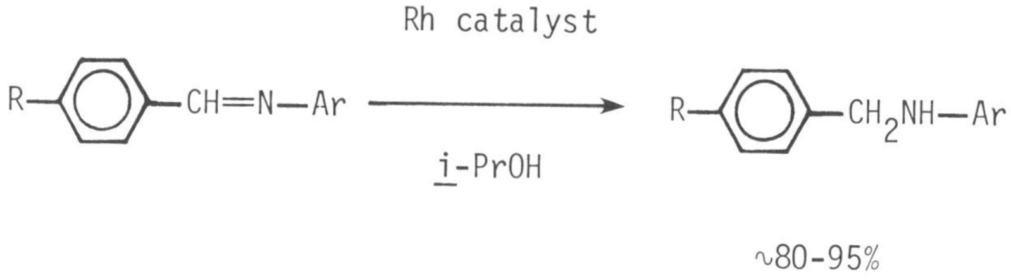
Rh catalyst
R
CH=N—Ar
i-PrOH
R
CH_2NH—Ar
~80-95%

R = H, Me, OMe, halogen, etc.
Ar = subst. Ph

VI.A.5-11 K. Yamada, M. Takeda, and T. Iwakuma, Tetrahedron Lett., 22, 3869 (1981).

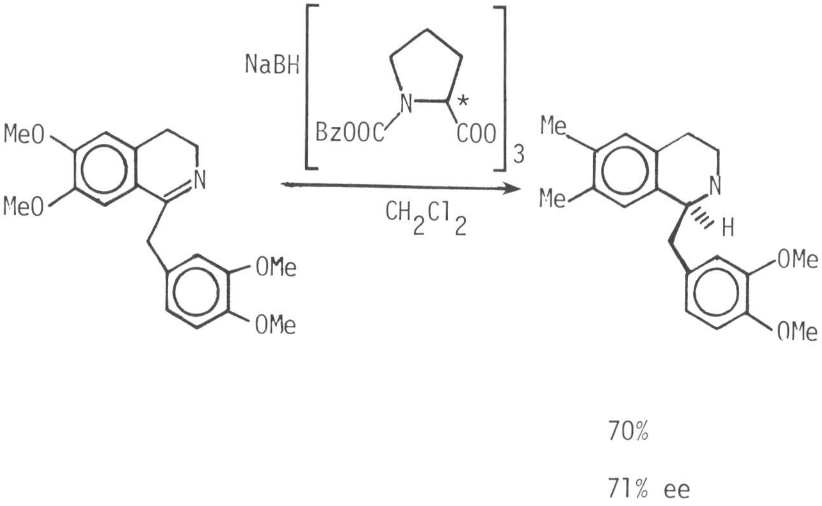
NaBH
N
*
BzOOC
COO
3
MeO
MeO
N
OMe
OMe
CH_2Cl_2
Me
Me
N
H
OMe
OMe
70%
71% ee

VI.A.5-12   J. Barluenga et al., J.C.S. Chem. Comm., 1178 (1981).

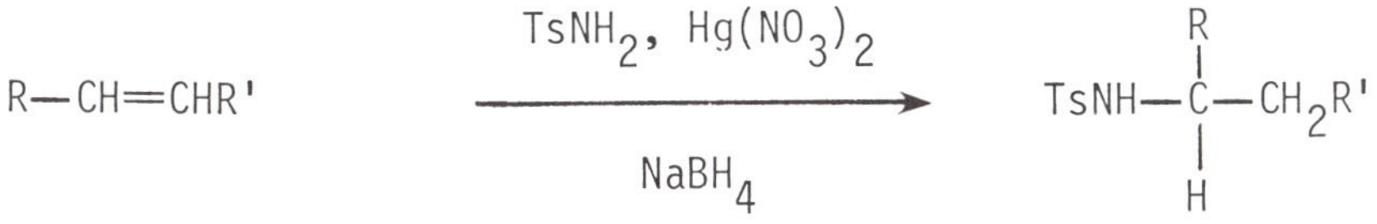

$$R-CH=CHR' \xrightarrow[\text{NaBH}_4]{\text{TsNH}_2, \text{Hg(NO}_3)_2} \text{TsNH}-\overset{\overset{\displaystyle R}{|}}{\underset{\underset{\displaystyle H}{|}}{C}}-CH_2R'$$

31-80%

R = 1° alkyl, Ph

R' = H, cyclic with R

VI.A.5-13   R. Fuks and M.V.D. Bril, Tetrahedron, 37, 2895 (1981).

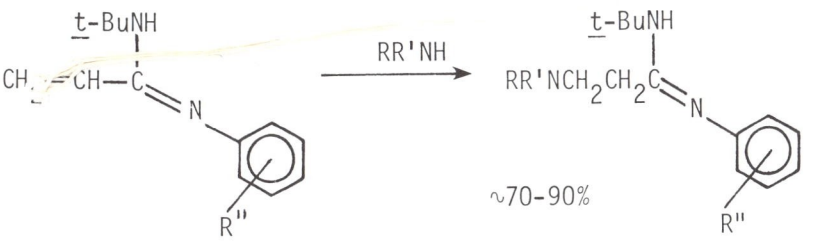

~70-90%

R,R' = alkyl, cyclic (piperidine, morpholine, pyrrolidine)

R" = Me, Cl, OMe, NO_2, etc.

VI.A.5-14   G. Bidan and M. Genies, Tetrahedron,37, 2297
(1981).

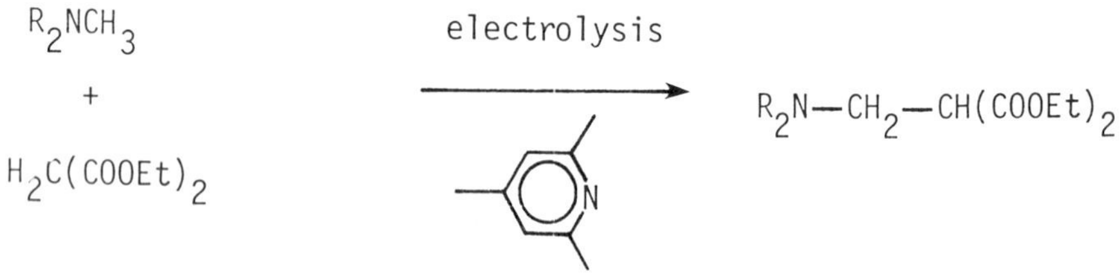

R_2NCH_3

\+

$H_2C(COOEt)_2$

electrolysis

$\xrightarrow{\hspace{3cm}}$

$R_2N—CH_2—CH(COOEt)_2$

∿50-70%

R = Me, Ph, Bz

VI.A.5-15   E.C. Taylor and J.S. Skotnicki, Synthesis, 606
(1981).

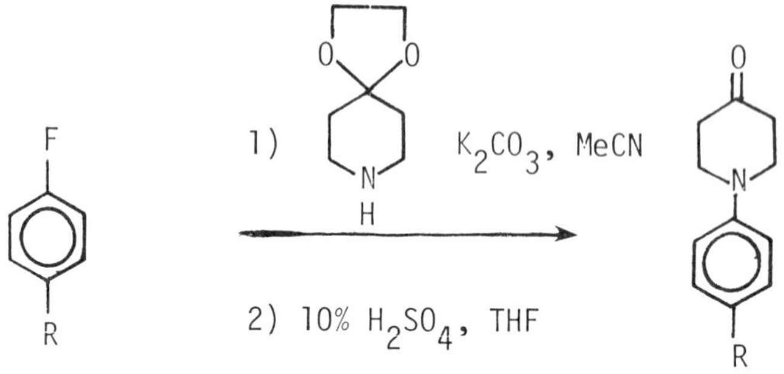

1) [diagram] K_2CO_3, MeCN

2) 10% H_2SO_4, THF

∿50-60% overall

R = -COOEt, -CN, -CHO, -COCH_3, -NO_2

VI.A.5-16   G.S. Poindexter, Synthesis, 541 (1981).

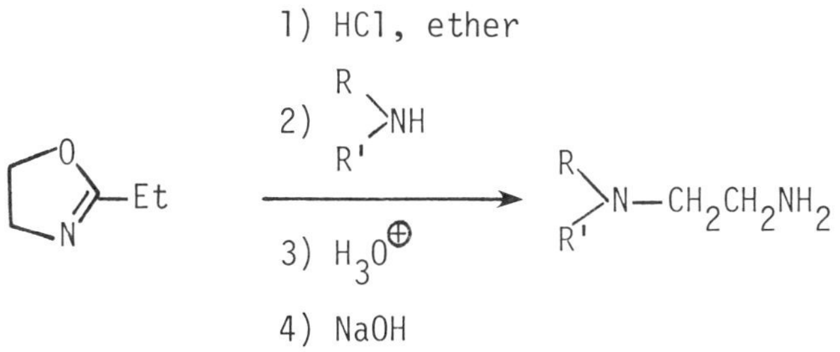

R,R' = H, Me, Et, cyclic, Ph   Widely varying yields.

VI.A.5-17   H. Yamamoto and K. Maruoka, J. Am. Chem. Soc., 103, 4186 (1981).

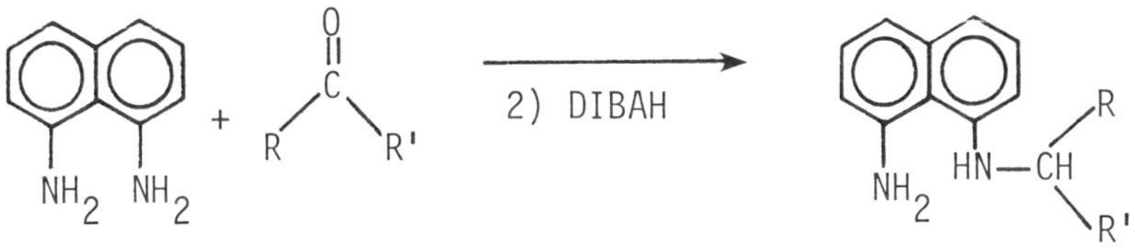

R,R' = alkyl, cyclic, H, Ph   84-98%

VI.A.5-18   G.E. Keck, R.R. Webb, and J.B. Yates, Tetrahedron 37, 4007 (1981).

Reveiw: "A Versatile Method for Carbon-Nitrogen Bond Formation via Ene Reactions of Acylnitroso Compounds"

VI. A. 6. Amino Acids and Derivatives

(see also: VI.A.4., VI.A.10.)

VI.A.6-1   F. Effenberger et al., Chem. Ber., 114, 173 (1981).

$$
\begin{array}{c}
\text{O} \\
\| \\
R-CH-C-OR \\
| \\
Br
\end{array}
\quad
\xrightarrow[\text{2) conc. HCl, }100^{\circ}]{\text{1) KNCO, ROH}}
\quad
\begin{array}{c}
R-CH-COOH \\
| \\
NH_3{\oplus}
\end{array}
$$

R = Amino acid side chains. An amino group may also be
introduced at the side of a halide in the side chain
(e.g. lysine).

VI.A.6-2   S. Yano et al., J. Am. Chem. Soc., 103, 2459
(1981).

$$
\begin{array}{c}
H_2N \\
\diagdown \\
\diagup \quad C(COO^{\ominus})_2 \\
H_3C
\end{array}
\quad
\xrightarrow{\text{chiral Co(III) complex}}
\quad
\begin{array}{c}
NH_2 \\
| \\
H_3C-CH-COO^{\ominus} \\
*
\end{array}
$$

88% yield

66% ee

VI.A.6-3   T. Shono, Y. Matsumura, and K. Tsubata, <u>Tetrahed-</u>
<u>ron Lett.</u>, <u>22</u>, 2411 (1981).

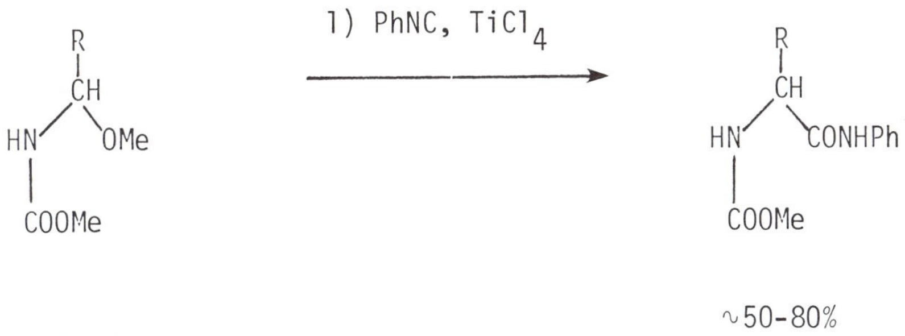

R = alkyl, cyclic

VI.A.6-4   U. Schollkopf <u>et al.</u>, <u>Synthesis</u>, 966 (1981).

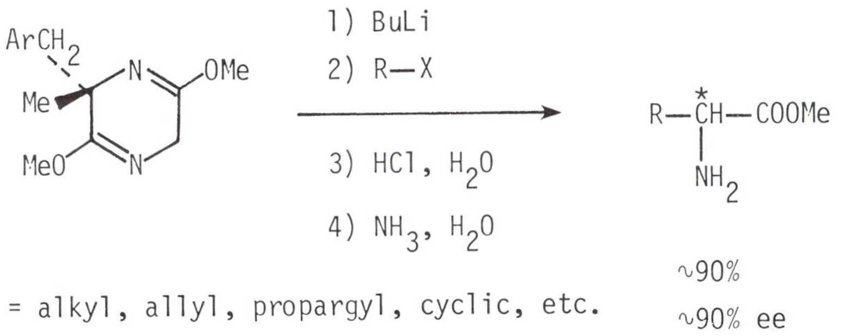

R = alkyl, allyl, propargyl, cyclic, etc.

∼90%
∼90% ee

VI.A.6-5   U. Schöllkopf, U. Groth, and C. Deng, <u>Angew. Chem.</u>
<u>Int. Ed.</u>, <u>20</u>, 798 (1981).

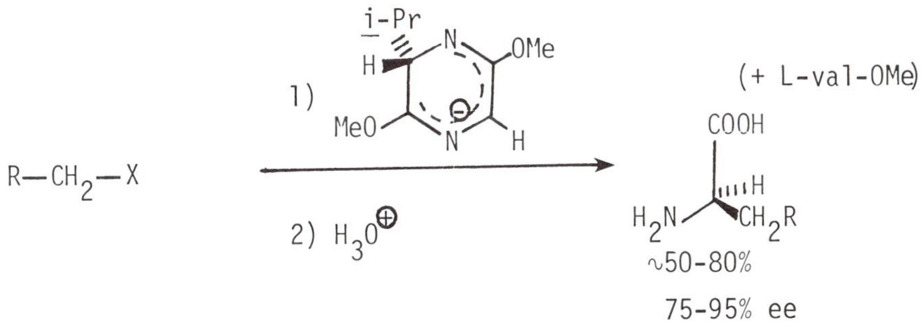

VI.A.6-6   P.A. MacNeil, N.K. Roberts, and B. Bosnich, <u>J. Am.</u>
<u>Chem. Soc.</u>, <u>103</u>, 2273 (1981).

$$\underset{R}{\overset{COOH}{\diagup}}\underset{NHCOPh}{\diagup} \xrightarrow[\text{Rh(I)·skewphos}]{H_2} R-CH-\overset{*}{C}HCOOH \atop NHCOPh$$

R = H, <u>i</u>-Pr, Ph, subst. Ph   ∿80-90% ee

$$\text{skewphos} = \overset{H}{\underset{Ph_2P}{\bigg\vert}}\overset{H}{\underset{PPh_2}{\bigg\vert}}$$

VI.A.6-7   J.K. Stille <u>et al.</u>, <u>J. Org. Chem.</u>, <u>46</u>, 2954 and
2960 (1981).

$$\underset{\substack{H \\ }}{\overset{R\quad R'}{\bigcirc}}\overset{NHAc}{\underset{COOH}{\diagup}} \xrightarrow[\text{EtOH/Et}_3\text{N}]{\text{chiral Rh(I)-polymer}} RCH_2-\overset{NHAc}{\underset{COOH}{\overset{|\,*}{C}H}}$$

R = OH, OAc   high yields,
R' = H, OMe   ∿80-90% ee

VI.A.6-8   W. Bergstein, A. Kleeman, and J. Martens,
<u>Synthesis</u>, 76 (1981).

$$\underset{R}{\overset{H}{\diagup}}C=C\underset{\underset{O}{\overset{|}{NH-C-R'}}}{\overset{COOH}{\diagup}} \xrightarrow[\text{chiral Rh catalyst}]{H_2} R-CH_2-\overset{*}{C}H-COOH \atop \underset{O}{\overset{|}{HN-C-R'}}$$

98-99%
∿90% ee

VI.A.7. Carbenes

(see also: I.D.)

VI.A.7-1 U.H.Brinker and J. Ritzer, J. Am. Chem. Soc., 103, 2116 (1981).

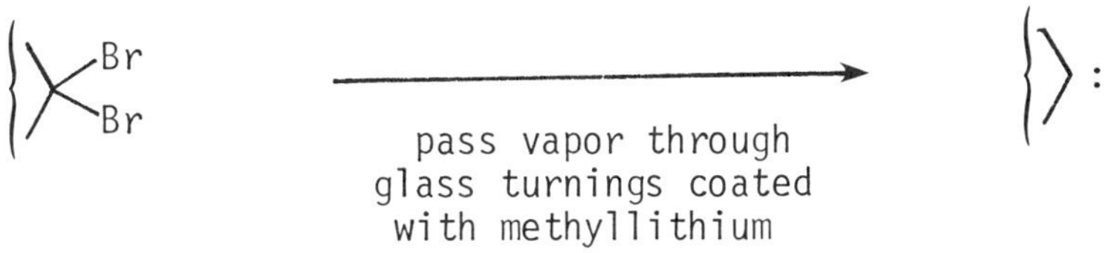
Br
Br
pass vapor through
glass turnings coated
with methyllithium

VI.A.7-2 M. Suda, Synthesis, 714 (1981).

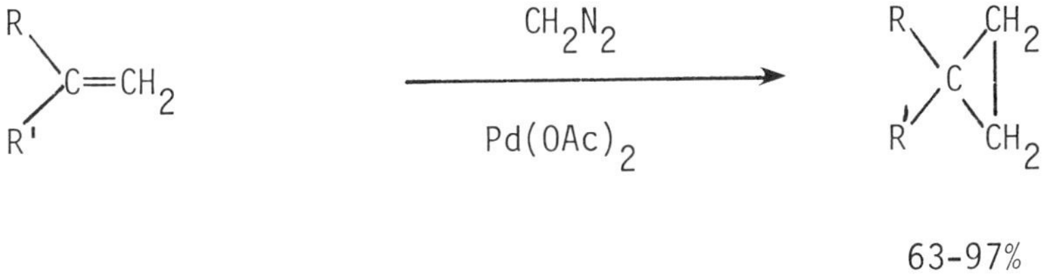
R
R'
C=CH2
CH2N2
Pd(OAc)2
R
R'
C
CH2
CH2
63-97%

R,R' = H, alkyl, $-CH_2OPh$, cyclic

VI.A.7-3 A.F. Noels et al., J.C.S. Chem Comm., 688 (1981).

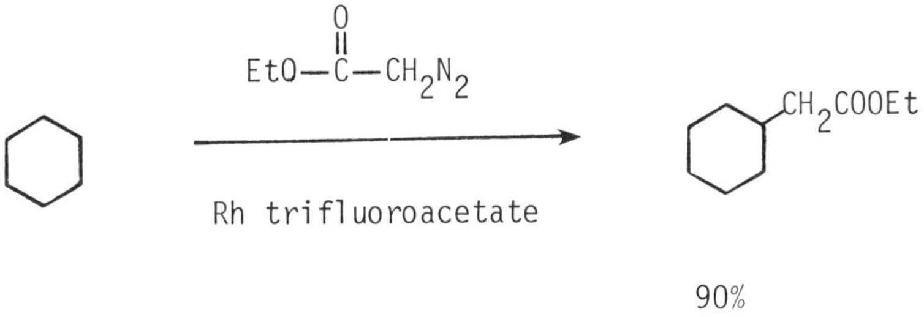
EtO—C(=O)—CH2N2
Rh trifluoroacetate
CH2COOEt
90%

VI.A.7-4   P.J. Stang, Israel J. Chem., 21, 119 (1981).

Review: "Small and Strained Ring Compounds via

Unsaturated Carbenes"

VI.A.7-5   M.P. Doyle et al., Tetrahedron Lett., 22, 1783
(1981); Synthesis, 787 (1981).

$$R^1-\underset{\underset{R^2}{|}}{C}=\underset{\underset{R^4}{|}}{C}-R^3 \quad \xrightarrow[\text{Rh, Cu catalysts}]{EtOOCCHN_2} \quad EtOOC-\triangle$$

(with triangle bearing R^1, R^2, R^3, R^4)

59-88%

Olefin = cyclohexene, styrene, ethyl vinyl ether,
          dihydropyran, etc.

VI.A.7-6   P.J. Stang and S.B. Christensen, J. Org. Chem.,
46, 823 (1981).

$$Me_2C=C\underset{SiMe_3}{\overset{OTf}{<}}$$

+

$$R'-\underset{\underset{}{\overset{S}{||}}}{C}-CHR_2$$

$$\xrightarrow[\text{glyme}]{Bu_4NF}$$

$$\underset{R_2C=C_{R'}}{\overset{H}{\underset{S}{\overset{}{>}}}C=CMe_2}$$

25-40%

Proceeds via $Me_2C=C:$

VI.A.8. Enamines

VI.A.8-1   N. DeKimpe and N. Schamp. Org. Prep. Proc. Int., 13, 241 (1981).

Review: "The Synthesis of β-Halogenated Enamines"

VI.A.8-2   S. Rajappa, Tetrahedron, 37, 1453 (1981).

Review: "Nitroenamines. Preparation, Structure, and Synthetic Potential"

VI.A.9. Epoxides

(see also: II.F.1.)

VI.A.9-1   W.C. Still and V.J. Novack, J. Am. Chem. Soc.,
103, 1283 (1981).

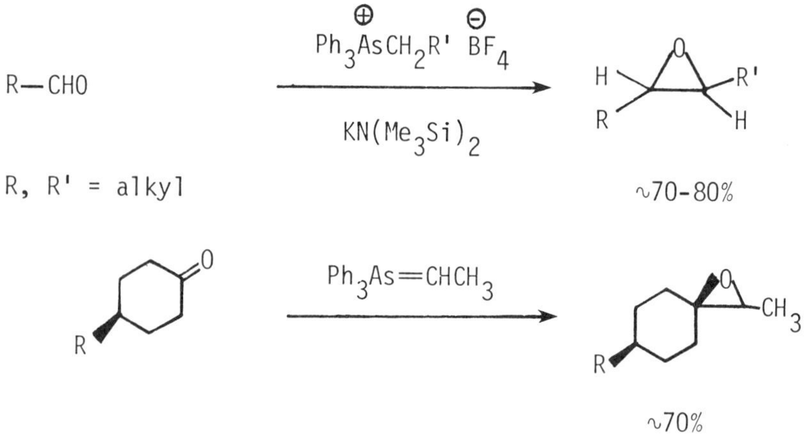

R—CHO $\xrightarrow[\text{KN(Me}_3\text{Si)}_2]{\text{Ph}_3\overset{\oplus}{\text{As}}\text{CH}_2\text{R' }\overset{\ominus}{\text{BF}}_4}$

R, R' = alkyl   ∿70-80%

$\xrightarrow{\text{Ph}_3\text{As}=\text{CHCH}_3}$

∿70%

VI.A.9-2   K. Takaki, M. Yasumura, and K. Negoro, Angew.
Chem. Int. Ed., 20, 671 (1981).

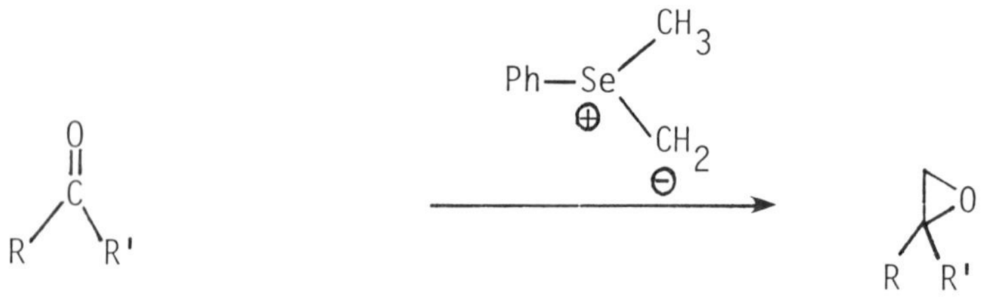

R = n-alkyl, subst. Ph
R' = H, Me, Et, -(CH$_2$)$_5$-

VI.A.9-3   C.N. Barry and S.A. Evans, Jr., J. Org. Chem., 46,
3361 (1981).

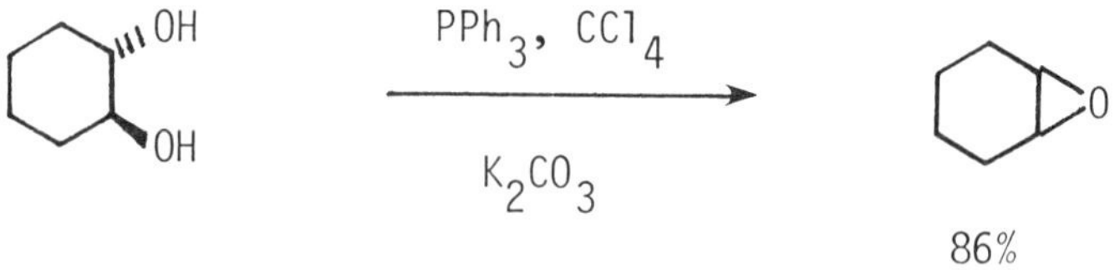

$\xrightarrow[\text{K}_2\text{CO}_3]{\text{PPh}_3, \text{ CCl}_4}$

86%

VI.A.10.   Esters

(see also: IV.E, V.D.)

VI.A.10-1   M. Yamaguchi et al., Bull. Chem. Soc. Japan, 54, 1470 (1981).

R—COOH + R'—OH $\xrightarrow{\hspace{3cm}}$ $R-\overset{\overset{\displaystyle O}{\|}}{C}-OR'$

(or R-COONa)   >90%

R = alkyl, subst. Ph

R' = 1°, 2° alkyl, Bz

VI.A.10-2   S. Kim and S. Yang, Synth. Comm., 11, 121 (1981).

R—COOH + R'OH $\xrightarrow{\hspace{3cm}}$ $R-\overset{\overset{\displaystyle O}{\|}}{C}-OR'$

                                                                    >90%

R, R' = alkyl, aryl, etc.

VI.A.10-3   G.A. Olah, S.C. Narang, and A.G. Luna, Synthesis, 790 (1981).

R—COOH $\xrightarrow[\text{2) R'OH, Et}_3\text{N}]{\text{1) SO}_2\text{ClF, Et}_3\text{N, CH}_2\text{Cl}_2}$ $R-\overset{\overset{\displaystyle O}{\|}}{C}-OR'$

                                                                    51-91%

R = subst. Ph, styryl
R' = Me, Et, Bz

VI.A.10-4   R.W. Johnson et al., Tetrahedron Lett., 22, 3715
(1981).

$$R-\overset{\overset{\text{O}}{\|}}{C}-O-\underset{}{\bigcirc}-OH \quad \xrightarrow[\text{R'OH}]{-2\ e^{\ominus}} \quad R-\overset{\overset{\text{O}}{\|}}{C}-OR'$$

~70-90%

R = pentyl, c-Hx, t-Bu
R' = Hx, c-Hx, t-Bu

VI.A.10-5   J.S. Grossert, W.M.N. Ratnayake, and T. Swee,
Can. J. Chem., 59, 2617 (1981).

$$R-\overset{\overset{\text{O}}{\|}}{C}-O-\overset{\overset{\text{O}}{\|}}{C}-CF_3 \quad \xrightarrow[\text{R' = Me, 2-naphthyl}]{\text{R'OH}} \quad R-\overset{\overset{\text{O}}{\|}}{C}-OR'$$

up to 90%

R = long chain alkyl
    (fatty acid)

VI.A.10-6   T. Mukaiyama et al., Chem. Lett., 563 (1981).

$$R'-\overset{\overset{\text{O}}{\|}}{C}-O-\underset{N}{\bigcirc}-\underset{N}{\bigcirc}$$

$$R-OH \quad \xrightarrow[\text{CsF}]{} \quad R-O-\overset{\overset{\text{O}}{\|}}{C}-R'$$

~80-90%
in most cases

R = alkyl, Bz, cholesteryl, etc.
R' = alkyl, Bz

Monoacylation of diols may be accomplished.

VI.A.10-7   R. Nakao, K. Oka, and T. Fukumoto, Bull. Chem.
Soc. Japan, 54, 1267 (1981).

$$R—COOH \xrightarrow[\text{Me}_3\text{SiCl}]{\text{R'OH}} R—\overset{\displaystyle O}{\overset{\|}{C}}—OR'$$

~90%

R = Me, Et, i-Pr, t-Bu
R' = Me, i-Pr, allyl, c-Hx

VI.A.10-8   G.A. Olah et al., Synthesis, 142 (1981).

$$R^1—\overset{\displaystyle O}{\overset{\|}{C}}—OR^2 \xrightarrow[\text{2) } R^3\text{OH}]{\text{1) Me}_3\text{SiI, I}_2} R^1—\overset{\displaystyle O}{\overset{\|}{C}}—OR^3$$

~70-98%

R^1 = Ph, t-Bu, styryl
R^2, R^3 = Me, Et, i-Pr, Bz

VI.A.10-9   N. Hashimoto, T. Aoyama, and T. Shioiri, Chem.
Pharm. Bull., 29, 1475 (1981).

$$R—COOH \xrightarrow[\text{MeOH}]{\text{Me}_3\text{SiCHN}_2} R—\overset{\displaystyle O}{\overset{\|}{C}}—OMe$$

R = alkyl, aryl   Quantitative in most cases.

VI.A.10-10   E. Fujita et al., Chem. Pharm. Bull., 29, 3202 (1981).

$$R—OH \xrightarrow[\text{BF}_3 \cdot \text{Et}_2\text{O}]{\text{Ac}_2\text{O}} R—OAc$$

R = alkyl, cycloalkyl

Acylates, alcohols in preference to phenols.

VI.A.10-11   G.H. Posner et al., Synthesis, 789 (1981); Tetrahedron Lett., 22, 5003 (1981); S.S. Rana, J.J. Barlow, and K.L. Matta, Tetrahedron Lett., 22, 5007 (1981).

Primary alcohols may be acetylated in the presence of secondary alcohols by using ethyl acetate in the presence of Woelm neutral alumina. Monoacetates are obtained in yields of about 50-70%.

VI.A.10-12   W.J. Greenlee and E.D. Thorsett, J. Org. Chem., 46, 5351 (1981).

$$\underset{\substack{|| \\ R—C—NH_2}}{O} \xrightarrow{\text{R'—OH, acidic resin}} \underset{\substack{|| \\ R—C—OR'}}{O}$$

∿80-90%

R = alkyl, aryl, heterocyclic
R' = Me, Et

VI.A.10-13   S. Kim and J.I.Lee, <u>J.C.S. Chem. Comm.</u>, 1231 (1981).

$$R-\overset{\overset{\text{O}}{\|}}{C}-S-\underset{N}{\bigcirc} \quad \xrightarrow{\text{LiCuR}_2', \ O_2} \quad R-\overset{\overset{\text{O}}{\|}}{C}-OR'$$

~70-90%

VI.A.10-14   M. Ueda, K. Seki, and Y. Imai, <u>Synthesis</u>, 991 (1981).

$$\bigcirc\hspace{-1em}\underset{N}{\overset{O}{\diagup}}\hspace{-0.5em}-S-\overset{\overset{\text{O}}{\|}}{C}-R \quad \xrightarrow{\text{R'OH, Et}_3\text{N}} \quad R-\overset{\overset{\text{O}}{\|}}{C}-OR'$$

82-98%

R = Me, Ph
R' = Ph, Bz

VI.A.10-15   M. Sekine, A. Kume, and T. Hata, <u>Tetrahedron Lett.</u>, <u>22</u>, 3617 (1981).

$$R-OH \quad \xrightarrow[\text{DBU, CH}_2\text{Cl}_2]{\overset{\displaystyle Ph-\overset{\overset{\text{O}}{\|}}{C}-\overset{\overset{\text{O}}{\|}}{P}(OEt)_2}{}} \quad R-O-\overset{\overset{\text{O}}{\|}}{C}-Ph$$

52-93%

R = alkyl

VI.A.10-16   W.H. Kruizinga, B. Strijtveen, and R.M. Kellogg, J. Org. Chem., 46, 4321 (1981).

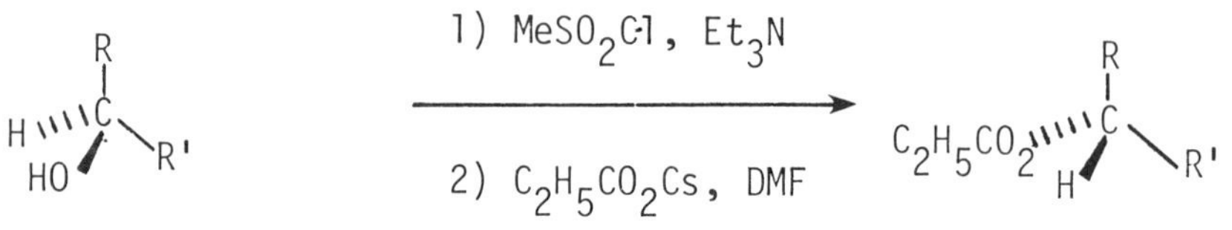

R,R' = alkyl, Ph, cyclic, etc.

48-100%

VI.A.10-17   K. Ogura, J. Watanabe, and H. Iida, Tetrahedron Lett., 22, 4499 (1981).

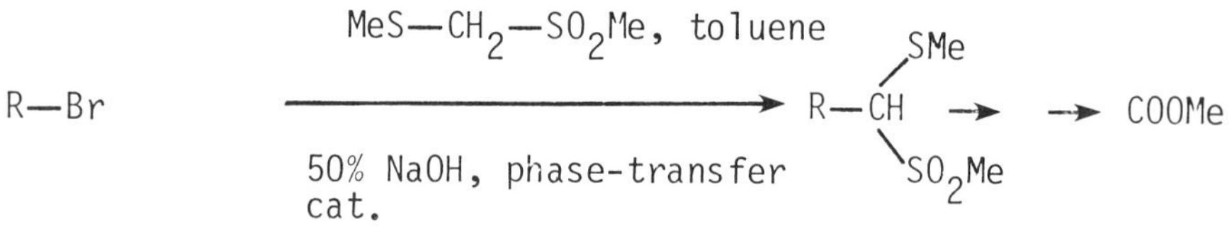

R = n-alkyl, Bz, cinnamyl

~80%

VI.A.10-18   S. Shinkai et al., Bull Chem. Soc. Japan, 54, 631 (1981).

Use of dimethylaminopyridine linked to polystyrene beads as a catalyst for ester synthesis by the DCC method. Rates are somewhat slower than for homogeneous DMAP, but yields are good.

VI.A.11.   Ethers

(see also: V.A.)

VI.A.11-1   D.E. Laycock and H. Alper, J. Org. Chem., $\underline{46}$, 289 (1981).

$$R_2C{=}S \quad \xrightarrow[\text{2) CO, Br}_2\text{, CH}_3\text{OH}]{\text{1) Cp}_2\text{Zr}\overset{H}{\underset{Cl}{<}}} \quad R_2CH{-}O{-}CH_3$$

R = alkyl, subst. Ph   ∿50-90%

VI.A.11-2   W.B. Motherwell et al., J.C.S. Chem. Comm., 503 (1981).

$$Ar{-}OH \quad \xrightarrow{\text{Ph}_4\text{BiOCOCF}_3} \quad Ar{-}O{-}Ph$$

Ar = subst Ph, naphthyl, estrone, etc.   51-77%

VI.A.11-3   G.A. Olah, A.P. Fung, and R. Malhotra, Synthesis, 474 (1981).

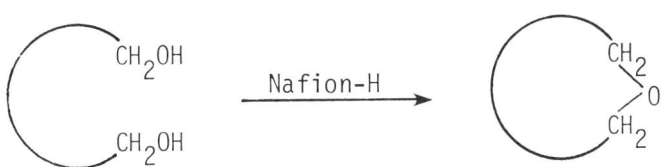

∿50-90%

Used for 5,6,7,8-membered rings

VI.A.11-4   C.N. Barry and S.A. Evans, Jr., <u>J. Org. Chem.</u>,
<u>46</u>, 3361 (1981).

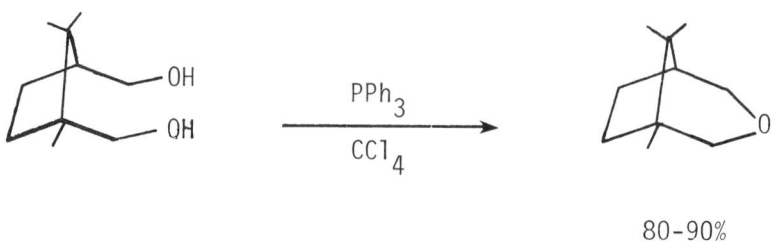

80-90%

VI.A.12.   Ketones and Aldehydes

(see also: I.A.2., II.A.1., III.F.1., II.F.2.)

VI.A.12-1   J.L. Fry and R.A. Ott, <u>J. Org. Chem.</u>, <u>46</u>, 602
(1981).

$$1) \text{ Et}_3\text{O}^{\oplus}\text{ BF}_4^{\ominus}$$

$$2) \text{ Et}_3\text{SiH}$$

$$\text{R—C} \equiv \text{N} \longrightarrow \text{R—CHO}$$

3) H_2O

R = alkyl, subst. Ph   ∿60-90%

VI.A.12-2   N. Kornblum and A.S. Erickson, <u>J. Org. Chem.</u>,
<u>46</u>, 1037 (1981).

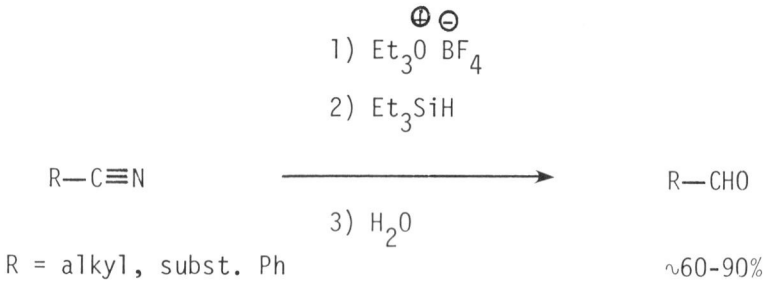

R = 3^0 alkyl, 3^0 benzylic, etc.

VI.A.12-3   D. Schinzer and C.H. Heathcock, <u>Tetrahedron Lett.</u>,
<u>22</u>, 1881 (1981).

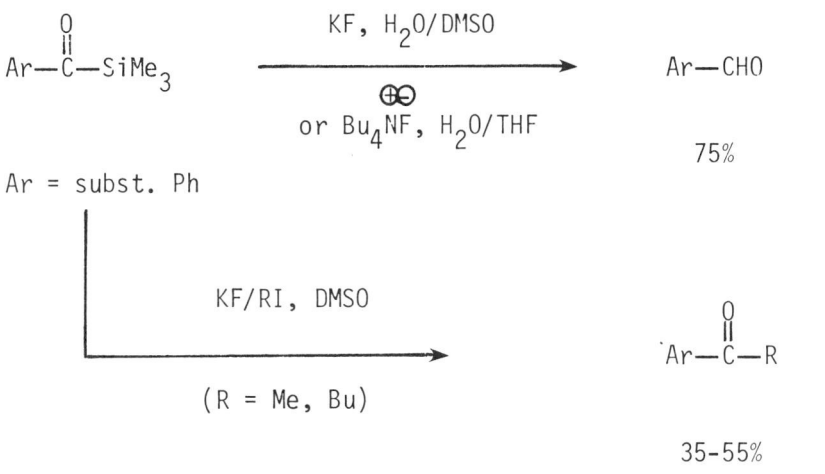

Ar = subst. Ph

VI.A.12-4   I. Degani, R. Fochi, and V. Regondi, <u>Tetrahedron
Lett.</u>, <u>22</u>, 1821 (1981).

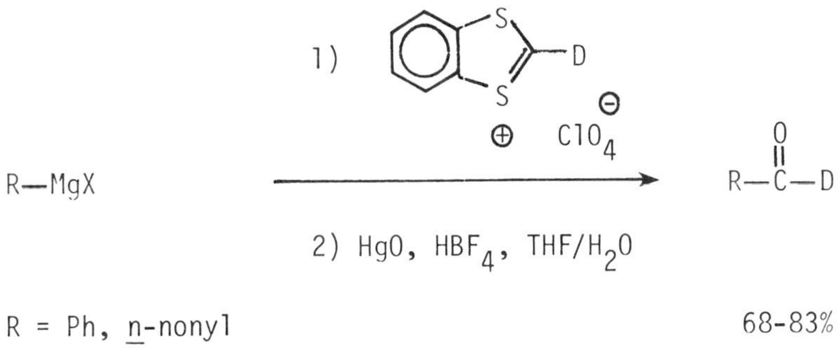

R—MgX

1) [benzodithiolium structure] —D
          ⊕   ClO₄⊖

2) HgO, HBF₄, THF/H₂O

R = Ph, <u>n</u>-nonyl

$$R-\overset{\overset{\displaystyle O}{\|}}{C}-D$$

68-83%

VI.A.12-5   G.A. Olah and M. Arvanaghi, Angew. Chem. Int. Ed., 20, 878 (1981).

$$RLi \quad + \quad \text{[piperidine-N-CHO]} \quad \xrightarrow[\text{2) } H_3O^{\oplus}]{} \quad R—CHO$$

~80-90%

R = alkyl, aryl, vinyl, acetylenic, etc.

VI.A.12-6   R.J.P. Corriu, J.J.E. Moreau, and M.P.Sat, J. Org. Chem., 46, 3372 (1981).

$$(CO)_4Fe \overset{Me_2}{\underset{Me_2}{<\!\!\!{Si \atop Si}}} \text{(benzene ring)}$$

1) hv,

$$R—CH—CN \atop R' \qquad \xrightarrow[\text{2) } H_3O^{\oplus}]{} \qquad R—CH—CHO \atop R'$$

R = H, alkyl, aryl, etc.   ~50-80%
R' H, Ph

VI.A.12-7   A. Pelter and J.M.Rao, J.C.S. Chem. Comm., 1149 (1981).

$$R_3B \quad \xrightarrow[\text{2) NaOH, } H_2O_2]{\text{1) } Li—C(SPh)_3} \quad R_2C{=}O$$

R = alkyl, cycloalkyl   72-82%

VI.A.12-8   S. Mukaiyama, J. Inanaga, and M. Yamaguchi, <u>Bull. Chem. Soc. Japan</u>, <u>54</u>, 2221 (1981).

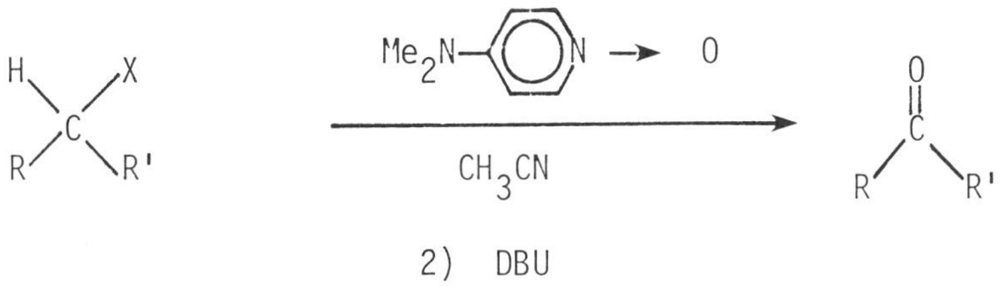

X = halogen
R, R' = H, Ph, alkyl, amino

VI.A.12-9   S. Kim and J.I. Lee, <u>J.C.S. Chem. Comm.</u>, 1231 (1981).

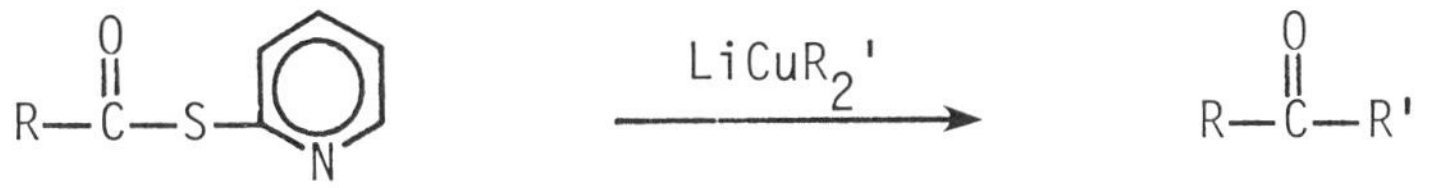

R = alkyl, Ph, Bz
R' = Me, <u>n</u>-Bu, <u>t</u>-Bu

70-99%

VI.A.12-10   V. Premasagar, V.A. Palaniswamy, and E.J. Eisenbraun, <u>J. Org. Chem.</u>, <u>46</u>, 2974 (1981).

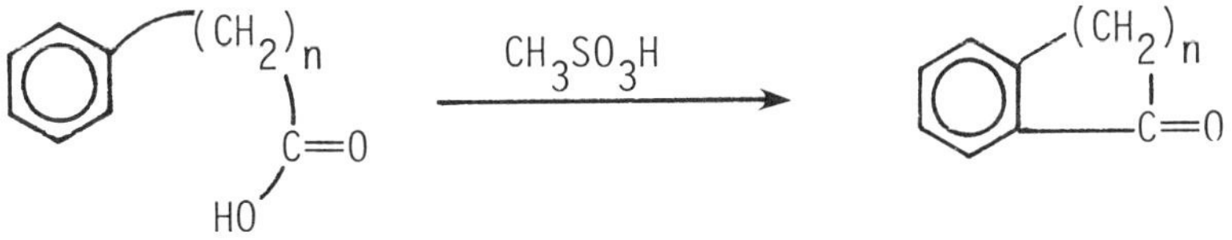

n = 2, 3

60-100%

VI.A.12-11   U. Niewöhner and W. Steglich, Angew. Chem. Int.
Ed., 20, 395 (1981).

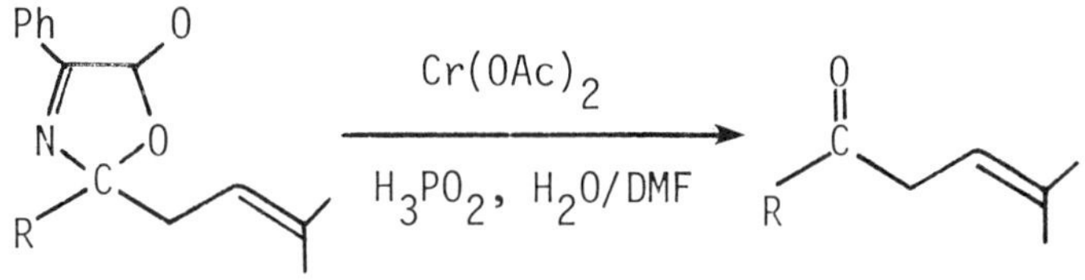

54-97%

R = alkyl, vinyl, subst. Ph, styryl, etc.

VI.A.12-12   I. Kuwajima et al., Bull Chem. Soc. Japan, 54,
3100 and 3510 (1981).

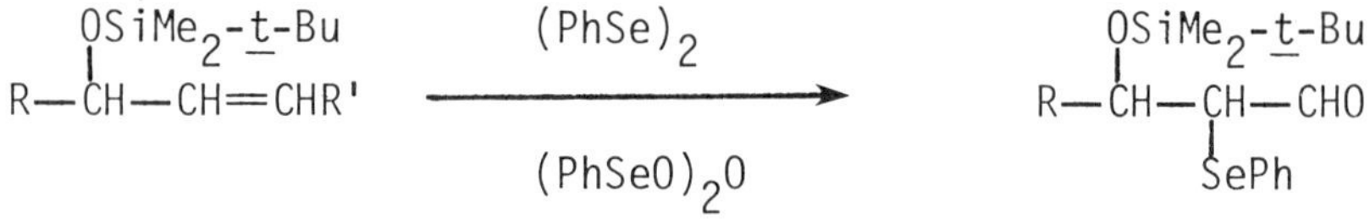

R = alkyl, Ph, cyclic   74-87%
R' = H, Me

VI.A.12-13   K.I. Paskevich, V.I. Saloutin, and I. Ya.
Postovskii, Russ. Chem. Rev., 50, 180 (1981).

Review: "Fluorine-Containing β-diketones"

VI.A.13.   Nitriles

VI.A.13-1   R.J. Card and J.L. Schmitt, J. Org. Chem., 46, 754 (1981).

$$R-CH_2OH \quad \xrightarrow[\text{Cu, } 300^0]{\text{NH}_3} \quad R-C\equiv N$$

$\sim$90%

R = 1^0 alkyl, subst. Ph

VI.A.13-2   R. Davis and K.G. Untch, J. Org. Chem., 46, 2985 (1981).

$$R-OH \quad \xrightarrow[\text{NaI, DMF/CH}_3\text{CN}]{\text{NaCN, Me}_3\text{SiCl}} \quad R-CN$$

$\sim$70-90%

R = 1^0, 2^0, 3^0 alkyl, cycloalkyl   in most cases

VI.A.13-3   M.T. Reetz and I. Chatziiosifidis, Angew. Chem. Int. Ed., 20, 1017 (1981).

$$R_3C-Cl \quad \xrightarrow[\text{SnCl}_4]{\text{Me}_3\text{SiCN}} \quad R_3C-CN$$

$\sim$60-80%

VI.A.13-4   D. Dauzonne, P. Demerseman, and R. Royer,
Synthesis, 739 (1981).

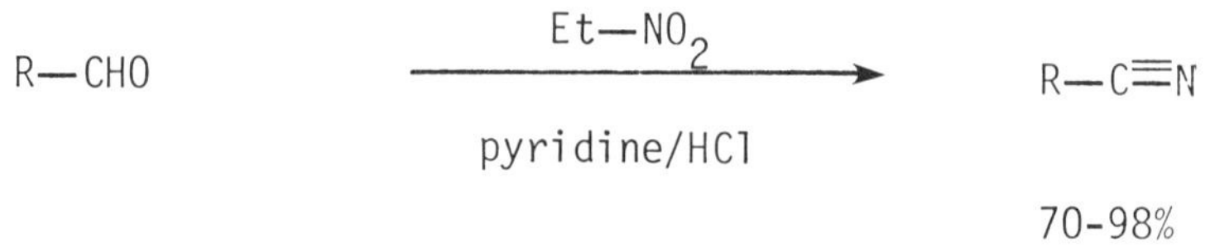

$$R-CHO \xrightarrow[\text{pyridine/HCl}]{Et-NO_2} R-C\equiv N$$

70-98%

R = subst. Ph, styryl, heterocyclic

VI.A.13-5   J. -P. Dulcere, Tetrahedron Lett., 22, 1599 (1981)

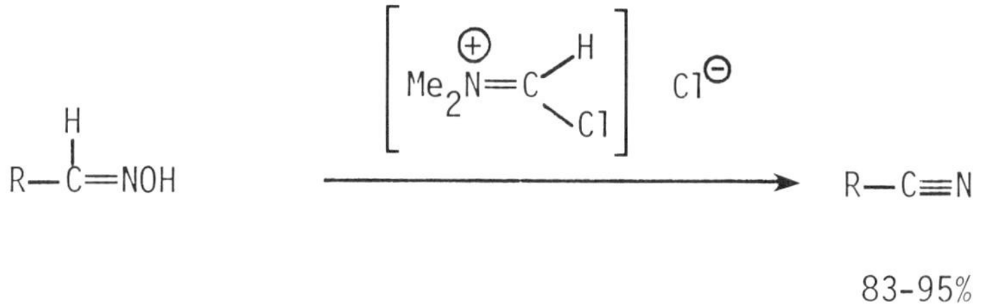

$$R-\overset{H}{\underset{}{C}}=NOH \xrightarrow{\left[Me_2N=C\overset{H}{\underset{Cl}{\diagdown}}\right] Cl^{\ominus}} R-C\equiv N$$

83-95%

R = 3° alkyl, containing allenes

VI.A.13-6   T. Keumi et al., Bull. Chem. Soc. Japan, 54,
1579 (1981).

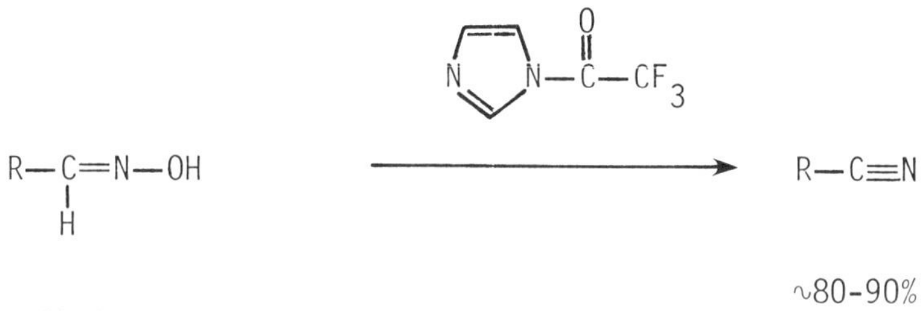

$$R-\overset{}{\underset{H}{C}}=N-OH \xrightarrow{\quad} R-C\equiv N$$

~80-90%

R = alkyl, styryl, subst. Ph, furyl

VI.A.13-7   P. Molina et al., Synthesis, 711 (1981).

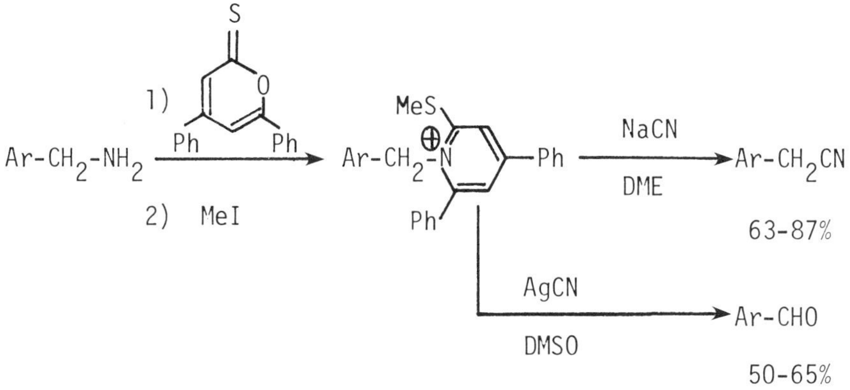

VI.A.14.   Nitro-compounds

VI.A.14-1   J.V. Crivello, J. Org. Chem., 46, 3056 (1981).

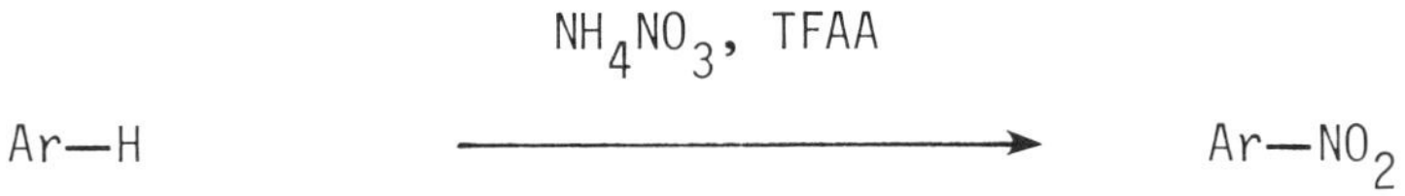

$$Ar\!-\!H \xrightarrow{\quad NH_4NO_3, \ TFAA \quad} Ar\!-\!NO_2$$

Ar = subst. Ph, naphthyl, etc.
Phenols are oxidized under these conditions

VI.A.14-2   R.P. Kozyrod and J.T. Pinhey, Tetrahedron Lett., 22, 783 (1981).

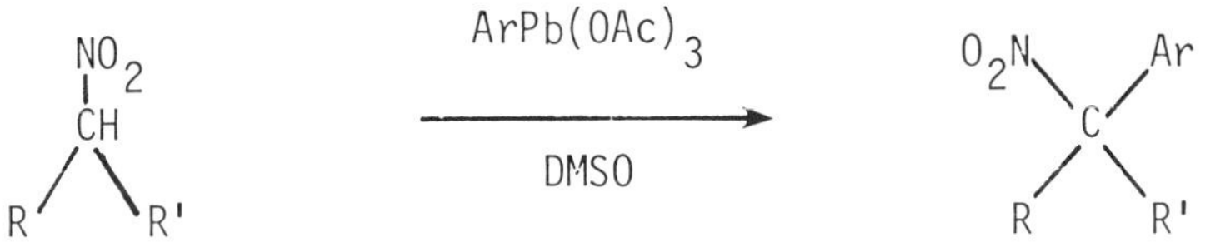

R, R' = Me, $-(CH_2)_4-$, $-(CH_2)_5-$ ~70%

Ar = Subst. Ph

VI.A.14-3   G.A. Olah, S.C. Narang, and A.P. Fung, J. Org. Chem., 46, 2706 (1981).

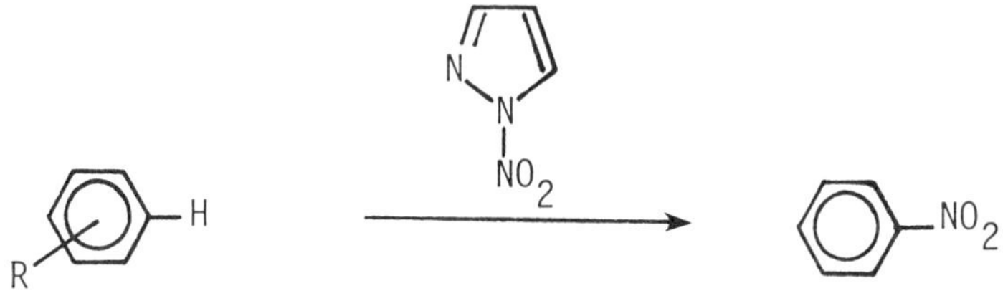

R = alkyl

~90%
usual (o,p) regioselectivity

VI.A.14-4   T. Sakakibara et al., J.C.S. Chem. Comm., 261 (1981).

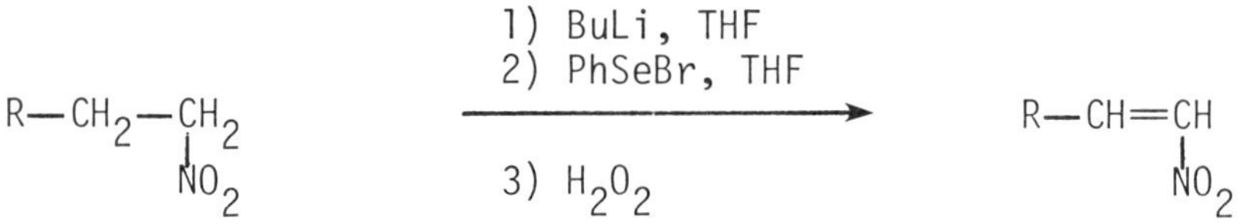

1) BuLi, THF
2) PhSeBr, THF

3) H₂O₂

~50-60% overall

R = alkyl

VI.A.15.   Nucleotides, etc.

   (see also: IV.I.1.a,b; V.F.)

VI.A.15-1   M. Hedayatullah, J. Het. Chem., 18, 339 (1981).

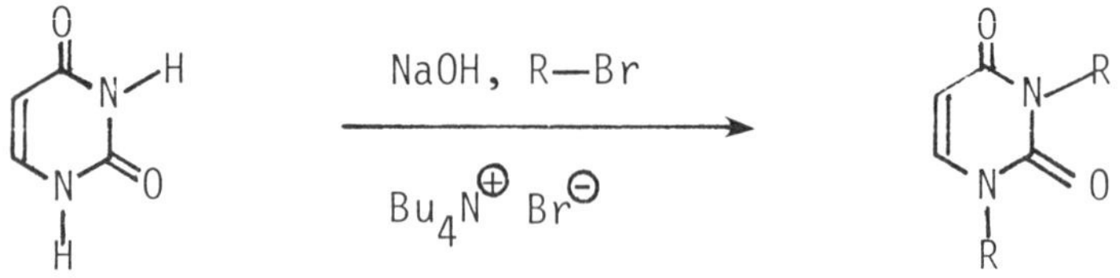

NaOH, R—Br

Bu₄N⊕ Br⊖

R = Me, Et, Bz   67-100%

VI.A.15-2   J. Yamawaki, T. Ando, and T. Hanafusa, <u>Chem.</u>
<u>Lett.</u>, 1143 (1981).

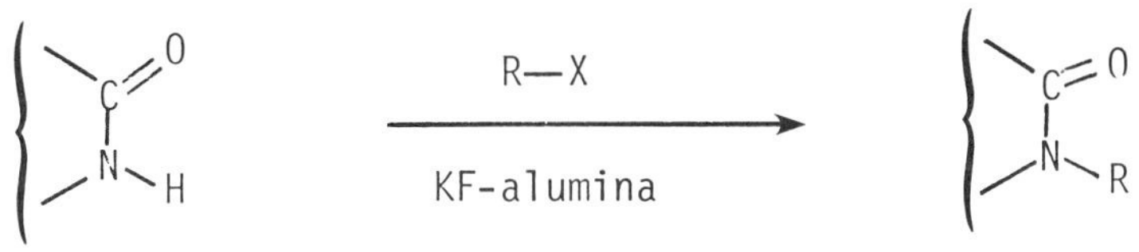

R—X = MeI, BzCl   ~80-90%

Uracil and xanthine derivatives may be methylated
similarly.

VI.A.15-3   A. Matsuda, Y. Kurasawa, and K.A. Watanabe,
<u>Synthesis</u>, 748 (1981).

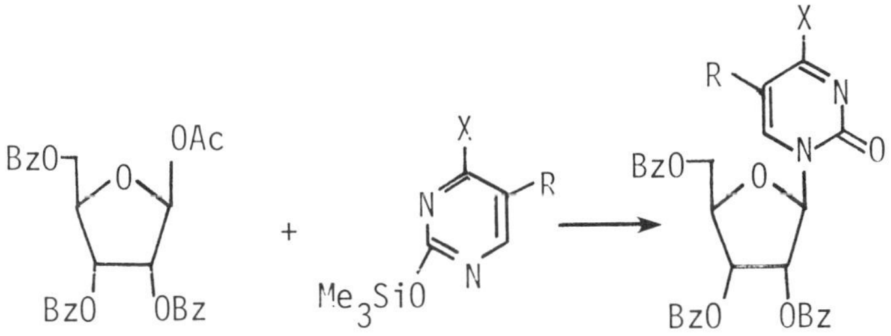

R = H, F, Me   73-85%
Z = OH, NH$_2$, NH—C(=O)—Ph—OMe

VI.A.15-4   H. Vorbruggen et al., Chem. Ber., 114, 1234 and
1279 (1981).

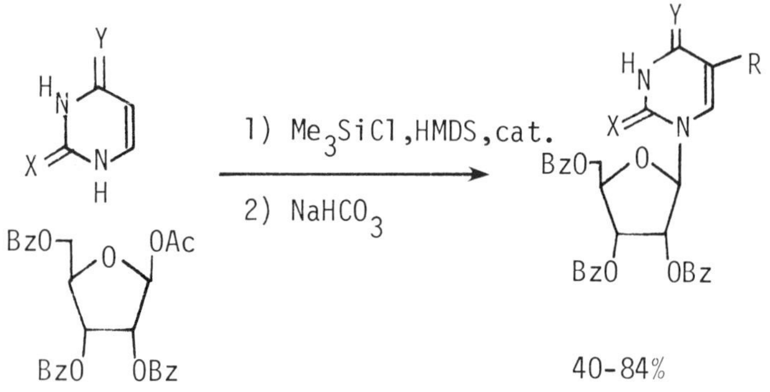

1) Me$_3$SiCl,HMDS,cat.

2) NaHCO$_3$

40-84%

X = O, S
Y = O, NH, NAc
R = H, OMe

VI.A.15-5   C.F.Bigge and M.P. Mertes, J. Org. Chem., 46,
1994 (1981).

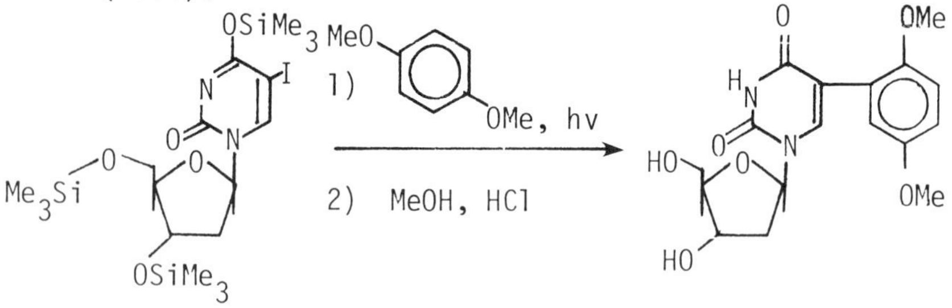

1)

OMe, hv

2)   MeOH, HCl

32%

VI.A.15-6   R.A. Lessor and N.J. Leonard, J. Org. Chem.,
46, 4300 (1981).

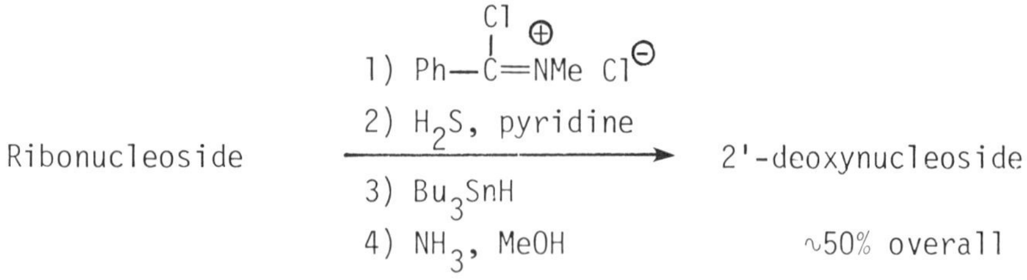

Ribonucleoside

1) Ph—C=NMe Cl$^\ominus$ (Cl, $\oplus$)

2) H$_2$S, pyridine

3) Bu$_3$SnH

4) NH$_3$, MeOH

2'-deoxynucleoside

∿50% overall

VI.A.15-7   M.J. Robins and J.S. Wilson, <u>J. Am. Chem. Soc.</u>,
<u>103</u>, 932 (1981).

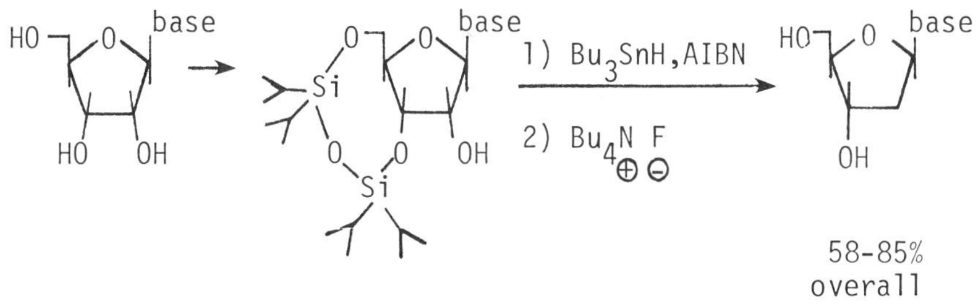

58-85%
overall

VI.A.16.   Olefins, Acetylenes

(see also: I.B., I.C., II.J., III.G.)

VI.A.16-1   D.J.S. Tsai and D.S. Matteson, <u>Tetrahedron Lett.</u>,
<u>22</u>, 2751 (1981).

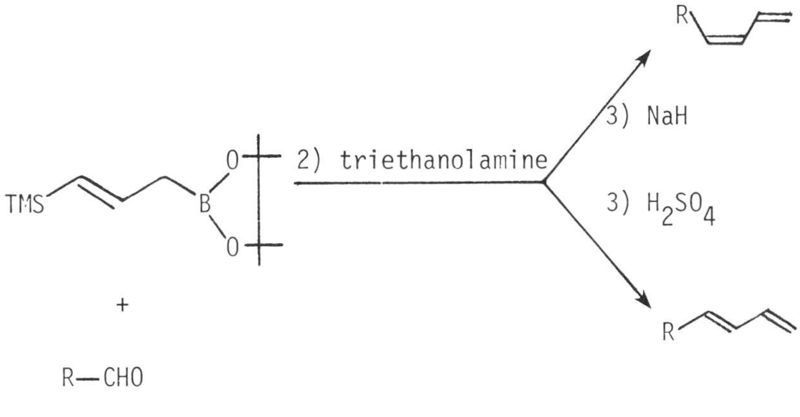

R = Ph, <u>n</u>-heptyl, AcO(CH$_2$)$_8$-   ∿70% overall

VI.A.16-2   W. Adam et al., J. Org. Chem., 46, 3359 (1981).

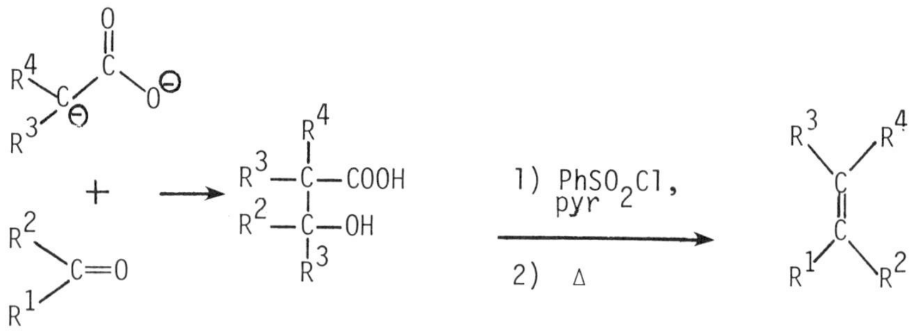

Widely varying yields.

Useful for sterically congested olefins.

VI.A.16-3   V.A. Curtis, F.J. Knutson, and R.J. Baumgarten, Tetrahedron Lett., 22, 199 (1981).

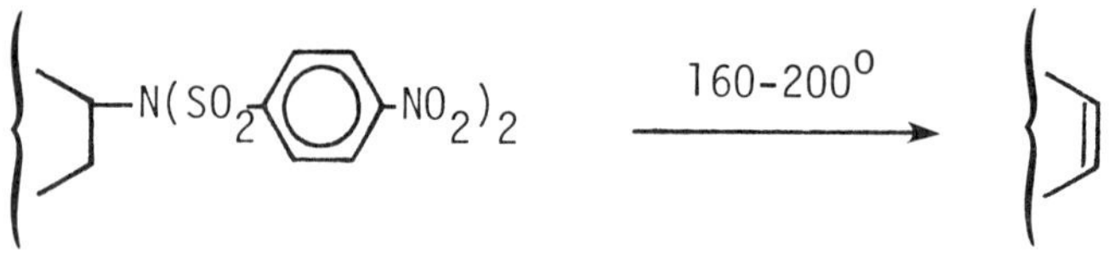

(cyclopentyl, cyclooctyl, etc.)   85-99%

VI.A.16-4   E.V. Dehmlow and M. Lissel, Tetrahedron, 37, 1653 (1981).

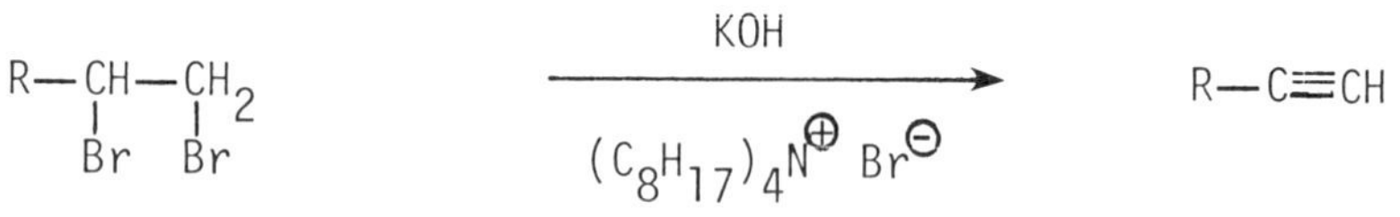

$$R-\underset{Br}{CH}-\underset{Br}{CH_2} \quad \xrightarrow[\;(C_8H_{17})_4N^{\oplus}\;Br^{\ominus}\;]{KOH} \quad R-C{\equiv}CH$$

~ 90%

R = alkyl, aryl, -CH(OEt)$_2$

VI.A.16-5   P.G. Karmarkar, A.A. Thakar, and M.S. Wadia,
Tetrahedron Lett., 22, 2301 (1981).

$$2 \ Ar \diagdown\diagup \!\!\!\!\!\!\!\diagdown NO_2 \quad \xrightarrow{\ H_2O_2, \ Et_3N\ } \quad Ar-C\equiv C-Ar$$

Ar = subst. Ph

~30%

VI.A.16-6   D.H. Wadsworth and B.A. Donatelli, Synthesis,
285 (1981).

$$\underset{\substack{\| \\ O}}{\overset{\displaystyle Ar \diagdown \quad \diagup Ar'}{C=C}} \quad \xrightarrow[\underline{o}\text{-dichlorobenzene}]{\ Al_2O_3, \ \nabla\ } \quad Ar-C\equiv C-Ar'$$

Ar = subst. Ph, naphthyl, etc.

~80-90%

VI.A.16-7   J.F. Normant and A. Alexakis, Synthesis, 841
(1981).

Reveiw: "Carbometallation (C-metallation) of Alkynes:

Stereospecific Synthesis of Alkenyl Derivatives"

VI.A.17.   Peptides

(see also: V.B., V.C., V.D., VI.A.4.)

VI.A.17-1   D.A. Buckingham et al., J. Am. Chem. Soc., 103, 7023 and 7025 (1981).

Use of Co(III)-chelated amino esters for peptide coupling:

$$\left[(en)_2 Co \overset{N H_2}{\underset{O=C}{\diagdown}} \overset{CHR}{\underset{OMe}{\diagup}} \right]^{+3} + H_2NCHR'COOMe \longrightarrow (en)_2 Co \overset{N H_2}{\underset{O=C}{\diagdown}} \overset{CHR}{\underset{NHCHR'COOMe}{\diagup}}$$

72-91%

This technique is used to synthesize tetrapeptides and an enkephalin.

VI.A.17-2   H. Takaku and M. Yoshida, J. Org. Chem., 46, 589 (1981).

Use of (8-Quinolinesulfonyl) tetrazole (I) as a coupling

agent in the phosphotriester approach to peptide synthesis.

I

VI.A.17-3   T. Teramoto et al., Tetrahedron Lett., 22, 1109 (1981).

Use of optically active N-hydroxytartrimides as activating

agents for enantioselective peptide synthesis.

VI.A.17-4   M. Wakselman and F. Acher, J.C.S. Chem. Comm.,
632 (1981).

Peptide synthesis using 1, a strained sultone, as a coupling

agent. Yields of dipeptides are about 60%

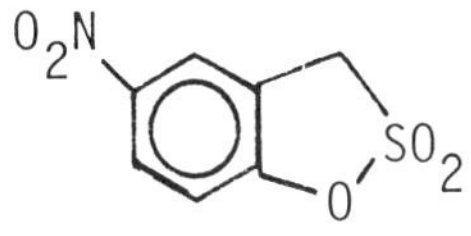

VI.A.17-5   T. Mukaiyama et al., Chem. Lett., 65 (1981).

Use of bis(o- or p-nitrophenyl) phenylphosphonate as a

coupling agent to synthesize leucine--enkephalin. Yields

for peptide couplings are 73-96%.

VI.A.17-6   R. Appel and E. Hiester, Chem. Ber., 114, 2649
(1981).

Use of Tris(dimethylamino) phosphane/Hexachloroethane

/1-Hydroxybenzotriazole as a condensing reagent for the

synthesis of arginine--containing peptides.

VI.A.17-7   K. Martinek, A.M. Semenov, and I.V. Berezin,
Doklady Chem., 254 420 (1981).

Use of chymotrypsin to catalyze synthesis of dipeptides in

a two-phase water/ethyl acetate system.

VI.A.17-8   T. Kunieda et al., Tetrahedron Lett., 22, 1257
(1981).

Use of $(PhO)_2\overset{\overset{O}{\|}}{P}$—N⟍O   as a carboxyl activating

group for peptide synthesis. No racemization is observed,

and protected dipeptides are formed in 83-98% yields.

VI.A.17-9   H. Chi-yi et al., Tetrahedron Lett., 22, 3467
(1981).

Use of 3-aminoacyl-tetrahydrothiazole-2-thione (TTT) as an

active amide for peptide synthesis.

$$P—NH—\overset{R}{\underset{|}{CH}}—\overset{\overset{O}{\|}}{C}—N⟍S$$

TTT amide

VI.A.17-10   E. Atherton, C.J. Logan, and R.C. Sheppard, _J.C.S. Perkin I_, 538 (1981).

"Use of base-labile N^{α}-9-fluorenylmethoxycarbonylamino

acids in combination with acid-labile t-butyl or p-alkoxy-

benzyl side-chain or carboxy-terminal (resin-linkage)

protecting groups enables solid-phase peptide synthesis to

be carried out under exceptionally mild reaction conditions."

VI.A.17-11   R. Colombo, _J.C.S. Chem. Comm._, 1012 (1981).

Non-acidic solid-phase peptide synthesis may be achieved

using base-labile N^{α}-fluoren-9-ylmethoxycarbonyl-substituted

amino acids, benzyl-based side-chain protecting groups, and

a p-nitrobenzhydrylamine resin. The final deprotection

step uses transfer hydrogenation with 1, 4-cyclohexadiene.

VI.A.17-12   F.H.C. Stewart, _Aust. J. Chem._, __34__, 2013 (1981).

Use of the 4-picolyl ester of 4-hydroxy-3-nitrophenylacetic

acid as the precursor of a series of protected amino acid

active esters used in peptide synthesis. Purification of

the protected peptides by acid washing was facilitated by

the presence of the basic 4-picolyl moiety.

VI.A.17-13   N. Hatanaka and I. Ojima, Chem. Lett., 231 and 1297 (1981).

A synthetic approach to oligopeptides through β-lactams:

VI.A.17-14   E. Giralt et al., Tetrahedron, 37, 2007 (1981).

Use of $ClCH_2$—Pab—resin as the solid polystyrene support in peptide synthesis allows for easier cleavage of the pure peptide and better acid stability of the support.

VI.A.17-15   E. Atherton et al., J.C.S. Chem. Comm., 1151 (1981).

"Copolymerization of dimethylacrylamide, ethylene bisacrylamide, and acryloylsarcosine methyl ester within the pores of rigid macroporous inorganic particles provides support suitable for continuous flow solid phase peptide synthesis and for other applications."

VI.A.17-16   R. Colombo, Tetrahedron Lett., 22, 4129 (1981).

Synthesis of protected peptide hydrazides on modified

poly(ethylene glycol) supports containing benzyl-, p-

benzyloxybenzyl, and tertiary-alkyl-oxycarbonylhydrazide

anchoring groups.

VI.A.17-17   M.K. Anwer and A.F. Spatola, Tetrahedron Lett.,
22, 4369 (1981).

Use of catalytic transfer hydrogenation, with ammonium

formate and $Pd(OAc)_2$, to remove peptides from solid phase

resins.

VI.A.17-18   J.A. vanNispen et al., Rec. Trav. Chim. Pays-
Bas, 100, 435 (1981).

Methanesulfonic acid may be used in place of liquid HF

to cleave small peptides from the standard Merrifield resin.

Advantages are fewer side reactions and a less dangerous

reagent.

VI.A.17-19   M. Bodanszky and J. Martinez, Synthesis, 333
(1981).

Review: "Side Reactions in Peptide Synthesis"

VI.A.18.   Vinyl Halides, Vinyl Ethers, Vinyl Esters

VI.A.18-1   H.P. On, W. Lewis, and G. Zweifel, Synthesis, 999 (1981).

$$\underset{H}{\overset{R}{>}}C=C\underset{X}{\overset{SiMe_3}{<}} \quad \xrightarrow{\text{NaOMe, MeOH}} \quad \underset{H}{\overset{R}{>}}C=C\underset{X}{\overset{H}{<}}$$

NaOMe, MeOH

$$\underset{H}{\overset{R}{>}}C=C\underset{SiMe_3}{\overset{X}{<}} \quad \longrightarrow \quad \underset{H}{\overset{R}{>}}C=C\underset{H}{\overset{X}{<}}$$

R = alkyl, Ph, etc; may containg THP ethers.   69-94%
X = Cl, Br

VI.A.18-2   G.W. Kabalka, E.E. Gooch, and H.C. Hsu, Synth. Comm., 11, 247 (1981).

$$R-C\equiv CH \quad \xrightarrow[\text{3) ICl, NaOAc}]{\begin{array}{l}\text{1) catecholborane}\\ \text{2) H}_2\text{O}\end{array}} \quad R-CH=CHI$$

68-88% (E isomer)

VI.A.18-3   S. Uemura et al., Bull. Chem. Soc. Japan, 54, 2843 (1981).

$$R-C\equiv C-R' \quad \xrightarrow{SO_2Cl_2} \quad \underset{Cl}{\overset{R}{>}}C=C\underset{R'}{\overset{Cl}{<}}$$

R, R' = Ph, H, alkyl   ∿50-90%

VI.A.18-4   D.R. Williams et al., Tetrahedron Lett., 22, 3745 (1981).

$$\begin{array}{c} R \\ R' \end{array}\!\!\!\!=\!O \quad \xrightarrow[\text{2) Zn, HOAc}]{\text{1) Br}_2\text{CHLi}} \quad \begin{array}{c} R \\ R' \end{array}\!\!\!\!=\!CHBr$$

62-95%

R, R' = H, alkyl, cyclic, Ph   (mixture of E and Z)

VI.A.18-5   G. Piancatelli et al., Tetrahedron Lett., 22, 1041 (1981).

$$\begin{array}{c} R \\ R' \end{array}\!\!\!C\!\!\begin{array}{c} OH \\ C\equiv CH \end{array} \quad \xrightarrow[\text{2) Pyridinium dichromate}]{\text{1) I}_2} \quad \begin{array}{c} R \\ R' \end{array}\!\!\!C\!\!=\!\!C\!\!\begin{array}{c} CHO \\ I \end{array}$$

R, R' = alkyl, cyclic, steroidal   30-66%

VI.A.18-6   S.V. Ley and A. J. Whittle, Tetrahedron Lett., 22, 3301 (1981).

$$\xrightarrow[\text{CH}_2\text{Cl}_2/\text{pyridine}]{\text{PhSeX}}$$

X = Cl, Br   ∿50-100%

VI.A.19.   Sulfur Compounds

(see also: II.E., III.C.)

VI.A.19-1   H. Hiemstra and H. Wynberg, J. Am. Chem. Soc., 103, 417 (1981).

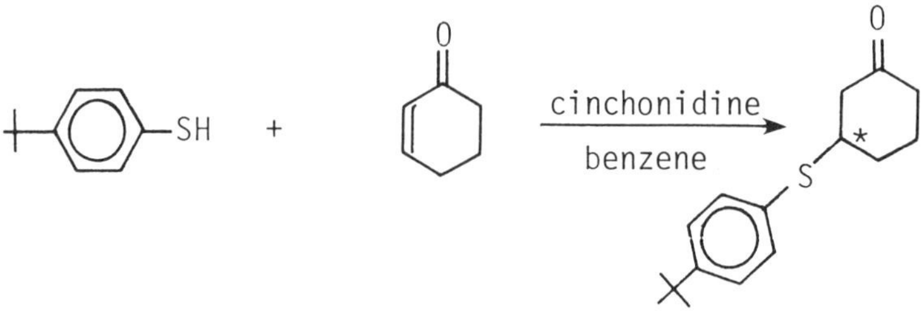

62% ee

Full paper, many examples

VI.A.19-2   H.J. Cristau et al., Synthesis, 892 (1981).

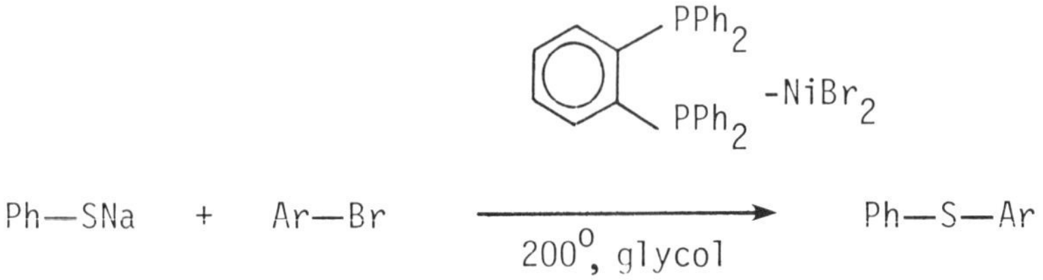

Ph—SNa   +   Ar—Br $\xrightarrow[200°, \text{glycol}]{}$ Ph—S—Ar

∿70-90%

Ar = subst. Ph, naphthyl, etc.

VI.A.19-3   F. Bottino, R. Fradullo, and S. Pappalardo,
J. Org. Chem., 46, 2793 (1981).

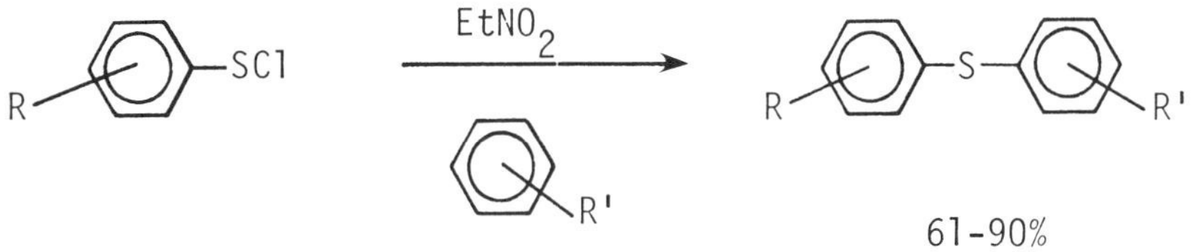

R = H, Me, Cl
R' = H, Me, OMe, Cl

VI.A.19-4   H. Suzuki and N. Sato, Chem. Lett., 267 (1981).

$$Ar-CH_2OH \xrightarrow[\text{benzene, } \Delta]{\text{PhSSPh, } P_2I_4} Ar-CH_2-SPh$$

77-95%

Ar = subst. Ph

VI.A.19-5   P. Jacob III and A.T. Shulgin, Synth. Comm., 11,
957 (1981).

$$ArLi \xrightarrow{R'SSR'} Ar-S-R'$$

Ar = Subst. Ph, pyridyl, furyl, etc.   ∿60-90%
R' = alkyl, aryl

VI.A.19-6   Y. Kikugawa, Chem. Lett., 1157 (1981).

$$
\underset{R}{\overset{O}{\underset{\displaystyle R'}{\parallel}}}C \quad + \quad R''SH \quad \xrightarrow[\text{CF}_3\text{COOH}]{\text{pyridine-borane}} \quad R-\overset{\displaystyle H}{\underset{\displaystyle R'}{\overset{|}{\underset{|}{C}}}}-SR''
$$

52-69%

R, R' = H, Ph, alkyl
R" = Bz, alkyl

_____

VI.A.19-7   Y. Tamura et al., Tetrahedron Lett., 22, 81
(1981).

$$
\text{ArH} \quad \xrightarrow[\text{TsOH}]{\overset{O}{\overset{\parallel}{\text{MeSCH}_2\text{COOEt}}}} \quad \overset{\text{SMe}}{\underset{}{\overset{|}{\text{Ar}-\text{CHCOOEt}}}}
$$

55-89%

Ar = subst. Ph, naphthyl, thiophene, etc.

_____

VI.A.19-8   T. Aida, T.H. Chan, and D.N. Harpp, Tetrahedron
Lett., 22, 1089 (1981).

$$
\underset{R}{R-CH-CH}\overset{S}{\underset{I}{\diagdown}}\underset{I\quad R'}{CH-CH-R'} \quad \xrightarrow{\text{Me}_3\text{N}} \quad \underset{R}{\overset{R}{\diagup}}C=C\overset{S}{\underset{H\ H}{\diagup\diagdown}}C=C\underset{R}{\overset{R}{\diagdown}}
$$

∿90%

R, R' = alkyl, cyclic

VI.A.19-9   A. de Groot and B.J.M. Jansen, Tetrahedron Lett., 22, 887 (1981).

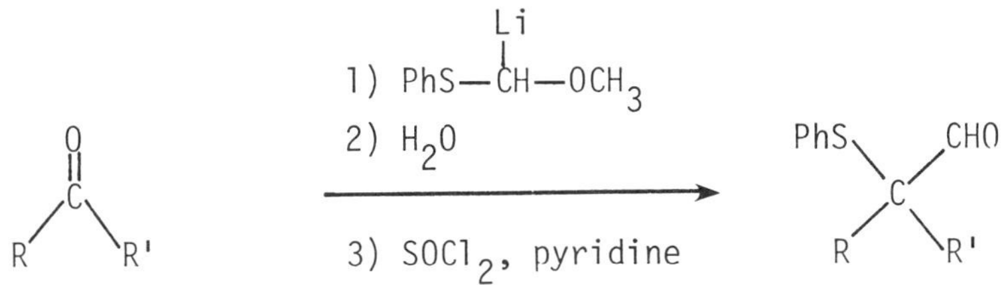

R, R' = Me, Et, Ph, -(CH$_2$)$_5$-   ∿80% overall

VI.A.19-10   S. Kim and S. Yang, Chem. Lett., 133 (1981).

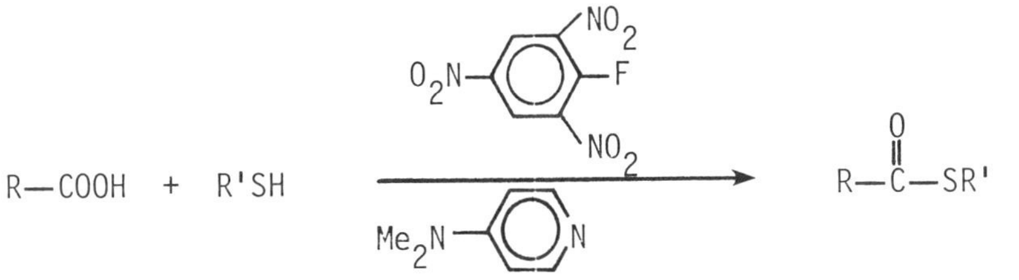

R—COOH + R'SH

R = alkyl, Ph, styryl

R' = 1°, 2°, 3° alkyl, Ph

80-95% in
most cases

VI.A.19-11   T. Kunieda, Y. Abe, and M. Hirobe, Chem. Lett., 1427 (1981).

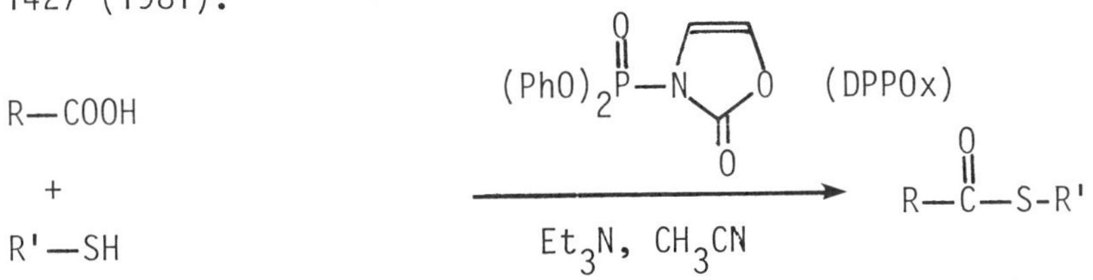

R—COOH

+

R'—SH

Et$_3$N, CH$_3$CN

∿90%

R = alkyl, Ph, Ph$_2$CH, etc.
R' = 1°, 2°, 3° alkyl, Ph, Bz

VI.A.19-12   S. Ohta and M. Okamoto, <u>Tetrahedron Lett.</u>, <u>22</u>, 3245 (1981).

$$R—COOH \xrightarrow[\text{2) R'SH, Mg(OEt)}_2]{\text{1) Im}_2\text{CO}} R—\overset{\overset{\displaystyle O}{\|}}{C}—SR'$$

~80-90%

R = alkyl, aryl, vinyl, esters, etc.

R' = Et, <u>t</u>-Bu, Ph, pyridyl, etc.

(Im = 1-imidazolyl)

VI.A.19-13   R.P. Volante, <u>Tetrahedron Lett.</u>, <u>22</u>, 3119 (1981).

$$R—OH \xrightarrow[\text{Ph}_3\text{P, CH}_3\text{COSH}]{\#NCO_2\underline{i}\text{-Pr})_2} R—S—\overset{\overset{\displaystyle O}{\|}}{C}—CH_3$$

>90%

R = alkyl, Bz, cinnamyl, 3β-cholesteryl

VI.A.19-14   M. Yamaguchi <u>et al.</u>, <u>Bull. Chem. Soc. Japan</u>, <u>54</u>, 943 (1981).

$$\xrightarrow[\text{DMAP, CH}_2\text{Cl}_2]{\text{R'SH}} R—\overset{\overset{\displaystyle O}{\|}}{C}—SR'$$

84-98%

R = alkyl, Ph, styryl

R' = Et, <u>t</u>-Bu, <u>i</u>-Pr, Ph

VI.A.19-15   E. Vedejs et al., J.Org. Chem., 46, 5253 (1981).

$$R-\overset{\overset{O}{\|}}{S}-\underset{\underset{Li}{|}}{CHR'}
\quad
\begin{array}{l}
1)\ ClPPH_2 \\
2)\ I_2 \\
\xrightarrow{\hspace{3cm}} \\
3)\ R''Li \\
4)\ O_2
\end{array}
\quad
R-\overset{\overset{O}{\|}}{S}-\overset{}{C}-R'$$

~50-70% overall

R = Me, Ph, t-Bu,
R'= alkyl, Ph, $\Big]$-$(CH_2)_4$-

VI.A.19-16   G. Cainelli et al., Synthesis, 302 (1981).

$$R-X \quad \xrightarrow[\text{hexane}]{\textcircled{P}\,\overset{\oplus}{}\,S\overset{\ominus}{}-\overset{\overset{O}{\|}}{C}-CH_3} \quad R-\overset{\overset{O}{\|}}{S}-\overset{}{C}-CH_3$$

~80-90%

X = Cl, Br, OTs
R = 1°, 2° alkyl, allyl, cyclohexyl, Bz, etc.

VI.A.19-17   M.K. Kaloustian and F. Khouri, Tetrahedron Lett., 22, 413 (1981).

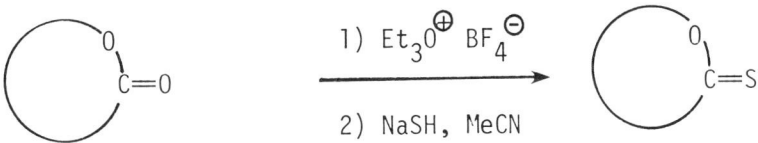

1) $Et_3O^{\oplus}\ BF_4^{\ominus}$

2) NaSH, MeCN

44-90%

VI.A.19-18   S.A. Benner, <u>Tetrahedron Lett.</u>, <u>22</u>, 1851 (1981).

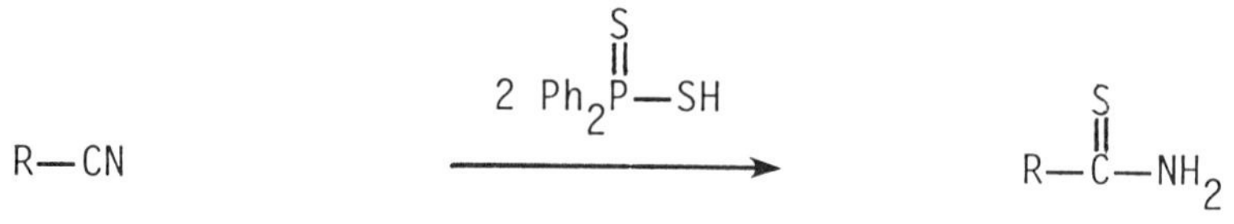

$$R-CN \xrightarrow{2\ Ph_2\overset{\overset{\displaystyle S}{\|}}{P}-SH} R-\overset{\overset{\displaystyle S}{\|}}{C}-NH_2$$

R = alkyl, cyclic, Bz, subst. Ph   73-91%

VI.B.1.   Ring Enlargement

VI.B.1-1   F. Huet, A. Lechevallier, and J. -M. Conia,
<u>Chem. Lett.</u>, 1515 (1981).

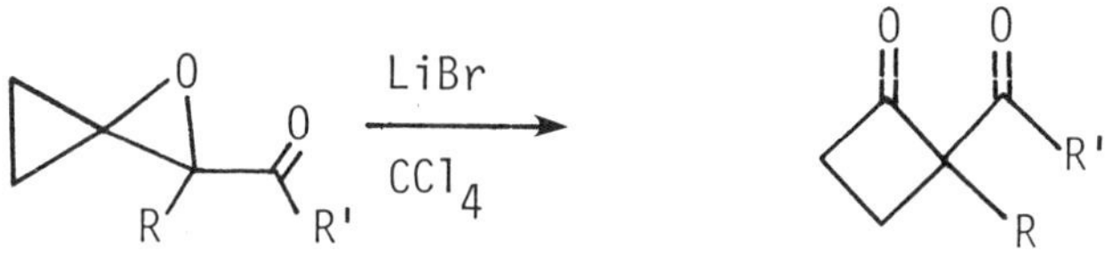

60-100%

R = H, Me
R' = Me, Ph

VI.B.1-2   M.L. Forcellese <u>et al.</u>, <u>J. Org. Chem.</u>, <u>46</u>, 3326
(1981).

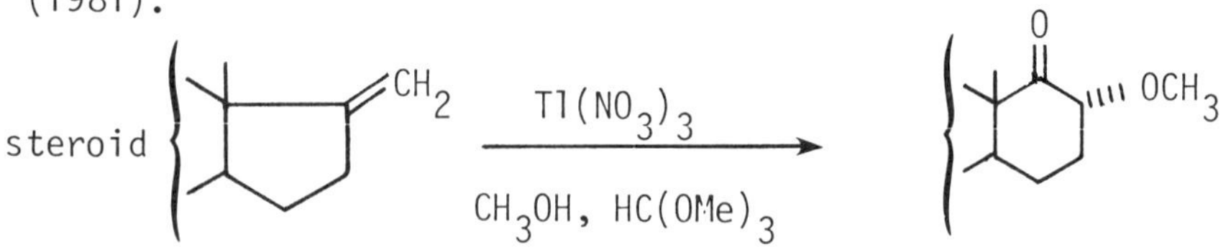

30-50%

VI.B.1-3   R. Okazaki et al., Angew. Chem. Int. Ed., 20, 799
(1981).

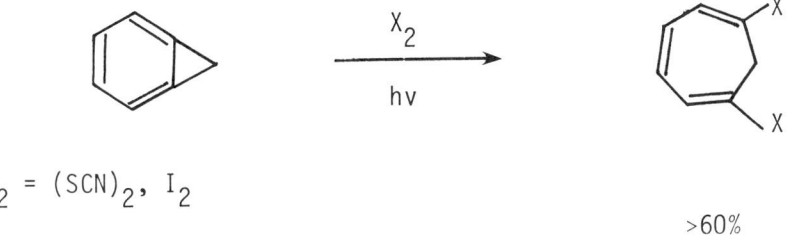

$X_2 = (SCN)_2, I_2$

>60%

VI.B.1-4   T. Shono et al., Chem. Lett., 551 (1981).

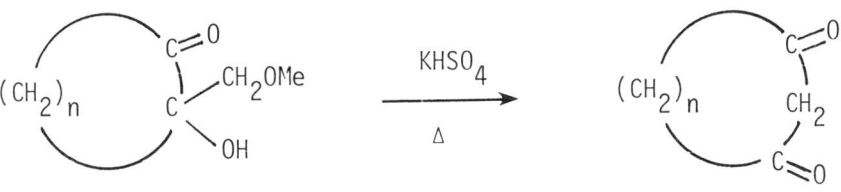

n = 2-6

50-75%

VI.B.1-5   P.A. Wender and S.M. Sieburth, Tetrahedron Lett.,
22, 2471 (1981).

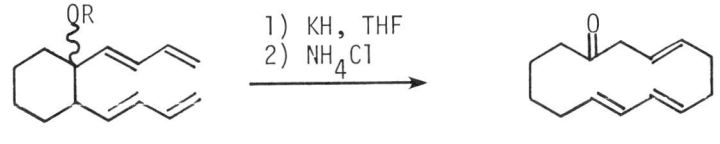

90%

VI.B.1-6   W. Tochtermann and P. Rosner, Chem. Ber., 114
3725 (1981).

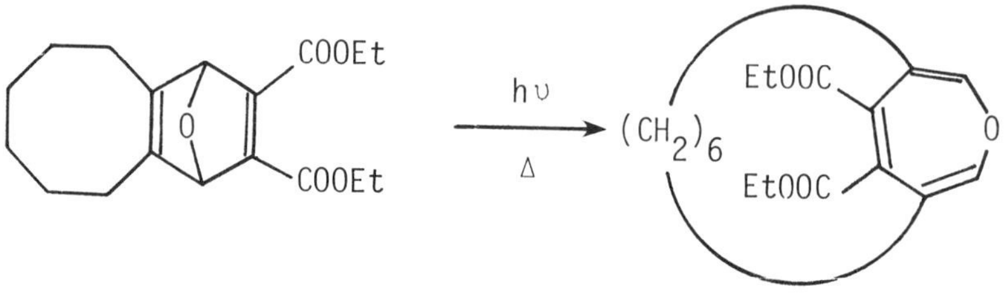

74%

VI.B.1-7   V. Bhat and R.C. Cookson, J.C.S. Chem. Comm.,
1123 (1981).

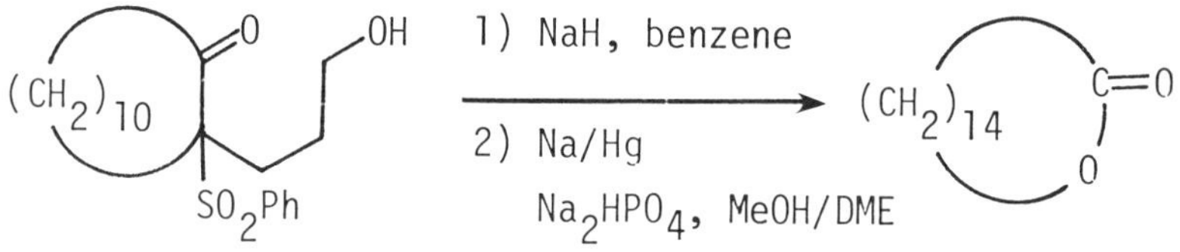

81%

VI.B.1-8   K.W. Blake and I. Gillies, J.C.S. Perkin I,
700 (1981).

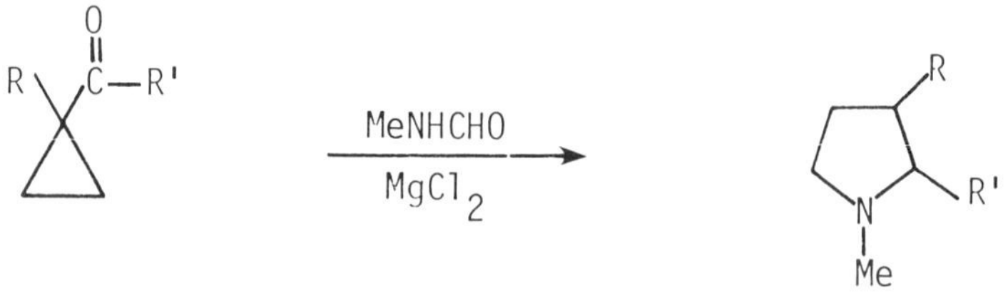

R = H, Ph
R' = H, alkyl, aryl

VI.B.1-9   J.N. Chatterjea et al., Liebigs Ann. Chem., 52
(1981).

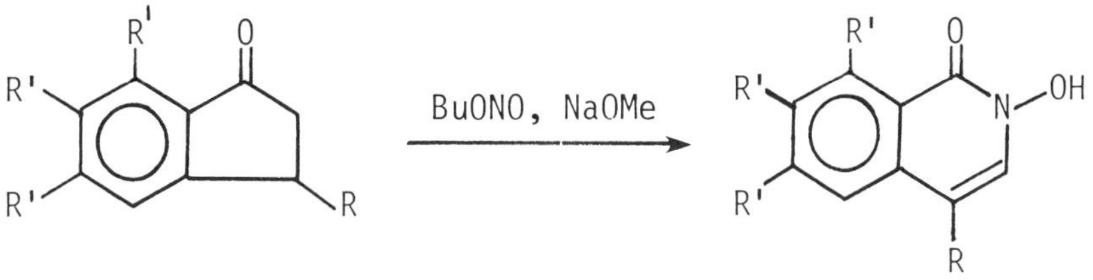

R = H, Me, Ph

R's = H, OMe

VI.B.1-10   D.S.C. Black and L.M. Johnstone, Angew. Chem.
Int. Ed., 20, 670 (1981).

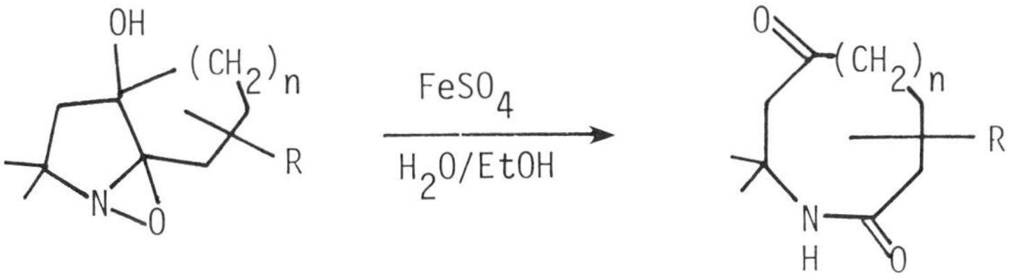

R = H, benzo, =CHPh, etc.

n = 1, 2

VI.B.1-11   P.A. Wender <u>et al.</u>, <u>Tetrahedron</u>, <u>37</u>, 3967 (1981).

Review: "Macroexpansion Methodology: Medium Ring

Synthesis Based on an Eight Unit Ring Expansion Process"

VI.B.2.   Ring Contraction

VI.B.2-1   M. Karpf and C. Djerassi, <u>J. Am. Chem. Soc.</u>,
<u>103</u>, 302 (1981).

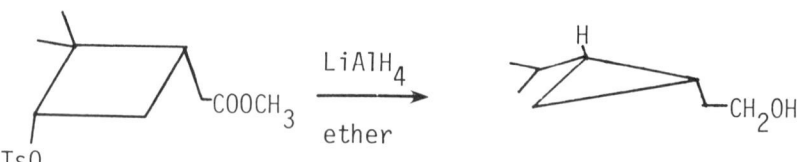

∿44%

VII

MISCELLANEOUS REVIEWS

VII-1   Lists of "Recent Reviews" published by the Journal of Organic Chemistry:

#7:   J. Org. Chem., 46, 1973 (1981).

#8:   J. Org. Chem., 46, 4634 (1981).

VII-2   A. Akelah and D.C. Sherrington, Chem. Rev., 81, 557 (1981); Synthesis, 413 (1981).

Review: "Application of Functionalized Polymers in

Organic Synthesis:

VII-3   D.C. Bailey and S.H. Langer, Chem. Rev., 81, 109 (1981).

Review: "Immobilized Transition-Metal Carbonyls and

Related Catalysts"

VII-4   C.M. Lukehart, Accounts Chem. Res., 14, 109 (1981).

Review: "Metalla-β-diketones and Their Derivatives"

_____

VII-5   J.R. Long, Aldrichimica Acta, 14, 63 (1981).

Review: "The Role of Silver Salts in Organic Processes"

_____

VII-6   B.B. Snider et al., Tetrahedron, 37, 3927 (1981).

Review: "Alkylaluminum Halides: Lewis Acid Catalysts
which are Bronsted Bases"

_____

VII-7   P. Heimbach, H. Schenkluhn, and K. Wisseroth, Pure
and Appl. Chem., 53, 2419 (1981).

Review: "Control in Transition Metal Catalyzed Organic
Synthesis"

VII-8   M. Pereyre and J.P.-Quintard, Pure and Appl. Chem.,
53, 2401 (1981).

Review: "Organotin Chemistry for Synthesis Applications"

VII-9   M.F. Semmelhack, Pure and Appl. Chem., 53, 2379
(1981).

Review: "Nucleophilic Addition to Diene- and Arene-metal

complexes"

VII-10   J. Tsuji, Pure and Appl. Chem., 53, 2371 (1981).

Review: "Palladium Catalysis in Natural Product Synthesis"

VII-11   B.M. Trost, Pure and Appl. Chem., 53, 2357 (1981);
Aldrichimica Acta, 14, 43 (1981).

Review: "Transition Metal Templates for Selectivity in

Organic Synthesis"

VII-12   E. Negishi, Pure and Appl. Chem., 53, 2333 (1981).

Review: "Bimetallic Catalytic Systems Containing Ti, Zr, Ni, and Pd. Their Applications to Selective Organic Synthesis."

VII-13   R.F. Heck, Pure and Appl. Chem., 53, 2323 (1981).

Review: "Palladium-catalyzed Syntheses of Conjugated Polyenes"

VII-14   M. Orchin, Accounts Chem. Res., 14, 259 (1981).

Review: "HCo(CO)$_4$, the Quintessential Catalyst"

VII-15   N.H. Andersen et al., Tetrahedron, 37, 4069 (1981).

Review: "The Use of Metalloids (-SiMe$_3$, -SnR$_3$) as Protected Carbanions"

VII-16   G.H. Posner et al., Pure and Appl. Chem., 53, 2307
(1981).

Review: "Asymmetric Synthesis using Organometallic

Reagents and Optically Pure Vinylic Sulfoxides"

VII-17   H. Wynberg, Rec. Trav. Chim. Pays-Bas, 100, 393
(1981).

Review: "Chance, Necessity, and Asymmetric Catalysis"

VII-18   H.C. Brown, P.K. Jadhav, and A.K. Mandal, Tetrahedron, 37, 3547 (1981).

Review: "Asymmetric Synthesis via Chiral Organoborane

Reagents"

VII-19   G. Solladié, Synthesis, 185 (1981).

Review: "Asymmetric Synthesis Using Nucleophilic

Reagents Containing a Chiral Sulfoxide Group"

VII-20   H.C. Brown and J.B. Campbell, Jr., Aldrichimica
Acta, 14, 3 (1981).

Review: "Synthesis and Applications of Vinylic Organo-

boranes"

_____

VII-21   A.H. Schmidt, Aldrichimica Acta, 14, 31 (1981).

Review: "Bromotrimethylsilane and Iodotrimethylsilane--

Versatile Reagents for Organic Synthesis"

_____

VII-22   I. Fleming, Chem. Soc. Rev., 10, 83 (1981).

Review: "Some Uses of Silicon Compounds in Organic

Synthesis"

_____

VII-23   S. Danishefsky, Accounts Chem. Res., 14, 400 (1981).

Review: "Siloxy Dienes in Total Synthesis"

VII-24   R. Noyori, S. Murata, and M. Suzuki, Tetrahedron, 37, 3899 (1981).

Review: "Trimethylsilyl Triflate in Organic Synthesis"

VII-25   K. Oka, Synthesis, 661 (1981).

Review: "Some Applications of Thionyl Chloride in Synthetic Organic Chemistry"

VII-26   G. Hamprecht, K. -H. König, and G. Stubenrauch, Angew. Chem. Int. Ed., 20, 151 (1981).

Review: "Alkylsulfamoyl Chlorides as Key Units in the Synthesis of Novel Biologically Active Compounds for Crop Protection"

VII-27   A.L. Lapidus and Y.Y. Ping, Russ. Chem. Rev., 50, 63 (1981).

Review: "Organic Syntheses Based on Carbon Dioxide"

VII-28   Yu. V. Belkin and N.A. Polezhaeva, Russ. Chem. Rev.,
50, 481 (1981).

Review: "The Chemistry of Stabilized Sulfonium Ylides"

VII-29   H.W. Moore and M.D. Gheorghin, Chem. Soc. Rev.,
10, 289 (1981).

Review: "Cyanoketenes: Synthesis and Cycloadditions"

VII-30   Y.G. Gololbov, I.N. Zhmurova, and L.F. Kasukhin,
Tetrahedron, 37, 437 (1981).

Review: "Sixty Years of Staudinger Reaction"

VII-31   T.A. Hase and J.K. Koskimies, Aldrichimica Acta,
14, 73 (1981).

Review: "A Compilation of References on Formyl and Acyl

Anion Synthons"

VII-32   H.P. Albicht and K. Issleib, <u>Z. Chem.</u>, <u>21</u>, 341 (1981).

Review: "C-Lithiated Phosphines as Synthons"

VII-33   W.T. Brady, <u>Tetrahedron</u>, <u>37</u>, 2949 (1981).

Review: "Synthetic Applications involving Halogenated Ketenes"

VII-34   J. Grinishaw and A.P. deSilva, <u>Chem. Soc. Rev.</u>, <u>10</u>, 181 (1981).

Review: "Photochemistry and Photocyclization of Aryl Halides"

VII-35   A.L.J. Beckwith, <u>Tetrahedron</u>, <u>37</u>, 3073 (1981).

Review: "Regio-selectivity and Stereo-selectivity in Radical Reactions"

VII-36   A.B. Smith, III and R.K. Dieter, Tetrahedron, 37,
2407 (1981).

Review: "The Acid-Promoted Decomposition of α-Diazo

Ketones"

VII-37   O. Mitsunobu, Synthesis, 1 (1981).

Review: "The Use of Diethyl Azodicarboxylate and Tri-

phenylphosphine in Synthesis and Transformation of

Natural Products"

VII-38   M. Balci, Chem. Rev., 81, 91 (1981).

Review: "Bicyclic Endoperoxides and Synthetic Applica-

tions"

VII-39   A. Williams and I.T. Ibrahim, Chem. Rev., 81, 589
(1981).

Review: "Carbodiimide Chemistry: Recent Advances"

VII-40   M. Mikolajczyk and P. Kielbasinski, <u>Tetrahedron</u>, <u>37</u>, 233 (1981).

Review: "Recent Developments in Carbodiimide Chemistry"

VII-41   G. Cainelli and G. Cardillo, <u>Accounts Chem. Res.</u>, <u>14</u>, 89 (1981).

Review: "Some Aspects of the Stereospecific Synthesis of Terpenoids by Means of Isoprene Units"

VII-42   A. Viola, J.J. Collins, and N. Filipp, <u>Tetrahedron</u>, <u>37</u>, 3765 (1981).

Review: "Intramolecular Pericyclic Reactions of Acetylenic Compounds"

VII-43   G. Illuminati and L. Mandolini, <u>Accounts Chem. Res.</u>, <u>14</u>, 95 (1981).

Review: "Ring Closure Reactions of Bifunctional Chain Molecules"

VII-44   G. Melloni, G. Modena, and U. Tonellato, Accounts
Chem. Res., 14, 227 (1981).

Review: "Relative Reactivities of Carbon-Carbon Double

and Triple Bonds toward Electrophiles"

VII-45   H.J. Schäfer, Angew. Chem. Int. Ed., 20, 911 (1981).

Review: "Anodic and Cathodic CC-Bond Formation"

VII-46   A. Padwa, T.J. Blacklock and W.F. Rieker, Israel
J. Chem., 21, 157 (1981).

Review: "Synthesis of Polycyclic Ring Systems via

Intramolecular [2 + 2]-cycloaddition Reactions of

Cyclopropene Derivatives"

VII-47   V.A. Galishev, V.N. Chistokletov, and A.A. Petrov,
Russ. Chem. Rev., 49, 880, (1980).

Review: "α,β-Unsaturated Heteroatomic Compounds in

1,3-dipolar Addition Reactions"

VII-48   V.I.M. Yuvenskii and B.K. Nefedov, Russ. Chem. Rev.,
50, 470 (1981).

Review: "Synthesis of Nitrogen-containing Compounds by

the Interaction of Nitro-compounds with Carbon Monoxide"

VII-49   D.H.R. Barton and W.B. Motherwell, Pure and Appl.
Chem., 53, 1081 (1981).

Review: "New and Selective Reactions and Reagents in

Natural Product Chemistry"

VII-50   D.H.R. Barton and W.B. Motherwell, Pure and Appl.
Chem., 53, 15 (1981).

Review: "New and Selective Reactions and Reagents in

Carbohydrate Chemistry"

VII-51   T. Kametani and H. Nemoto, Tetrahedron, 37, 3 (1981).

Review: "Recent Advances in the Total Synthesis of

Steroids via Intramolecular Cycloaddition Reactions"

AUTHOR INDEX

Modena, G. - 125, 470

Moiseenkov, A. M. - 145

Molina, P. - 232, 433

Moore, H. W. - 182, 466

Moreno-Manas, M. - 20, 80, 241

Moriarty, R. M. - 252, 254

Morita, T. - 290

Moro-Oka, Y. - 241

Morton, T. H. - 146

Mosher, H. S. - 71

Motherwell, W. B. - 425

Moulines, J. - 353

Muchowski, J. M. - 47

Mukaiyama, T. - 31, 42, 45, 50,

   57, 68, 70, 85, 108, 176,

   318, 381, 401, 420, 441

Muller, P. - 251, 268

Muller, R. K. - 113

Mulzer, J. - 18, 75

Murahashi, S. I. - 336

Murai, A. - 93, 171

Murata, I. - 354

Murphy, W. S. - 207, 300

Nair, V. - 72

Nakai, T. - 196, 198, 199

Nakajima, R. - 245

Nakamura, A. - 39, 69

Nakamura, E. - 119

Nakanashi, S. - 346

Nakano, T. - 326

Nakao, R. - 421

Nakata, T. - 279

Nakatsuka, S. I. - 28

Narasimhan, N. S. - 92, 344

Narula, A. P. S. - 222

Narula, A. S. - 3

Nedelec, L. - 119

Negishi, E. - 30, 132, 137,

   144, 246, 462

Negoro, K. - 418

Nekrasova, G. V. - 57

Neuenschwander, M. - 148, 165

Newcomb, M. - 7

Newman, M. S. - 219

Newton, R. F. - 85, 165

Nicalaou, K. C. - 105, 175,

   271, 347

Niewohner, U. - 430

Niwa, M. - 172

Noels, A. F. - 108, 415